EXPLORATIONS IN BASIC BIOLOGY

Sixth Edition

EXPLORATIONS IN BASIC BIOLOGY

Sixth Edition

STANLEY E. GUNSTREAM

Pasadena City College

Pasadena, California

PRENTICE HALL
Englewood Cliffs, New Jersey 07632

Editors: Kristin Watts Peri/Robert Pirtle
Production Manager: Paul Smolenski
Production Supervisor: John Travis
Cover designed by Ellen Pettengell-Proof Positive/Farrowlyne Associates, Inc.
Drawings by Mary Dersch
Photo Research by Diane Austin

This book was set in Century Schoolbook by Compset, Inc.

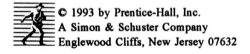

© 1993 by Prentice-Hall, Inc.
A Simon & Schuster Company
Englewood Cliffs, New Jersey 07632

Printed in the United States of America
10 9 8 7 6 5 4

ISBN   0-02-348523-2

Prentice-Hall International (UK) Limited, *London*
Prentice-Hall of Australia Pty. Limited, *Sydney*
Prentice-Hall Canada Inc., *Toronto*
Prentice-Hall Hispanoamericana, S.A., *Mexico*
Prentice-Hall of India Private Limited, *New Delhi*
Prentice-Hall of Japan, Inc., *Tokyo*
Simon & Schuster Asia Pte. Ltd., *Singapore*
Editora Prentice-Hall do Brasil, Ltda., *Rio de Janeiro*

CONTENTS

PREFACE

The sixth edition of *Explorations in Basic Biology,* like earlier editions, is designed for use in the laboratory component of introductory general biology courses. It is compatible with any modern biology textbook. The exercises provide a variety of options for either one or two semester courses and one, two or three quarter courses. The exercises are appropriate for three-hour laboratory sessions, but they are also adaptable for a two-hour laboratory format.

Explorations in Basic Biology is designed to simplify the work of instructors and to enhance learning by students.

MAJOR FEATURES

1. The thirty-nine exercises provide a *wide range of options* for the instructor, and the range of activities within an exercise further increases the available options.
2. Each exercise is basically *self-directing,* which allows students to work independently without continuous assistance by the instructor.
3. Each exercise, and its major subunits, are *self-contained* so that the instructor may arrange the sequence of exercises, or the activities within an exercise, to suit his or her preferences. In addition, portions of an exercise may be deleted without negatively impacting the continuity of the exercise.
4. Each exercise begins with a list of **Objec-** **tives** that outlines the minimal learning responsibilities of the student.
5. The text of each exercise starts with a discussion of **background information** that is necessary to (a) understand the subject of the exercise, and (b) prepare the student for the activities that follow. The inclusion of the background information minimizes the need for introductory explanations and assures that all lab sections receive the same background information.
6. Over 250 **illustrations** are provided to enhance students' understanding of both background information and laboratory procedures. New **key terms** are in bold print for easy recognition by students.
7. Before beginning the laboratory activities, students are asked to demonstrate their understanding of the background information by labeling illustrations and completing the portion of the Laboratory Report that covers this material.
8. The required **Materials** (equipment and supplies) are listed for each activity in the exercise. This list helps the student to obtain the needed materials and guides the laboratory technician in setting up the laboratory. The exercises utilize standard equipment and materials that are available in most biology departments.
9. Activities to be performed by students are identified by an **Assignment** heading that clearly distinguishes activities to be performed from the background informa-

tion. The assignment sections are numbered sequentially within each exercise and on the Laboratory Report to facilitate identification and discussion.

10. Laboratory procedures are clearly described in a step-wise manner within each Assignment section.

11. A **Laboratory Report** is provided for each exercise to guide and reinforce students' learning. The laboratory reports not only provide a place for students to record observations, impressions, collected data, and conclusions, but they also provide a convenient means of assessing student understanding.

MAJOR CHANGES IN THE SIXTH EDITION

The sixth edition has been improved to make it even more useful for both instructors and students. Some of these improvements have originated with suggestions from adopters.

1. Three new exercises have been added to increase the flexibility of the manual and options for instructors.

   a. *Exercise 1, Orientation,* is designed to prepare students for success in the laboratory by focusing on understanding biological terminology, performing simple metric measurements and calculations, and developing observational skills by constructing a dichotomous key to categorize samples of leaves.

   b. *Exercise 24, Sensory Perception in Humans,* includes sensory perception activities removed from Exercise 23, Neural Control, and provides additional activities in the study of human senses.

   c. *Exercise 39, Animal Behavior,* provides an opportunity to investigate innate behavior in flatworms, sowbugs and *Drosophila* flies.

2. Several exercises have been substantially improved.

   a. *Exercise 9, Terrestrial Plants,* now includes a study of external plant structure, monocot and dicot differences, and flower structure for those instructors who do not emphasize plant histology.

   b. *Exercise 14, Dissection of the Fetal Pig,* has been rewritten and includes new illustrations to make the dissection easier for students.

   c. *Exercise 21, Excretion,* has been enhanced by a series of "unknown" simulated urine samples which students must analyze and match with common human disorders.

   d. *Exercise 23, Neural Coordination,* has had the sensory perception activities transferred to Exercise 24, and these have been replaced with additional human reflexes for investigation.

   e. *Exercise 33, Heredity,* has been augmented with simple "hands on" analysis of progeny in monohybrid and dihybrid crosses without involving Chi Square analysis, which is reserved for analyses performed later in the exercise.

   f. *Exercise 37, Ecological Relationships,* has been modified by simplifying the ecosystem analysis section to make the analyses easier for students.

3. Improvements have been made in nearly every exercise. These include updating information, improving the clarity of explanations and directions, reorganizing the material, improving illustrations, and rewording questions on the Laboratory Reports.

   • The *assignment numbers* within each exercise now are the same as the numbers of the section headings on the Laboratory Reports. This will make it easier for instructors to correlate the assignments with the Laboratory Reports when directing students.

   • The *tables and graphs* in which students record data have been moved to the Laboratory Reports to keep all student responses in one place for easier assessment by instructors.

   • Additional material has been added to some exercises to provide *more options* for instructors in tailoring the exercises for their courses.

   • A greater emphasis is placed on the *color-coding* of anatomical features in illustrations as a pedagogical activity. Many students seem to profit from this activity.

4. Over 25 *new or modified illustrations* are included to facilitate learning by students.

5. *Eight pages of color plates* are provided that illustrate representative monerans, protists, fungi and algae. These plates emphasize microscopic organisms and structures to help students in their study of these organismic groups.
6. The headings for the Objectives and Assignment sections of each exercise are now in *color* for easy recognition by students.

HELPING STUDENTS PREPARE FOR THE LABORATORY

Learning may be enhanced by having students carefully prepare for each laboratory session. The following activities should be done *before* students come to the laboratory session. These activities will prepare students for their laboratory study and also allow them to use the entire laboratory time for observations and experiments.

1. *Review* the objectives of the assigned exercise to learn their minimal learning responsibilities.
2. *Read* the background information to understand the topic to be studied in the laboratory and to learn the key terms in bold print.
3. *Label* any illustrations requiring labeling and *color-code* the major features of the illustrations.
4. *Complete* the items on the Laboratory Report that deal with the background information.

ACKNOWLEDGMENTS

Many adopters across the country have provided helpful suggestions for improving the sixth edition. Special recognition is due those whose critical reviews of the fifth edition provided a basis for the present revision: Glen R. Parsons, University of Mississippi; Francis X. Lobo, Marywood College; Turner M. Spencer, Thomas Nelson Community College; Edward V. Koprowski, California State University–Northridge; and Robert G. Futrell, Jr., Rockingham Community College.

I appreciate the useful suggestions made by colleagues at Pasadena City College, especially Judy Greenlee, Wendie Johnston and Mel Stehsel. In addition, Dr. Greenlee's review of the initial draft of the revision provided valuable insights and helpful suggestions.

The skills and helpfulness of the editorial and production personnel at Macmillan, especially Kristin Watts Peri, John Travis, Chris Migdol and Diane Austin, have eased the work of revision. Once again, original drawings by Mary Dersch have enhanced the illustrations in the sixth edition. I am grateful for their assistance.

Adopters are encouraged to communicate to the author or publisher helpful suggestions that will improve the usefulness of this manual for their courses.

TO THE STUDENT

Laboratory study is a major part of a course in biology. It provides an opportunity for you to study biological organisms and processes and to correlate your findings with the textbook and lectures. You will examine biological specimens, conduct experiments, collect and analyze data, and form conclusions. In this way, you will experience the process of biology, and it is this process that distinguishes science from other academic disciplines.

Many students consider the laboratory to be the most interesting and enjoyable part of the course. Of course, your enjoyment will depend on your success in the laboratory. And, your success in the laboratory depends largely on your attitude, motivation, and how you prepare for and carry out the laboratory activities.

PREPARATION

You can increase your chances of success in the laboratory by being well-prepared for each laboratory session. Completing the following activities *before* you arrive in the laboratory will strengthen your preparation.

1. Read the assigned laboratory exercise giving special attention to the *Objectives,* which identify your minimal learning responsibilities, and the *introductory background information* that you must understand in order to carry out the laboratory assignments.
2. As you read the exercise, learn the *key terms,* which are in bold print, and de-

velop a general understanding of the *procedures* to be followed in the laboratory assignments.
3. Label and color-code the *illustrations.* Coloring the major parts of the illustrations will help you learn structural characteristics and relationships.
4. Complete the items on the *laboratory report* that deal with the background information. This will not only help you learn this material, but it will allow you to devote your laboratory time to the required activities.
5. Bring your textbook to the laboratory for use as a reference.

LABORATORY ACTIVITIES

Once you are in the laboratory, the following suggestions will help you to use your time most effectively.

1. Remove the *laboratory report* for the assigned exercise from the manual so you will not have to flip pages when completing it. Keep your completed laboratory reports in a three-ring binder for future reference.
2. Read the directions in each *assignment* completely through, and be sure that you understand them, before starting a laboratory activity. Follow the directions carefully and in sequence, unless directed otherwise by your instructor.
3. Work carefully and thoughtfully. Remember that *thinking* is a vital part of your ac-

tivities. You will not have to rush if you are well prepared.

4. Complete the laboratory report thoughtfully. Just haphazardly filling in the blanks is not enough. Each laboratory report is structured to guide your learning.

5. When helpful, discuss your procedures, observations and results with your lab partners. This will improve your understanding. If you need help, ask your instructor.

LABORATORY SAFETY AND HOUSEKEEPING

These guidelines are to be followed for the safety and benefit of both you and your classmates.

1. Use laboratory equipment with care. Clean and return it to its correct location after use. Report any problems to your instructor.

2. Inform your instructor of any spills or breakage to see if there are special procedures to be followed in cleaning up.

3. Report any injuries, even minor ones, to your instructor immediately.

4. Tie back long hair and roll up loose sleeves when using open flames.

5. Do not smoke, eat, drink, or apply cosmetics in the laboratory.

6. Wash glassware after use. Return each item to its correct location after cleaning.

7. Clean your workstation at the end of the laboratory session.

PART I

SOME FUNDAMENTALS

1

ORIENTATION

Laboratory study is an important part of a course in biology. It provides opportunities for you to observe and study biological organisms and processes and to correlate your findings with the textbook and lectures. It allows the conduction of experiments, the collection of data, and the analysis of data to form conclusions. In this way, you experience the process of science, and it is this process that distinguishes science from other disciplines.

PROCEDURES TO FOLLOW

Your success in the laboratory depends on how you prepare for and carry out the laboratory activities. The procedures that follow are expected to be used by all students.

Preparation

Before coming to the laboratory, complete the following activities to prepare yourself for the laboratory session.
1. Read the assigned exercise to understand (a) the objectives, (b) the meaning and spelling of new terms, (c) the introductory background information, and (d) the procedures to be followed.
2. Label and color-code illustrations as directed in the manual, and complete the items on the laboratory report related to the background information.
3. Bring your textbook to the laboratory for use as a reference.

Working in the Laboratory

The following guidelines will save time and increase your chances of success in the laboratory.
1. Remove the laboratory report from the manual so that you can complete it without flipping pages. Laboratory reports are three-hole punched so you can keep completed reports in a binder.
2. Follow the directions explicitly and in sequence unless directed otherwise.
3. Work carefully and thoughtfully. You will not have to rush if you are well prepared.
4. Discuss your procedures and observations with other students. If you become confused, ask your instructor for help.
5. Answer the questions on the laboratory report thoughtfully and completely. They are

provided to guide the learning process. Just filling in the blanks is not acceptable.

Laboratory Safety and Housekeeping

1. Use equipment with care. Report any problems to your instructor.
2. Inform your instructor of any breakage or spills, and ask for assistance in proper cleanup and disposal procedures.
3. Immediately report any injuries, even minor ones, to your instructor.
4. Tie back long hair and roll up loose sleeves when using open flames.
5. Clean glassware and equipment before and after use. Return each item to its proper location at the end of the session, and clean your workstation.
6. Do not smoke, eat, drink, or apply cosmetics in the laboratory.

BIOLOGICAL TERMS

One of the major difficulties encountered by beginning students is learning biological terminology. Each exercise has new terms emphasized in bold print so that you do not overlook them. Be sure to know their meanings prior to the laboratory session.

Biological terms are composed of a root word, and either a prefix or a suffix, or both. The **root word** provides the main meaning of the term. It may occur at the beginning or end of the term, or it may be sandwiched between a **prefix** and a **suffix**. Both prefix and suffix modify the meaning of the root word. The parts of a term are often joined by adding *combining vowels* that make the term easier to pronounce. The following examples illustrate the structure of biological terms.

1. The term *endocranial* becomes *endo/crani/al* when separated into its components. *Endo* is a prefix meaning within; *crani* is the root word meaning skull; *al* is a suffix meaning pertaining to. Therefore, the literal meaning of endocranial is "pertaining to within the skull."
2. The term *arthropod* becomes *arthr/o/pod* when separated into its components. *Arthr*

is a prefix meaning joint; *o* is a combining vowel; *pod* is the root word meaning foot. Therefore, the literal meaning of arthropod is "jointed foot."

Once you understand the structure of biological terms, learning the terminology becomes much easier. *Appendix A contains the meaning of common prefixes, suffixes, and root words. Use it frequently to help you master new terms.*

Assignment 1

Using Appendix A, **complete item 1 on Laboratory Report 1 that begins on page 353**.

UNITS OF MEASUREMENT

In making measurements, scientists use the **International System of Units (SI)**, which is commonly called the **metric system**. The metric system is the only system of measurement used in many countries of the world. It is an easy and convenient system, once it is learned. Some of the common units are shown in Table 1.1. Study the table to become familiar with the names and values of the units. Note that within each category the units vary by powers of 10, which allows easy conversion from one unit to another. The English equivalents are given for some units for comparison and so that you can convert values from one system to the other. If you study the table, you will be able to recognize the relative value of the prefixes in the names of the units, which are summarized below.

kilo- = unit × 1000
deci- = unit ÷ 10
centi- = unit ÷ 100
milli- = unit ÷ 1000
micro- = unit ÷ 1,000,000

A major advantage of the metric system is that units within a category may be converted from one to another by multiplying or dividing by the correct power of 10. The following examples show how this is done.

1. To convert 3.25 meters to millimeters

$$3.25 \text{ m} \times \frac{1000 \text{ mm}}{1 \text{ m}} = 3250 \text{ mm}$$

TABLE 1.1
Common Metric System Units

Category	Symbol	Unit	Value	English Equivalent
Length	km	kilometer	1000 m	0.62 mi
	m	meter*	1 m	39.37 in.
	dm	decimeter	0.1 m	3.94 in.
	cm	centimeter	0.01 m	0.39 in.
	mm	millimeter	0.001 m	0.04 in.
	μm	micrometer	0.000001 m	0.00004 in.
Mass	kg	kilogram	1000 g	2.2 lb
	g	gram*	1 g	0.04 oz
	dg	decigram	0.1 g	0.004 oz
	cg	centigram	0.01 g	0.0004 oz
	mg	milligram	0.001 g	
	μg	microgram	0.000001 g	
Volume	l	liter*	1 l	1.06 qt
	ml	milliliter	0.001 l	0.03 oz
	μl	microliter	0.000001 l	

*Denotes the base unit.
Source: Table 1.1: *Human Biology: Laboratory Explorations.*

2. To convert 76 micrograms to milligrams

$$76 \ \mu g \times \frac{1 \ mg}{1000 \ \mu g} = 0.076 \ mg$$

Temperature

Scientists use the **degree Celsius** (°C) as the base unit for temperature measurements, whereas the degree **Fahrenheit** (°F) is used by most nonscientists in the United States. A simple comparison of the two systems is shown below.

	°C	°F
Boiling point of water	100	212
Freezing point of water	0	32

You can convert from one system to the other by using these formulas:

Celsius to Fahrenheit

$$°F = \left(\frac{9}{5}\right)(°C) + 32$$

Fahrenheit to Celsius

$$°C = \frac{5}{9}(°F - 32)$$

Materials

Balance
Beaker, 250 ml
Graduated cylinder, 25 ml
Millimeter ruler, clear plastic

Assignment 2

1. *Complete item 2a on the laboratory report.*
2. Perform the procedures that follow and record your calculations and data in item 2 on the laboratory report.
   a. Determine the diameter of a penny in millimeters. Then convert your answer to centimeters, meters, and inches.
   b. Using a balance, determine the weight (mass) of a 250-ml beaker in grams. Then, convert your answer to milligrams and ounces. If you need help in using the balance, see your instructor.
   c. Using, a graduated cylinder, beaker, and balance, determine the weight (mass) of 20 ml of water. When measuring the volume of water, be sure to read the top of the water column at the bot-

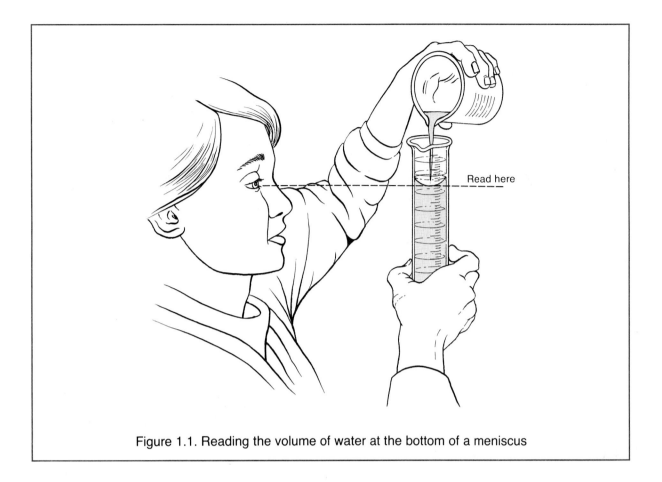

Figure 1.1. Reading the volume of water at the bottom of a meniscus

tom of the meniscus as shown in Figure 1.1.

3. **Complete item 2 on the laboratory report**.

OBSERVATIONS

In the laboratory, you will be asked to make many observations, and these observations require both seeing and thinking. The process of biology involves making careful observations to collect data, analyzing the data, and forming conclusions.

For example, careful observations are necessary to identify biological organisms. Each species (kind of organism) has certain characteristics that are similar to related organisms and characteristics that are different from related organisms. The greater the similarity between organisms, the closer is their relationship. Biologists use this principle of similarity to determine the closeness of relationship between organisms. You have subconsciously used this same principle in recognizing the various breeds of dogs as dogs and in distinguishing between dogs and cats.

When classifying organisms, biologists develop a **dichotomous key**, based on the characteristics of the organisms, to separate a variety of organisms according to species. A dichotomous key is constructed so that a single characteristic is considered at each step, an either/or decision, to separate the organisms into *two groups*. A series of these either/or decisions ultimately separates each type of organism from other types. For example, if we were to develop a dichotomous key to distinguish dogs, cats, monkeys, gorillas, and humans (while ignoring all other organisms), it might look like this:

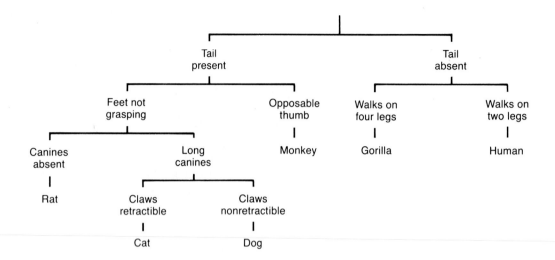

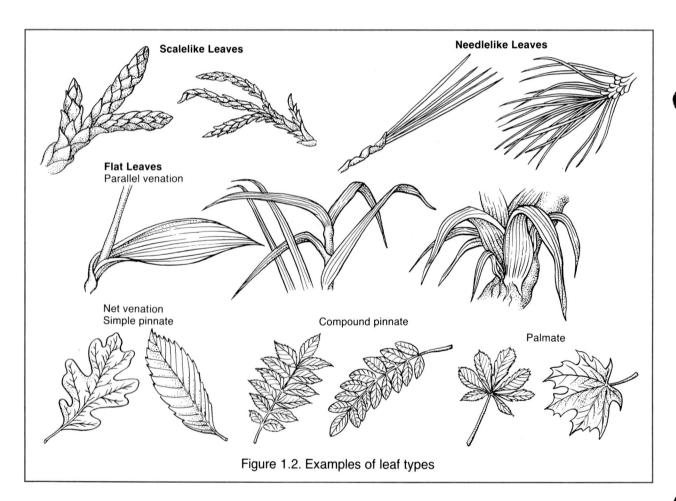

Figure 1.2. Examples of leaf types

This dichotomous key is purely artificial, but it illustrates how one is constructed. *Note that only one distinguishing characteristic is considered at each step.*

Materials

Leaves of various types

Assignment 3

1. There are several sets of 8 to 10 different types of leaves provided in the laboratory for your examination. Each leaf is numbered. Your objective is to develop a dichotomous key to categorize the leaves of one set according to their observable characteristics: To develop such a key, you must first observe the leaves carefully to determine their similarities and distinguishing characteristics. You are not trying to identify the names of the plants from which the leaves were obtained. Instead, you are to record the number of the leaf that fills the position at the end of each final branch of your key.

In making your key, consider leaf characteristics such as presence or absence of a petiole (leaf stalk), the arrangement of leaf veins (parallel or netlike), general shape (needlelike, scalelike, or flat, thin leaves), simple or compound leaves, shape of the leaf margin (e.g., smooth, toothed, indented), and so forth. Figure 1.2 illustrates certain characteristics of leaves to help you get started.

There is no "one correct way" to make this dichotomous key. It can be made in several ways, but it will challenge your ability to recognize the distinguishing characteristics of the leaves.

2. **Construct your dichotomous key in item 3 of the laboratory report and complete the laboratory report.**

2

THE MICROSCOPE

A **microscope** is a precision instrument and an essential tool in the study of cells, tissues, and minute organisms. It must be handled and used carefully at all times. Most of the microscopic observations in this course will be made with a **compound microscope**, but a **dissecting microscope** will be used occasionally. A microscope consists of a lens system, a controllable light source, and a mechanism for adjusting the distance between the objective lens and the object to be observed.

To make the observations required in this course, you must know how to use a microscope effectively. This exercise provides an opportunity for you to develop skills in microscopy.

THE COMPOUND MICROSCOPE

The major parts of the compound microscope are shown in Figure 2.1. Refer to this figure as you read this text, and *label the parts* indicated on the figure. Your microscope may be somewhat different than the one illustrated.

The **base** rests on the table and, in most microscopes, contains a built-in **light source** and a **light switch**. Some microscopes have a light intensity (voltage) control knob on the base, usually associated with the light switch. The **arm** rises from the base and supports the stage, lens system, and control mechanisms. The **stage** is the flat surface on which microscope slides are placed for viewing. **Stage clips**, or a **mechanical stage**, hold the slide in place.

Most microscopes have a **condenser** located below the stage. It concentrates the light on the object and may be raised or lowered by the **condenser control knob**. Usually, the condenser should be raised to its highest position. An **iris diaphragm** is built into the base of the condenser. The **iris diaphragm control lever** (a rotatable wheel in some microscopes) varies the amount of light entering the condenser and the lens system.

The **body tube** is supported by the arm and has an **ocular lens** at the upper end and a **revolving nosepiece** with the attached **objective lenses** at the lower end. The nosepiece is rotated to bring different objectives into viewing position. The objectives usually click into viewing position.

Student microscopes usually have three objec-

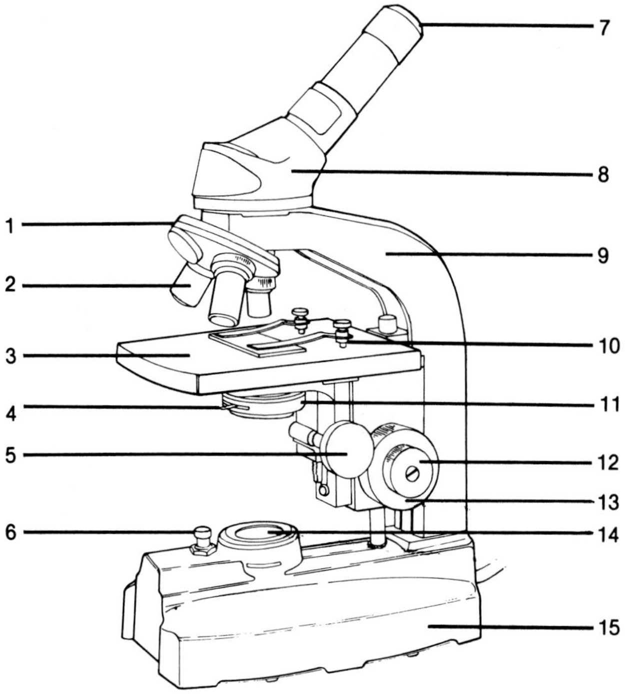

7	_____ Arm
	_____ Base
	_____ Body tube
8	_____ Coarse-focusing knob
	_____ Condenser
9	_____ Condenser control knob
	_____ Fine-focusing knob
10	_____ Iris diaphragm lever
11	_____ Light source
12	_____ Light switch
13	_____ Objective
14	_____ Ocular
	_____ Revolving nosepiece
15	_____ Stage
	_____ Stage clamp

Figure 2.1. Compound microscope

tives. The shortest is the **scanning objective**, which has a magnification of 4×. The **low-power objective** with a 10× magnification is intermediate in length. The **high-power** (high-dry) **objective** is the longest and usually has a magnification of 40×, but may have a 43× or 45× magnification in some microscopes. In this manual, the high-power objective is often called the 40× objective. Some microscopes have an **oil-immersion objective** (100× magnification) that is a bit longer than the high-power objective.

There are two focusing knobs. The **coarse-focusing knob** has the larger diameter and is used to bring objects into rough focus when using the 4× and 10× objectives. The **fine-focusing knob** has a smaller diameter and is used to bring objects into fine focus. It is the *only* focusing knob used with the high-power and oil-immersion objectives.

Magnification

The magnification of each lens is fixed and inscribed on the lens. The ocular usually has a 10× magnification. The powers of the objectives may vary but usually are 4×, 10× and 40×. The **total magnification** is calculated by multiplying the power of the ocular by the power of the objective.

Resolving Power

The quality of a microscope depends on its ability to **resolve** (distinguish) objects. Magnification without resolving power is of no value. Modern microscopes increase both magnification and resolution by a careful matching of light source and precision lenses. Most microscopes have a blue light filter located in either the condenser or the light source since resolving power increases as the wavelength of light decreases.

Student microscopes usually can resolve objects that are 0.5 μm or more apart. The best light microscopes can resolve objects that are 0.1 μm or more apart.

Contrast

Sufficient **contrast** must be present among the parts of an object for the parts to be distinguishable. Contrast results from the differential absorption of light by the parts of the object. Sometimes, stains must be added to a specimen to increase the contrast. A reduction in the amount of light improves contrast when viewing unstained specimens.

Focusing

A microscope is focused by increasing or decreasing the distance between the specimen on the slide and the objective lens. The focusing procedure used depends on whether your microscope has a **movable stage** or **movable body tube** (see Figure 2.2). Both procedures are described below. Use the one appropriate for your microscope. As a general rule, you should start focusing with the low-power (10×) objective unless the large size of the object requires starting with the 4× objective.

Focusing with a Movable Body Tube

1. Rotate the 10× objective into viewing position.
2. While using the coarse-focusing knob and *looking from the side* (not through the ocular), lower the body tube until it stops or until the objective is about 3 mm from the slide.
3. While looking through the ocular, slowly raise the body tube by turning the coarse-focusing knob toward you until the object

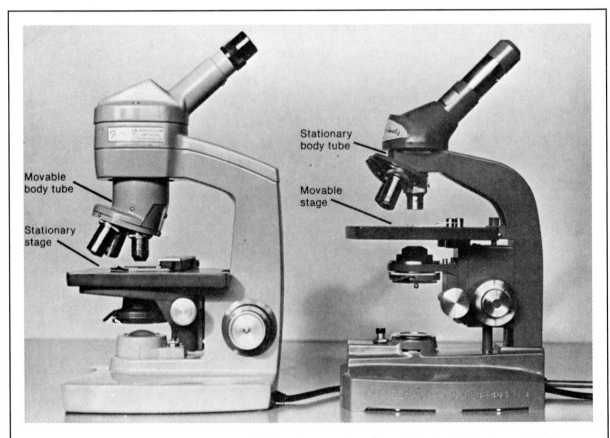

Figure 2.2. Comparison of two microscope designs

becomes visible. Use the fine-focusing knob to bring the object into sharp focus.

Focusing with a Movable Stage

1. Rotate the $10\times$ objective into viewing position.
2. While using the coarse-focusing knob and *looking from the side* (not through the ocular), raise the stage to its highest position or until the slide is about 3 mm from the objective.
3. While looking through the ocular, slowly lower the stage by turning the coarse-focusing knob away from you until the object comes into focus. Use the fine-focusing knob to bring the object into sharp focus.

Switching Objectives

Your microscope is **parcentric** and **parfocal**. This means that if an object is centered and in sharp focus with one objective, it will be centered and in focus when another objective is rotated into the viewing position. However, slight adjustments to recenter and refocus (with the fine-focusing knob) may be necessary. As you switch objectives from $4\times$ to $10\times$ to $40\times$ to increase magnification, the (1) working distance, (2) diameter of the field, and (3) light intensity are *reduced* as magnification increases. Note this relationship in Figure 2.3.

Slide Preparation

Specimens to be viewed with a compound microscope are placed on a **microscope slide** and are usually covered with a **cover glass**. Specimens may be mounted on slides in two different ways. A **prepared slide** (permanent slide) has a permanently attached cover glass, and the specimen is usually stained. A **wet-mount slide** (temporary slide) has the specimen mounted in a liquid, usually water, and covered with a cover glass. In this course, you will observe commercially prepared permanent slides and wet-mount slides that you will make. Wet-mount slides are prepared as shown in Figure 2.4.

Care of the Microscope

You should carry a microscope upright in front of you, not at your side. Use one hand to support

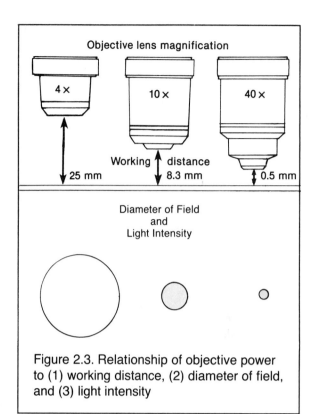

Figure 2.3. Relationship of objective power to (1) working distance, (2) diameter of field, and (3) light intensity

the base and the other to grasp the arm. See Figure 2.5. Develop the habit of cleaning the lenses prior to using the microscope. Use only special lint-free lens paper. If the lens paper does not clean the lenses, inform your instructor. If any liquid gets on the lenses during use, wipe it off immediately and clean the lenses with lens paper.

When you are finished using the microscope, perform these steps:

1. Remove the slide. Clean and dry the stage.
2. Clean the lenses with lens paper.
3. Rotate the nosepiece so that no objective projects beyond the front of the stage.
4. Raise the stage to its highest position *or* lower the body tube to its lowest position in accordance with the type of microscope you are using.
5. Unplug the light cord and loosely wrap it around the arm below the stage. Add a dustcover, if present.
6. Return the microscope to the correct cabinet cubicle.

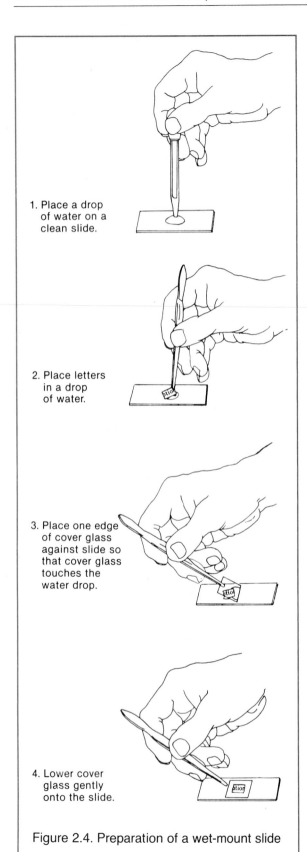

1. Place a drop of water on a clean slide.

2. Place letters in a drop of water.

3. Place one edge of cover glass against slide so that cover glass touches the water drop.

4. Lower cover glass gently onto the slide.

Figure 2.4. Preparation of a wet-mount slide

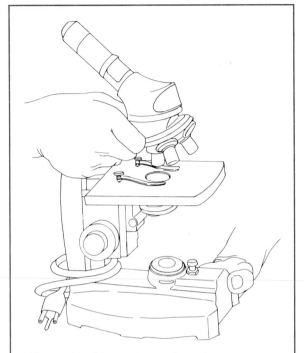

Figure 2.5. How to carry the microscope.
Grasp the arm of the microscope with one hand and support the base with the other.

Assignment 1

1. Label Figure 2.1.
2. Obtain the microscope assigned to you. Carry it as described above and place it on the table in front of you. Locate the parts shown in Figure 2.1. Clean the lenses with lens paper. Try the knobs and levers to see how they work.
3. Raise the condenser to its highest position and keep it there. Plug in the light cord and turn on the light. If your microscope has a voltage control knob, adjust it to an intermediate position to prolong the life of the bulb.
4. Rotate the 4× objective into viewing position and look through the ocular. The circle of light that you see is called the **field of view** or simply the **field**.
5. While looking through the ocular, open and close the iris diaphragm and note the change in light intensity. Repeat for each objective and note that light intensity decreases as the power of the objective increases. Therefore, you will need to adjust

the light intensity when you switch objectives. *Remember to use reduced light intensity when you are viewing unstained and rather transparent specimens.*

6. If your microscope has a voltage control knob, repeat item 5 while leaving the iris diaphragm open but changing the light intensity by altering the voltage.

7. ***Complete item 1 on Laboratory Report 2 that begins on page 357.***

Developing Microscopy Skills

The following microscopic observations are designed to help you develop skill in using a compound microscope.

Materials

Dissecting instruments
Kimwipes
Medicine droppers
Metric ruler, clear plastic
Microscope slide and cover glass
Newspaper
Water in dropping bottle
Pond water
Prepared slide of fly wing

Assignment 2

1. Obtain a microscope slide and cover glass. If they are not clean, wash them with soap and water, rinse, and dry them. Use a paper towel to dry the slide, but use Kimwipes to blot the water from the fragile cover glass.

2. Use scissors to cut three letters from a newspaper with the letter *i* as the middle letter.

3. Prepare a wet-mount slide of the letters as shown in Figure 2.4. Use a paper towel to soak up any excess water. If too little water is present, add a drop at the edge of the cover glass and it will flow under the cover glass.

4. Place the slide on the stage with the letters over the stage aperture, the circular opening in the stage. Secure it with either a mechanical stage or stage clamps. See Figure 2.6. The slide should be parallel to the edge of the stage nearest you with the letters oriented so that they may be read with the naked eye.

5. Rotate the 10× objective into viewing position and bring the letters into focus using the focusing procedure described above that is appropriate for your microscope.

6. Rotate the 4× objective into the viewing position. Center the letter *i* and bring it into sharp focus. Can you see all of the *i*? Can you see the other letters? What is different about the orientation of the letters when viewed with the microscope instead of the naked eye?

7. Move the slide to the left while looking through the ocular. Which way does the image move? Practice moving the slide while viewing through the ocular until you can quickly place a given letter in the center of the field.

8. Center the *i* and bring it into sharp focus. Rotate the 10× objective into position. Is the *i* centered and in focus? If not, center it and bring it into sharp focus. How much of the *i* can you see?

9. Rotate the 40× objective into position. Is it centered and in focus? All that you can see at this magnification are "ink blotches" that compose the *i*. If you do not see this, center the *i* and bring it into focus with the *fine-focusing knob. Never use*

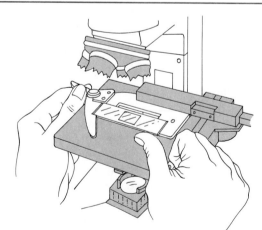

Figure 2.6. Placing the slide in the mechanical stage.

The retainer lever is pulled back to place the slide against the stationary arm. Then, the retainer lever is released to secure the slide.

the coarse-focusing knob with the high-power objective.

10. Practice steps 1 through 4 until you can quickly center the dot of letter *i* and bring it into focus with each objective. **Remember**, you are *never* to start observations with the high-power objective. Instead, start at a lower power and work up to the 40× objective.

11. ***Complete item 2 on the laboratory report***.

12. Remove the slide and set it aside.

Depth of Field

When you view objects with a microscope, you obviously are viewing the objects from above. The vertical distance within which structures are in sharp focus is called the **depth of field**, and it decreases as magnification increases. You will learn more about depth of field by performing the observations that follow.

Assignment 3

1. Obtain a prepared slide of a fly wing. Observe the tiny spines on the wing membrane with each objective starting with the 4× objective. Can you see all of a spine at each magnification?

2. Using the 4× objective, locate a large spine at the base of the wing where the veins converge and center it in the field. Can you see all of it?

3. Rotate the 10× objective into position and observe the spine. Can you see all of it? Practice focusing up and down the length of the spine, and note that you can see only a portion of the spine at each focusing position.

4. Rotate the 40× objective into position and observe the spine. At each focusing position, you can see only a thin "slice" of the spine. To determine the spine's shape, you have to focus up and down the spine using the *fine-focusing knob*.

5. ***Complete item 3 on the laboratory report***.

The preceding observations demonstrate that when viewing objects with a greater depth (thickness) than the depth of field, you see only a two-dimensional plane "optically cut" through the object. To discern an object's three-dimen-

sional shape, a series of these images must be "stacked up" in your mind as you focus through the depth of the object.

Diameter of Field

When using each objective, you must know the diameter of the field to estimate the size of observed objects. Estimate the diameter of field for each magnification of your microscope as described in the section that follows.

Assignment 4

1. Place the clear, plastic ruler on the microscope stage as shown in Figure 2.7. The edge of the ruler should extend across the

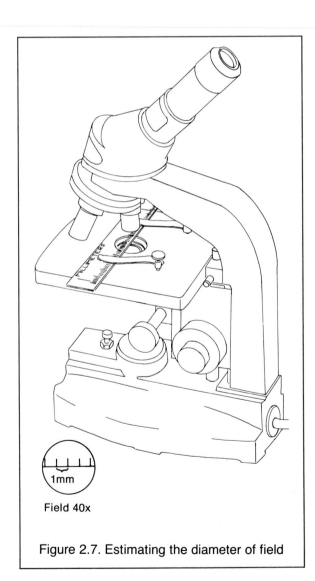

Field 40x

Figure 2.7. Estimating the diameter of field

diameter of the field. Focus on the metric scale with the 4× objective, and adjust the ruler so that one of the millimeter marks is at the left edge of the field. Estimate and record the diameter of field at 40× by counting the spaces and portions thereof between the millimeter marks.

2. Use these equations to calculate the diameter of field at 100× and 400×:

$$\text{Diam. (mm) at } 100\times = \frac{40\times}{100\times} \times \frac{\text{diam. (mm) at } 40\times}{1}$$

$$\text{Diam. (mm) at } 400\times = \frac{40\times}{400\times} \times \frac{\text{diam. (mm) at } 40\times}{1}$$

3. Return to your slide of newspaper letters, and estimate the diameter of the dot of the letter *i* and the length of the letter *i* including the dot.

4. ***Complete item 4 on the laboratory report.***

Application of Microscopy Skills

In this section, you will use the skills and knowledge gained in the preceding portions of the exercise.

Assignment 5

1. Prepare a wet-mount slide of two crossed hairs, one blond and the other brunette. Obtain 1-cm lengths of hair from cooperative classmates.
2. Using the 4× objective, center the crossing point of the hairs in the field and observe. Are both hairs in sharp focus?
3. Examine the crossed hairs at 100× magnification. Are both hairs in sharp focus? Determine which hair is on top by using focusing technique. If you have trouble with this, see your instructor.
4. Examine the crossed hairs at 400× magnification. Are both hairs in focus? Move the crossing point to one side and focus on the blond hair. Using focusing technique, observe surface and optical midsection views of the hair as shown in Figure 2.8. Note that you cannot see both views at the same time. What does this tell you about the depth of field in relationship to the diameter of the hair?
5. ***Complete items 5a to 5e on the laboratory report.***
6. Make slides of the pond water samples and

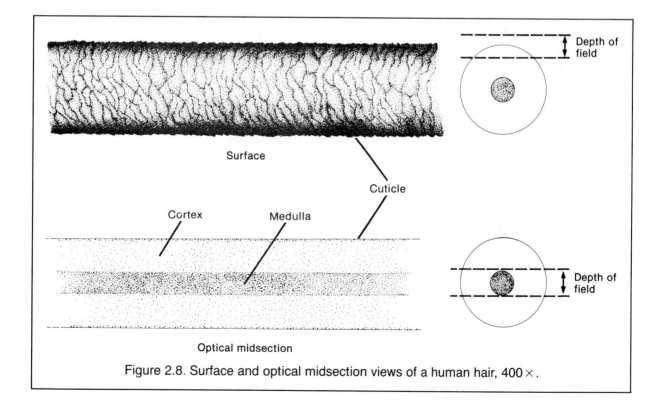

Figure 2.8. Surface and optical midsection views of a human hair, 400×.

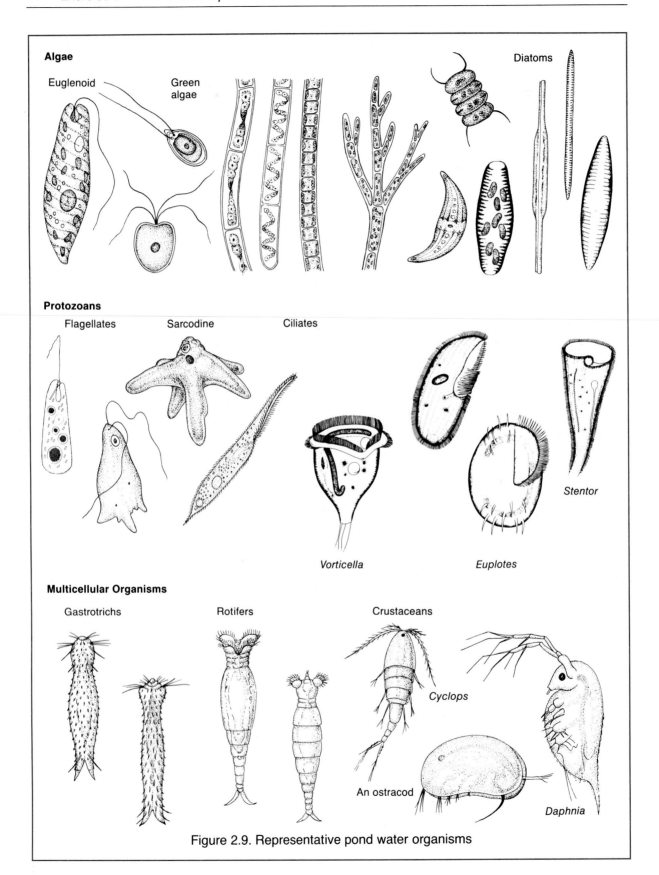

Algae

Euglenoid

Green algae

Diatoms

Protozoans

Flagellates

Sarcodine

Ciliates

Vorticella

Euplotes

Stentor

Multicellular Organisms

Gastrotrichs

Rotifers

Crustaceans

Cyclops

An ostracod

Daphnia

Figure 2.9. Representative pond water organisms

examine them microscopically. The object is to sharpen your microscopy skills rather than to identify the organisms. However, you may see organisms like those shown in Figure 2.9. Note the size, color, shape, and motility of the organisms. **Draw a few of the organisms in the space for item 5f on the laboratory report**.

7. Prepare your microscope for return to the cabinet as described previously, and return it.

THE DISSECTING MICROSCOPE

A **dissecting microscope** is used to view objects that are too large or too opaque to observe with a compound microscope. The two oculars enable stereoscopic observations and usually are 10× in magnification. Most student models have two objectives that provide 2× and 4× magnification so that total magnification is 20× and 40×. Some models have a zoom feature that enables observations at intermediate magnifications. Objects are usually viewed with reflected light instead of transmitted light, although some dissecting microscopes provide both types of light sources.

The parts of a dissecting microscope are shown in Figure 2.10. Note the single focusing knob and the two oculars. The oculars may be moved inward or outward to adjust for the distance between the pupils of your eyes. One ocular has a focusing ring that may be adjusted to accommodate for differences in visual acuity in your eyes.

Materials

Coin
Desk lamp
Metric ruler, clear plastic

Assignment 6

1. Obtain a dissecting microscope from the cabinet, and locate the parts shown in Figure 2.10.
2. Place a coin or other object on the stage, illuminate it with a desk lamp, and exam-

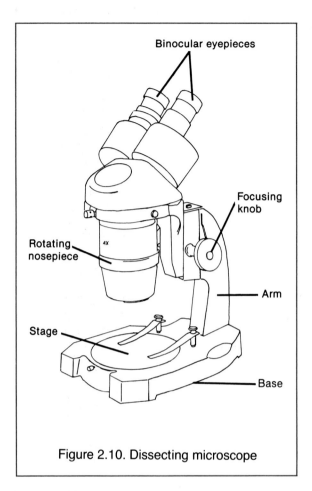

Figure 2.10. Dissecting microscope

ine it with both objectives. To accommodate for differences in acuity in your eyes, focus first with the focusing knob while viewing through the ocular that *cannot* be individually adjusted. Then use the **focusing ring** on the other ocular to bring the object into sharp focus for that eye.

3. Practice focusing until you are good at it. Move the coin on the stage, noting the direction in which the image moves. Examine the grooves and ridges that produce your unique fingerprint.
4. Determine the diameter of field at each magnification.
5. *Complete items 6 and 7 on the laboratory report*.

THE CELL

Living organisms exhibit two fundamental characteristics that are absent in nonliving things: **self-maintenance** and **self-replication**. The smallest unit of life that exhibits these characteristics is a single living cell. Thus, a single cell is the *structural and functional unit of life*:
1. All organisms are composed of cells.
2. All cells arise from preexisting cells.
3. All hereditary components of organisms occur in cells.

Two different types of cells occur in the biotic world: prokaryotic cells and eukaryotic cells. Prokaryote cells are rather primitive and occur only in bacteria and cyanobacteria. Eukaryotic cells compose all other organisms. Your study in this exercise will emphasize the structure of eukaryotic cells.

Prokaryotic Cells

Prokaryotic cells lack a nucleus and membrane-bound organelles. A single circular chromosome is located in an irregular **DNA (deoxyribonucleic acid) region** in the interior of the cell. A small amount of **cytoplasm** containing **ribosomes** lies between the DNA region and the **cell membrane**. A supportive and protective **cell wall** is secreted just exterior to the cell membrane. See Figure 3.1 and Plate 1, which is located following page 52.

Materials

Prepared slides of bacteria and cyanobacteria

Assignment 1

1. Examine the prepared slides of bacteria and cyanobacteria at 400× total magnification. Can you detect the cellular components?
2. *Complete item 1 on Laboratory Report 3 that begins on page 361.*

Eukaryotic Cells

As you read the following descriptions of cell anatomy, locate and label the organelles on Figures 3.2 and 3.3 and note their functions in Table 3.1. The structures of the cellular organelles are shown as viewed with an electron microscope; most organelles are too small to be seen with your microscope.

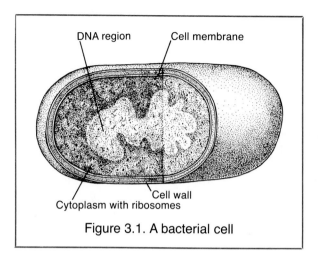

DNA region   Cell membrane

Cell wall
Cytoplasm with ribosomes

Figure 3.1. A bacterial cell

TABLE 3.1
Functions of the Major Cellular Organelles

Organelle	Function
Cell membrane	Controls the passage of materials into and out of the cell: maintains the integrity of the cell.
Chromosomes	Contain the genetic code, in DNA molecules, that controls the life processes of the cell.
Centrioles	Form microtubule systems, including the spindle in dividing animal cells (absent in plant cells).
Endoplasmic reticulum	Forms channels for the movement of materials throughout the cell and surfaces for chemical reactions.
Golgi apparatus	Stores, modifies, and packages materials for export from the cell.
Lysosomes	Contain strong digestive enzymes that are released to digest old and damaged cells (absent in plant cells).
Microfilaments	Provide support for the cytoplasm.
Microtubules	Provide support for the cytoplasm; form the spindle in dividing animal cells and flagella in flagellated cells.
Mitochondria	Sites of cellular respiration that releases energy from nutrients to form adenosine triphosphate (ATP).
Nuclear envelope	Controls the passage of materials between the nucleus and the cytoplasm.
Nucleolus	Assembles precursors (RNA and protein) of ribosomes prior to their export to the cytoplasm.
Nucleus	Control center of the cell because it contains DNA that controls cellular functions and inheritance.
Plastids	Chloroplasts are sites of photosynthesis. Chromoplasts contain pigments that give colorations to flowers and fruits. Leukoplasts are often sites of starch storage.
Secretory vesicles	Carry substances to the cell membrane for release from the cell.
Vacuoles	Contain a variety of substances.

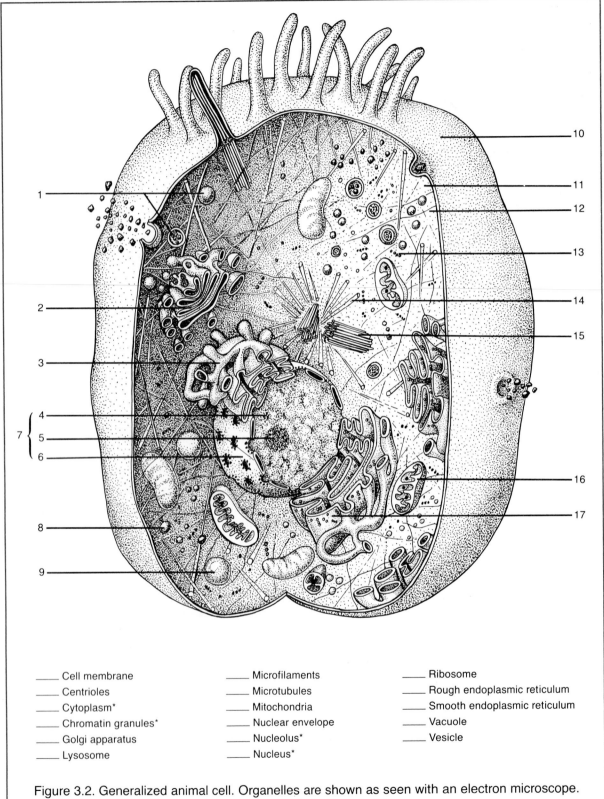

____ Cell membrane
____ Centrioles
____ Cytoplasm*
____ Chromatin granules*
____ Golgi apparatus
____ Lysosome

____ Microfilaments
____ Microtubules
____ Mitochondria
____ Nuclear envelope
____ Nucleolus*
____ Nucleus*

____ Ribosome
____ Rough endoplasmic reticulum
____ Smooth endoplasmic reticulum
____ Vacuole
____ Vesicle

Figure 3.2. Generalized animal cell. Organelles are shown as seen with an electron microscope. Organelles visible with your microscope are noted with asterisks.

The Animal Cell

Animal cells are surrounded by a **cell (plasma) membrane** that is composed of two back-to-back phospholipid layers and associated proteins. Intracellular membranes have the same structure. The bulk of the cell consists of **cytoplasm**, the semifluid or gel-like substance in which the nucleus and other **organelles** are embedded. It is supported by a lattice formed of very fine **microfilaments** and slightly larger **microtubules** that form the cytoskeleton.

The **nucleus** is a large, spherical organelle that contains the **chromosomes**. In nondividing cells, the chromosomes are uncoiled and elongated so that only bits and pieces of them may be seen as **chromatin granules** at each focal plane when viewing the nucleus with a microscope. In dividing cells, the chromosomes coil tightly and appear as dark-staining rod-shaped structures. The spherical, dark-staining structure in the nucleus is the **nucleolus**, which is composed of rubonucleic acid (RNA) and protein. The nucleus is surrounded by a **nuclear envelope** that is composed of two adjacent membranes perforated by pores. The pores enable the movement of materials between the nucleus and cytoplasm.

The outer membrane of the nuclear envelope is continuous with the **endoplasmic reticulum (ER)**, a series of folded membranes that permeate the cytoplasm. **Smooth ER** lacks ribosomes; **rough ER** is studded with ribosomes (shown as dots in the figures). **Ribosomes** are tiny organelles consisting of RNA and protein that may occur singly, in clusters, or in chains. They are located either free in the cytoplasm or on the rough ER. The **Golgi apparatus** is a stack of membranes associated with the ERs usually near the nucleus.

A pair of **centrioles**, short cylindrical bodies composed of microtubules, are oriented perpendicular to each other near the nucleus. **Mitochondria** are elongated structures formed of a larger, folded, inner membrane surrounded by a smaller, nonfolded membrane. Mitochondria are more abundant in cells with a high metabolic rate. **Vacuoles**, small fluid-filled spaces enveloped by a membrane, may be scattered in the cytoplasm. **Lysosomes** are small vacuoles that contain powerful digestive enzymes. **Secretory vesicles** are very small vacuoles that carry materials to the cell membrane for export.

The Plant Cell

Plant cells contain all of the organelles found in animal cells, except centrioles and lysosomes. Mature plant cells are characterized by the presence of cell walls, plastids, and a central vacuole. A rigid **cell wall**, formed of cellulose, is located just exterior to the cell membrane. **Plastids** are enveloped by a double membrane and are classified according to the pigments that they contain. **Chloroplasts** contain chlorophyll and carotenes and are the only plastids shown in Figure 3.3. **Chromoplasts** contain various red, orange, or yellow pigments. **Leukoplasts** contain no pigments and are colorless.

Immature plant cells have numerous small vacuoles, but in mature cells they combine to form a large **central vacuole** that constitutes much of the cell volume.

Assignment 2

1. Label Figures 3.2 and 3.3 and color-code the organelles.
2. ***Complete item 2 on the laboratory report***.

MICROSCOPIC STUDY

In this section, you will prepare wet-mount slides for the study of cell structure.

Materials

Dissecting instruments
Kimwipes
Medicine droppers
Microscope slides and cover glasses
Toothpicks, flat
Dropping bottles of:
   iodine (I_2 + KI) solution
   methylene blue, 0.01%
   sodium chloride, 0.9%
Amoeba proteus culture
Elodea shoots
Onion bulb, red

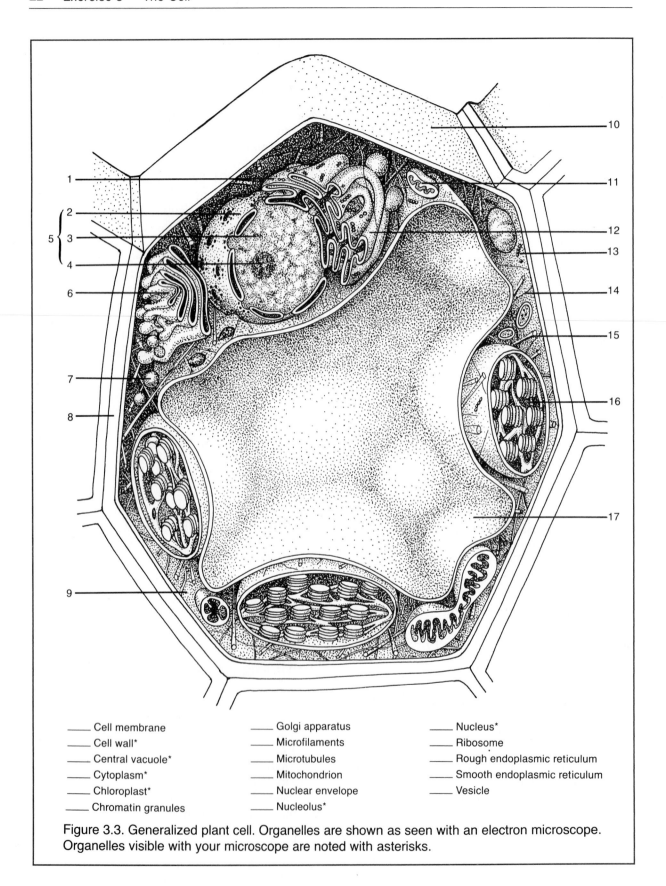

Figure 3.3. Generalized plant cell. Organelles are shown as seen with an electron microscope. Organelles visible with your microscope are noted with asterisks.

_____ Cell membrane
_____ Cell wall*
_____ Central vacuole*
_____ Cytoplasm*
_____ Chloroplast*
_____ Chromatin granules

_____ Golgi apparatus
_____ Microfilaments
_____ Microtubules
_____ Mitochondrion
_____ Nuclear envelope
_____ Nucleolus*

_____ Nucleus*
_____ Ribosome
_____ Rough endoplasmic reticulum
_____ Smooth endoplasmic reticulum
_____ Vesicle

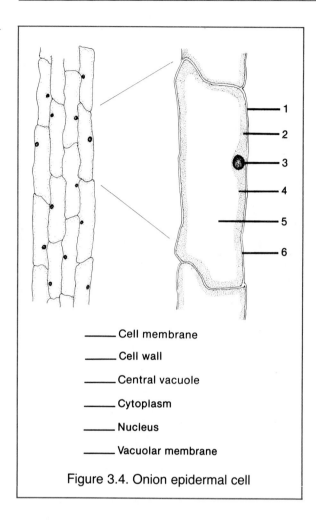

_____ Cell membrane

_____ Cell wall

_____ Central vacuole

_____ Cytoplasm

_____ Nucleus

_____ Vacuolar membrane

Figure 3.4. Onion epidermal cell

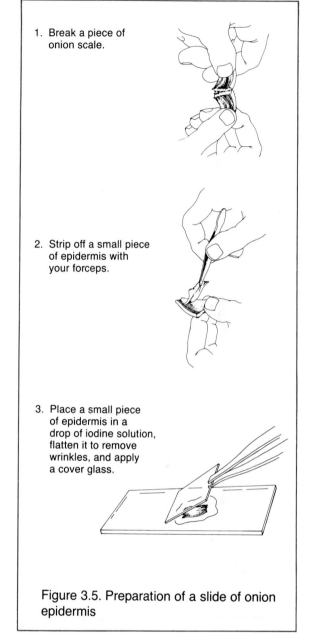

1. Break a piece of onion scale.

2. Strip off a small piece of epidermis with your forceps.

3. Place a small piece of epidermis in a drop of iodine solution, flatten it to remove wrinkles, and apply a cover glass.

Figure 3.5. Preparation of a slide of onion epidermis

Onion Epidermal Cells

Epidermal cells of an onion scale show many of the features found in nongreen plant cells, and are excellent subjects to use in beginning your study of cells.

Assignment 3

1. Label Figure 3.4 using information from the previous section.
2. Prepare a wet-mount slide of the inner epidermis of an onion scale as shown in Figure 3.5. Use a drop of iodine solution as the mounting fluid and add a cover glass.
3. Examine the cells at $100\times$ and $400\times$. Note the arrangement of the cells. Locate the parts shown in Figure 3.4.
4. Prepare a wet-mount slide of the red epidermis from the outer surface of the onion

scale. The color is due to **anthocyanin**, a water-soluble pigment in the central vacuole. Note the size of the vacuole and the location of the cytoplasm and nucleus.
5. **_Complete item 3 on the laboratory report_**.

Elodea Leaf Cells

Elodea is a water plant with simple leaves composed of cells that are easy to observe and that exhibit the characteristics of green plant cells.

Assignment 4

1. Label Figure 3.6.
2. Prepare a wet-mount slide of an *Elodea* leaf in this manner:
   a. Use forceps to remove a young leaf from near the tip of the shoot and mount it in a drop of water.
   b. Add a cover glass and observe at 40× and 100×.
3. Note the arrangement of the cells and the "spine" cells along the edge of the leaf. Locate a light green area for study. The thickness of the leaf is composed of more than one layer of cells. Switch to the 40× objective and focus through the thickness of the leaf. Determine the number of cell layers present.

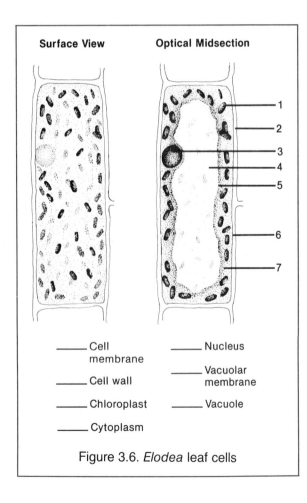

Surface View	Optical Midsection

_____ Cell membrane   _____ Nucleus

_____ Cell wall   _____ Vacuolar membrane

_____ Chloroplast   _____ Vacuole

_____ Cytoplasm

Figure 3.6. *Elodea* leaf cells

4. Using the 40× objective, focus through the depth of a cell and locate as many parts shown in Figure 3.6 as possible. The nucleus is spherical and slightly darker than the cytoplasm.
5. Focus carefully to observe surface and optical midsection views. See Figure 3.6. Note the basic shape of a cell.
6. Examine a spine cell at the edge of the leaf with reduced illumination to locate the nucleus, vacuole, and cytoplasm uncluttered with chloroplasts.
7. ***Complete item 4 on the laboratory report***.

Human Epithelial Cells

The epthelial cells lining the inside of your mouth are easily obtained for study, and they exhibit some of the characteristics of animal cells.

Assignment 5

1. Label Figure 3.7.
2. Prepare a wet-mount slide of human epithelial cells as shown in Figure 3.8, using 0.9% sodium chloride (NaCl) as the mounting fluid.
3. Add a cover glass and observe at 100× and 400×. What cell structures are present? Note the differences between these cells and the plant cells.
4. ***Complete item 5 on the laboratory report***.

The *Amoeba*

The common freshwater protozoan *Amoeba* exhibits many characteristics of animal cells, and it is large enough for you to see the cellular structure rather well. The nucleus, cytoplasm, vacuoles, and cytoplasmic granules are readily visible. The flowing movement of the *Amoeba* allows it to capture minute organisms that are digested in **food vacuoles**. The *Amoeba* also has **contractile vacuoles** that maintain its water balance by collecting and pumping out excess water.

Ectoplasm, the clear outer portion of the cytoplasm, is located just interior to the cell mem-

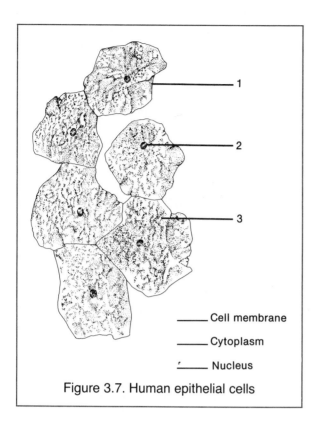

_____ Cell membrane

_____ Cytoplasm

_____ Nucleus

Figure 3.7. Human epithelial cells

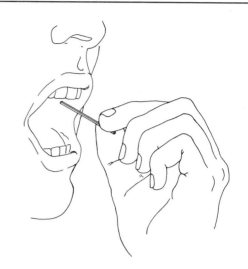

1. Gently scrape inside of mouth with a toothpick to obtain epithelial cells.

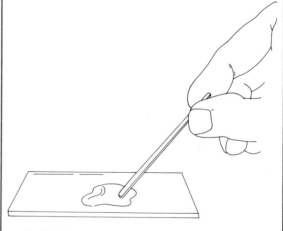

2. Swirl the toothpick in a drop of 0.9% NaCl on a clean slide.

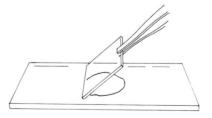

3. Gently apply a cover glass.

Figure 3.8. Preparation of a slide of human epithelial cells

brane. Most of the cytoplasm consists of the granular **endoplasm**. It contains the organelles and may be either gel-like, the **plasmagel**, or fluid, the **plasmasol**. Reversible changes between plasmagel and plasmasol result in the flowing **amoeboid movement** of the *Amoeba*. Your white blood cells also exhibit amoeboid movement as they slip through capillary walls and wander among the body tissues engulfing disease-causing organisms and cellular debris.

Assignment 6

1. Use a medicine dropper to obtain fluid *from the bottom* of the culture jar and place a drop on a clean slide. Do *not* add a cover glass.
2. Use the 10× objective to locate and observe an *Amoeba*. Note the manner of movement and locate the nucleus, cytoplasm, and vacuoles. The nucleus is spherical and a bit darker than the granular cytoplasm. Contractile vacuoles appear as

spherical bubbles in the cytoplasm, while food vacuoles contain darker food particles within the vacuolar fluid. Compare your specimen with that shown in Figure 3.9 and Plate 2.5. Observe the way the *Amoeba* moves while remembering that some of your white blood cells move in a similar way.

3. Note the appearance of the cytoplasm. Does the appearance of the inner and outer portions differ? Observe the nucleus and vacuoles. Gently add a cover glass to your slide and observe an *Amoeba* at 400×. When the pressure of the cover glass becomes too great, the cell membrane will rupture, and the contents of the cell will spill out, killing the *Amoeba*.

4. ***Complete item 6 on the laboratory report***.

THE ARRANGEMENT OF CELLS

Organisms may exhibit unicellular, colonial, or multicellular forms of body organization. **Unicellular organisms** consist of a single cell that performs all of the functions of the organism. **Colonial organisms** are composed of a group of similar cells, and each cell functions more or less independently. **Multicellular organisms** usually are composed of vast numbers of different types of cells with each type performing a different function.

In most multicellular organisms, cells are arranged to form tissues, and different tissues are grouped together to form organs. A **tissue** is a group of similar cells performing a similar function. Muscle and nerve tissues are examples. An **organ** is composed of several different types of tissues that function together to carry out the particular functions of the organ. The heart and brain are examples. In most animals, coordinated groups of organs compose an **organ system** and work together to carry out the functions of the organ system. The circulatory and skeletal systems are examples.

In this section, you will examine prepared slides of multicellular organisms. There are three basic types of prepared slides: whole mounts (w.m.) contain the entire organism; longitudinal sections (l.s.) are thin sections cut through the organism along its longitudinal axis; cross sections or transverse sections (x.s.) are thin sections of the organism cut perpendicular to the longitudinal axis.

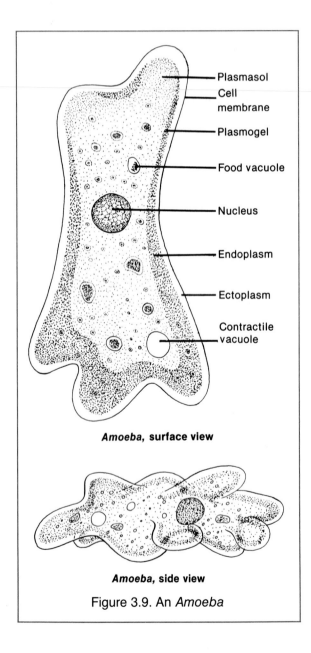

Amoeba, **surface view**

- Plasmasol
- Cell membrane
- Plasmogel
- Food vacuole
- Nucleus
- Endoplasm
- Ectoplasm
- Contractile vacuole

Amoeba, **side view**

Figure 3.9. An *Amoeba*

Materials

Prepared slides of:
    earthworm (*Lumbricus*), x.s.
    buttercup (*Ranunculus*) root, x.s.

Assignment 7

1. Examine the prepared slides of earthworm, x.s., and buttercup root, x.s. (Handle prepared slides by the edges or label to avoid getting fingerprints on the cover glass.) Each tissue is composed of similar cells that are characteristic for that tissue.
2. ***Complete the laboratory report***.

4

CHEMICAL ASPECTS

OBJECTIVES

On completion of the laboratory session, you should be able to:
1. Determine the number of protons, neutrons, and electrons in an atom of an element from data shown in a Periodic Table of the Elements.
2. Diagram a shell model of a simple atom.
3. Describe the formation of ionic and covalent bonds in accordance with the octet rule.
4. Diagram or make ball-and-stick models of simple molecules.
5. Describe and perform simple tests for carbohydrates, fats, and proteins.
6. Define all terms in bold print.

Life at its most fundamental level consists of complex chemical reactions, so having some understanding of the chemicals in living organisms and the nature of chemical reactions is important in your study of biology.

Chemical substances are classified into two groups: elements and compounds. An **element** is a substance that cannot be broken down by chemical means into any simpler substance. Table 4.1 lists the most common elements found in living organisms. An **atom** is the smallest unit of an element that retains the properties (characteristics) of the element.

Two or more elements may combine to form a **compound**. Water (H_2O) and table salt (NaCl) are simple compounds; carbohydrates, fats, and proteins are complex compounds. The smallest unit of a compound that retains the properties of

a compound is a **molecule**. It is formed of two or more atoms joined by **chemical bonds**.

ATOMIC STRUCTURE

Atoms are composed of three basic components: **protons**, **neutrons**, and **electrons**. Protons and neutrons have a mass of one atomic unit and are located in the central region of an atom, the nucleus. Protons possess a positive electrical charge of one ($+1$), but neutrons have no charge. Electrons have a negative electrical charge of one (-1), almost no mass, and orbit at near the speed of light around the atomic

TABLE 4.1
Common Elements in the Human Body

Element	Symbol	Percentage*
Oxygen	O	65
Carbon	C	18
Hydrogen	H	10
Nitrogen	N	3
Calcium	Ca	2
Phosphorus	P	1
Potassium	K	0.35
Sulfur	S	0.25
Sodium	Na	0.15
Chlorine	Cl	0.15
Magnesium	Mg	0.05
Iron	Fe	0.004
Iodine	I	0.0004

*By weight.

28

nucleus. The attraction between positive and negative charges is what keeps the electrons spinning about the nucleus.

Atoms of a given element differ from those of all other elements in the number and arrangement of protons, neutrons, and electrons, and these differences determine the properties of each element. Chemists have used these differences to construct a **Periodic Table of the Elements**.

Table 4.2 is a simplified periodic table of the first 20 elements. Each element is identified by its **symbol**. Note the location of the **atomic number**, which indicates the number of protons in each atom. Since a neutral atom possesses the same number of protons as electrons, the atomic number also indicates the number of electrons in each atom of the element. Note that the elements are sequentially arranged in the table according to atomic number.

TABLE 4.2
Simplified Periodic Chart of the Elements—1 Through 20*

I	II	III	IV	V	VI	VII	VIII
1 H hydrogen 1.0							2 He helium 4.0
3 Li lithium 7.0	4 Be beryllium 9.0	5 B boron 11.0	6 C carbon 12.0	7 N nitrogen 14.0	8 O oxygen 16.0	9 F fluorine 19.0	10 Ne neon 20.2
11 Na sodium 23.0	12 Mg magnesium 24.3	13 Al aluminum 27.0	14 Si silicon 28.1	15 P phosphorus 31.0	16 S sulfur 32.1	17 Cl chlorine 35.5	18 Ar argon 40.0
19 K potassium 39.1	20 Ca calcium 40.1						

Atomic Number

Atomic Symbol

Atomic Mass

*The dark line separates metals on the left from nonmetals.

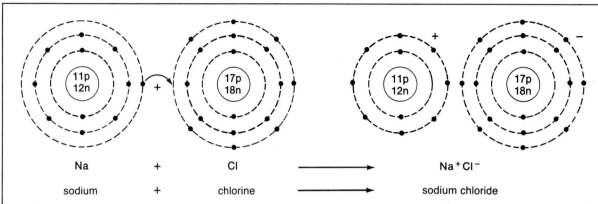

Figure 4.1. The formation of sodium chloride, an ionic reaction. Chlorine gains an electron from sodium with the result that each atom has eight electrons in the outer shell.

The atom's **mass number** (commonly called atomic weight) equals the sum of protons plus neutrons in each atom. Because the number of neutrons may vary slightly in atoms of the same element, the atomic mass reflects the *average* mass of the atoms. Atoms of an element that vary slightly in mass due to a difference in the number of neutrons are called **isotopes**.

The **shell model** is a useful, although not quite accurate, representation of atomic structure. The energy levels occupied by electrons are represented by circles (shells) around the nucleus. Electrons are shown as dots on the circles.

Examine Figure 4.1. Note the arrangement of the electrons in sodium (Na) and chlorine (Cl). The innermost shell can contain a maximum of only two electrons. Atoms with two or more shells can contain a maximum of eight electrons in the outermost shell.

Assignment 1

1. Use the simplified Periodic Table of the Elements in Table 4.2 to determine the number of protons, electrons, and neutrons in the first 20 elements.
2. Compare the number of electrons in the outer shell of the elements shown with their position in Table 4.2.
3. *Complete item 1 on Laboratory Report 4 that begins on page 365*.

REACTIONS BETWEEN ATOMS

The **octet rule** states that atoms react with each other to achieve eight electrons in their outermost shells. An exception to this is that the first shell is filled by only two electrons.

The octet rule is achieved by (1) losing or gaining electrons or (2) sharing electrons. Atoms with one, two, or three electrons in the outer shell donate (lose) these electrons to other atoms so that the adjacent shell, filled with electrons, becomes the outer shell. Atoms with six or seven electrons in the outer shell receive (gain) electrons from adjacent atoms to fill their outer shells. Other atoms usually attain an octet by sharing electrons.

Ionic Bonds

Typically, an atom has a net charge of zero, because it has the same number of protons (positive charges) and electrons (negative charges). If an atom gains electrons, it becomes negatively charged since the number of protons is constant. Similarly, the loss of electrons causes an atom to be positively charged. An atom with an electrical charge is called an **ion**.

Ionic bonds are formed between two ions by the attraction of opposite charges that have resulted from gaining or losing electrons. Examine Figure 4.1. Sodium and chlorine unite by an ionic bond to form sodium chloride (NaCl). Note that with the transfer of the single electron in the outer shell of sodium to chlorine, each atom has eight electrons in its outer shell. This transfer causes sodium to have a positive charge ($+1$) and chlorine to have a negative charge (-1). The opposite charges hold the ions together to form the molecule. This type of bonding occurs between **metals** (electron donors) and **nonmetals** (electron recipients).

Covalent Bonds

When nonmetals react with nonmetals, electrons are shared instead of being passed from one atom to another. This sharing of electrons usually forms a **covalent bond**. The electrons are shared in pairs, one from each atom. The electrons spend some time in the outer shell of each atom, so that the outer shell of each atom meets the octet rule.

Examine Figure 4.2. Note how one atom of oxygen (O) and two atoms of hydrogen (H) share their electrons to form one molecule of water (H_2O). Since oxygen is sharing a single pair of electrons with each hydrogen, each hydrogen is joined with oxygen by a single covalent bond. Compare the shell model of the molecule with the **structural formula**.

TABLE 4.3
Bonding Capacities

Element	Number of Bonds	Potential Bonds
Carbon	4	$-\overset{\textstyle|}{\underset{\textstyle|}{C}}-$
Nitrogen	3	$\overset{|}{\underset{/\ \backslash}{N}}$
Oxygen	2	$-O-$
Hydrogen	1	$H-$

Table 4.3 shows the covalent bonding capacity of the four elements that form about 96% of the human body. The bonding pattern shows each potential covalent bond represented by a single line. A common practice is to draw structural formulas of molecules with covalent bonds using lines to represent the bonds. One line indicates a single bond (one pair of shared electrons), two

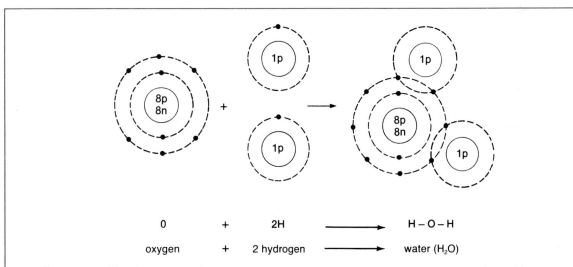

Figure 4.2. The formation of water, a covalent reaction. A pair of electrons is shared between each hydrogen atom and the oxygen atom. Shared electrons are counted as belonging to each atom.

lines indicate a double bond (two pairs of shared electrons), and so forth.

Assignment 2

1. Study Figures 4.1 and 4.2 to understand the characteristics of ionic and covalent bonds.
2. ***Complete item 2 on the laboratory report.***

ACIDS, BASES, AND pH

An **acid** is a substance that releases **hydrogen ions** (H^+) when dissolved in water. The greater the degree of dissociation, the greater is the strength of the acid. Hydrogen chloride is a strong acid.

$$HCl \longrightarrow H^+ + Cl^+$$

A **base** is a substance that releases **hydroxide ions** (OH^-) when dissolved in water. The greater the degree of dissociation, the stronger is the base. Sodium hydroxide is a strong base.

$$NaOH \longrightarrow Na^+ + OH^-$$

Chemists use a **pH scale** to indicate the strength of acids and bases. Figure 4.3 shows the scale, which ranges from 0 to 14 and indicates the proportionate concentration of hydrogen and hydroxide ions at the various pH values. When the concentration of hydrogen ions increases, the concentration of hydroxide ions decreases, and vice versa. A change of one (1.0) in the pH number indicates a 10-fold change in the hydrogen ion concentration because the pH scale is a logarithmic scale.

Pure water has a pH of 7, the point where the concentration of hydrogen and hydroxide ions are equal. Observe this in Figure 4.3. Acids have a pH less than 7; bases have a pH greater than 7. The greater the concentration of hydrogen ions, the stronger the acid and the *lower* the pH number. The greater the concentration of hydroxide ions, the stronger the base and the *higher* the pH number.

Living organisms are sensitive to the concentrations of hydrogen and hydroxide ions and must maintain the pH of their cells within narrow limits. For example, your blood is kept very close to pH 7.4. Organisms control the pH of cel-

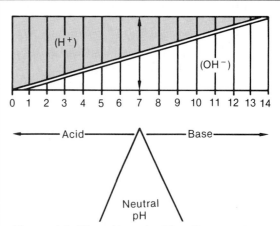

Figure 4.3. The pH scale. The diagonal line indicates the proportionate amount of hydrogen (H^+) and hydroxide (OH^-) ions at each pH value.

lular and body fluids by buffers. A **buffer** is a compound or a combination of compounds that can combine with or release hydrogen ions to keep the pH of a solution relatively constant.

Materials

Beakers, 100 ml
Graduated cylinder, 50 ml
pH test papers, wide and narrow ranges
Buffer solution, pH 7, in 1000-ml flask
Distilled water, pH 7, in 1000-ml flask
Dropping bottles of:
   after-shave lotion
   Alka-Seltzer® solution
   bromthymol blue
   detergent solution
   hydrochloric acid, 1.0%
   household ammonia
   lemon juice
   white vinegar

Assignment 3

1. ***Complete items 3a and 3b on the laboratory report.***
2. Use pH test papers to determine the pH of the various solutions. Place 1 drop of the solution to be tested on a small strip of pH

test paper while holding it over a paper towel. Compare the color of the pH paper with the color code on the dispenser to determine the pH. Use the wide-range test paper first to determine the approximate pH. Then, use a narrow-range test paper for the final determination.

3. Place 25 ml of (a) distilled water, pH 7, and (b) buffer solution, pH 7, in separate, labeled beakers. Measure and record the pH of each.

4. Add 6 drops of bromthymol blue to each beaker. Mix by swirling the liquid. The fluid in each beaker should be a pale blue. Bromthymol blue is a pH indicator that is blue at pH 7.6 and yellow at pH 6.0.

5. Place the beaker of distilled water and bromthymol solution on a white piece of paper. Add 1.0% HCl drop by drop to the solution and mix thoroughly by swirling the liquid between each drop. Count and record the number of drops required to turn the solution yellow, which indicates a pH of 6.

6. Use the preceding method to determine the number of drops of 1.0% HCl required to lower the buffer solution from pH 7 to pH 6.

7. ***Complete items 3c and 3d on the laboratory report***.

IDENTIFYING BIOLOGICAL MOLECULES

Carbohydrates, lipids, and proteins are complex organic compounds found in living organisms. Carbon atoms form the basic framework of their molecules. In this section, you will analyze plant and animal materials for the presence of these compounds, which constitute the basic organic foods of heterotrophs. Work in groups of two to four students. If time is limited, your instructor will assign certain portions to separate groups, with all groups sharing the data.

Carbohydrates

Carbohydrates are formed of carbon, hydrogen, and oxygen. Their molecules are characterized by the H—C—OH grouping in which the ratio of hydrogen to oxygen is 2:1. **Mono-**

saccharides are simple sugars containing three to seven carbon atoms, and they serve as building blocks for more complex carbohydrates. The combination of two monosaccharides forms a **disaccharide** sugar, and the union of many monosaccharides forms a **polysaccharide**. Six carbon sugars, such as glucose, are primary energy sources for living organisms. Starch in plants and glycogen in animals are polysaccharides used for nutrient storage. See Figure 4.4.

Use the iodine test for starch and Benedict's test for reducing sugars (most 6-carbon and some 12-carbon sugars) to identify the presence of these carbohydrates.

Iodine Test for Starch

1. Place one dropper (1 ml) of the liquid to be tested in a clean test tube. (A "dropper" means *one dropper full* of liquid.)

2. Add 3 drops of iodine solution to the liquid in the test tube and shake gently to mix.

3. A gray to blue-black coloration of the liquid indicates the presence of starch in that order of increasing concentration.

Benedict's Test for Reducing Sugars

1. Place 1 dropper (1 ml) of the liquid to be tested in a clean test tube.

2. Add 3 drops of Benedict's solution to the liquid in the test tube and shake gently to mix.

3. Heat the tube to near boiling in a water bath for 2–3 min. (A water bath consists of a beaker that is half-full of water, heated on a hot plate or over a burner. The test tube is placed in the beaker so that it is heated by the water in the beaker rather than directly by the heat source.)

4. A light green, yellow, orange, or brick-red coloration of the liquid indicates the presence of reducing sugars in that order of increasing concentration.

Lipids

Lipids include an array of oily and waxy substances, but the most familiar are the neutral fats (triglycerides). They are composed of carbon, oxygen, and hydrogen and have fewer oxygen atoms than carbohydrates. A fat molecule consists of three long fatty acid chains joined to

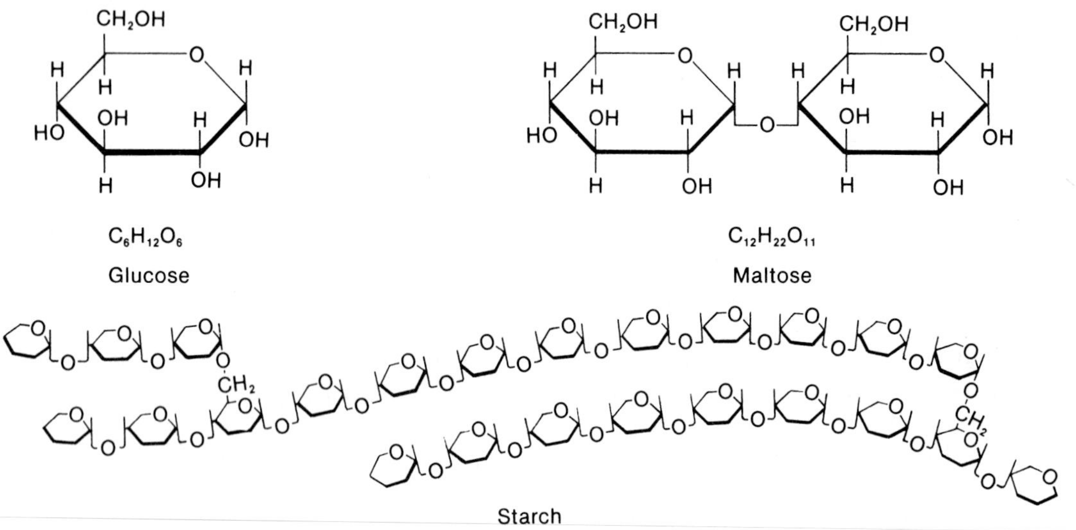

Figure 4.4. Carbohydrates

a single glycerol molecule. See Figure 4.5. Note the difference in the carbon-to-carbon bonding between saturated and unsaturated fats. Fats serve as an important means of nutrient storage in organisms and are not soluble in water.

The presence of lipids may be determined by a paper spot test and Sudan IV, a dye that selectively stains lipids.

Paper Spot Test

1. Place a drop of the liquid to be tested on a piece of paper. Blot it with a paper towel and let it dry. If the substance to be tested is a solid, it can be rubbed on the paper. Remove the excess with a paper towel. Animal samples may need to be gently heated.
2. After the spot has dried, hold the paper up to the light and look for a permanent translucent spot on the paper indicating the presence of lipids.

Sudan IV Test

1. Place a dropper of distilled water in a clean test tube and add 3 drops of Sudan IV. Shake the tube from side to side to mix well.

2. Add 10 drops of the substance to be tested and mix again. Let the mixture stand for 5 min.
3. Lipids will be stained red and will rise to the top of the water or be suspended in tiny globules.

Proteins

Proteins are usually large, complex molecules composed of many **amino acids** joined together by **peptide bonds**. Their molecules contain nitrogen in addition to carbon, oxygen, and hydrogen. Proteins play important roles by forming structural components of cells and tissues and by serving as enzymes. See Figure 4.6.

Proteins may be identified by a Biuret test that is specific for peptide bonds.

Biuret Test

1. Place a dropper of the solution to be tested in a clean test tube.
2. Add a dropper of 10% sodium hydroxide (NaOH) and mix by shaking the tube from side to side. *Caution*: NaOH is a caustic substance that can cause burns. If you get

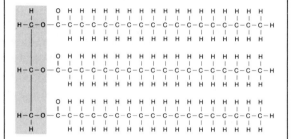

B

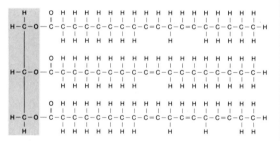

C

Figure 4.5. Fats.
A. The bonding of fatty acids to glycerol to form a fat.
B. Glycerol tristearate, a saturated fat. C. Linseed oil, an unsaturated fat. R represents the remainder of the molecular.

Figure 4.6. Formation of peptide bonds.
Proteins are formed of a large number of amino acids joined together by peptide bonds. R represents the remainder of the amino acid.

Materials

Hot plate
Beaker, 250 ml
Dissecting instruments
Medicine dropper
Mortar and pestle
Test tubes
Test-tube holder
Test-tube rack
Dropping bottles of:
   albumin, 1.0%
   Benedict's solution
   copper sulfate, 0.05%
   corn oil
   egg white, 1.0%
   glucose, 0.1%
   iodine (I_2 + KI) solution
   sodium hydroxide, 10%
   soluble starch, 0.1%
   Sudan IV dye

Assignment 4

1. To evaluate the results of the chemical tests that you will perform on the substances provided, it is necessary to prepare a **set of standards** that demonstrates both positive and negative results for each test. The standards are prepared by performing each test on (a) a substance giving a positive reaction and (b) a substance (water) giving a negative reaction.

it on your skin or clothing, wash it off *immediately* with lots of water.

3. Add five drops of 0.05% copper sulfate (CuSO$_4$) to the tube and shake from side to side to mix. Observe after 3 min.

4. A violet color indicates the presence of proteins.

2. Prepare a set of standards as shown on the chart in item 4a of the laboratory report and record your results in the chart. The test tubes used must be clean; if in doubt wash them. Follow these procedures.

   a. Label eight clean test tubes for the substances listed in the "Substances Tested" column of the chart. Note that the paper-spot test does not require test tubes.

   b. Add to the test tubes a dropper (about 1 ml) of the substances to be tested.

   c. Add the reagents for the tests to the appropriate test tubes and perform each test.

   d. Perform the paper-spot test on brown paper.

   e. ***Record your results in the chart in item 4a of the laboratory report***.

3. ***Complete items 4b and 4c on the laboratory report***.

4. Perform each of the tests on substances (unknowns) listed on the chart in item 4d on the laboratory report or provided by your instructor. In some cases, it will be necessary to macerate the material to obtain liquid for testing. Use this procedure.

   a. Place a ½-in. cube of the substance to be tested in a mortar. Add 5 ml of distilled water. Use a pestle to crush and grind the cube into a pulpy mass.

   b. Pour off the liquid and use it as the test solution.

5. Compare your results with the standards and ***record your results in the chart in item 4d and complete the laboratory report***.

DIFFUSION AND OSMOSIS

Materials constantly move into and out of cells. The **selective permeability** of the cell membrane permits some materials to pass through it, but prevents others. Furthermore, the materials that can pass through the membrane may change from moment to moment.

Materials move through a cell membrane by two different processes. **Passive transport**, commonly called diffusion, results from the normal, random motion of molecules and does not require an expenditure of energy by the cell. **Active transport** requires the use of energy by the cell, does not depend on molecular motion, and may move materials either with or against a **concentration gradient**. In a similar way, materials that would normally diffuse out of a cell may be prevented from doing so by **active retention**.

The Scientific Method

In this exercise, you will gain experience in the methodology of science that will help you understand how science differs from other forms of inquiry. Basically, science involves systematic observation and experimentation. See Figure 5.1.

Scientists systematically collect data by making either direct or indirect **observations**. A testable **hypothesis**, a tentative explanation, is then made that possibly accounts for the data. A controlled experiment is designed to test the hypothesis and determine the effect of a single **independent variable** (variable being tested) on the **dependent variable** (phenomenon being studied).

A **controlled experiment** consists of (1) an **experimental setup** that is exposed to the independent variable and (2) a **control setup** that is *not* exposed to the independent variable. All other variables are controlled, that is, kept the same in both setups.

Data (results) from the experiment are collected and analyzed to see if they support or refute the hypothesis. Then, a conclusion, and usually a new hypothesis, is made. Scientific knowledge advances through the testing, rejection, and refinement of hypotheses.

Whenever a hypothesis is repeatedly supported by experimental data gathered by many different scientists, it is often restated as a theory. For example, the cell theory states that all organisms are composed of cells. Note that a **scientific theory** has been repeatedly supported

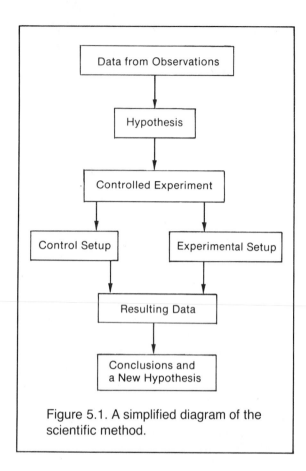

Figure 5.1. A simplified diagram of the scientific method.

by experimental data collected by numerous scientists.

BROWNIAN MOVEMENT

Molecules of liquids and gases are in constant, random motion. A molecule moves in a straight-line path until it bumps into another molecule, and then it bounces off into a different straight-line course. Since this motion is temperature dependent, the higher the temperature, the greater the molecular movement. This movement cannot be observed directly. However, tiny particles suspended in a liquid may be observed with a microscope as they are moved in a random fashion due to bombardment by molecules composing the liquid. This vibratory movement is indirect evidence of molecular motion and is called **Brownian movement** after Robert Brown who first described it in 1827.

Materials

Dissecting needle
Medicine dropper
Microscope slide and cover glass
Powdered carmine dye
Dropping bottle of detergent-water solution

Assignment 1

1. Place a drop of water-detergent solution on a clean slide.
2. Dip a *dry* tip of a dissecting needle into powdered carmine, and, while holding the tip above the drop of fluid, tap the needle with your finger to shake only a few dry particles into the drop. Add a cover glass.
3. Examine the dye particles at 400× to observe Brownian movement.
4. ***Complete item 1 on Laboratory Report 5 that begins on page 369.***

DIFFUSION

The constant, random motion of molecules is what enables diffusion to occur. **Diffusion** is the net movement of the same kind of molecules from an area of their higher concentration to an area of their lower concentration. Thus, the molecules move down a **concentration gradient**. Molecules that initially are unequally distributed in a liquid or gas tend to move by diffusion throughout the medium until they are equally distributed. Diffusion is an important means of distributing materials within cells and of passively moving substances through cell membranes.

Diffusion and Temperature

In the first experiment you will test this hypothesis: Temperature has no effect on the rate of diffusion. To test the hypothesis, you will perform a **controlled experiment** to determine whether the rate of diffusion is the same at two different temperatures. The water temperature in each beaker will serve as the control for the other. Note that there is only one independent variable, the difference in temperature, and that all other conditions are the same. Having only one independent variable in an experiment is desirable to obtain clear-cut results.

Materials

Hot plate
Beakers, 250 ml
Beaker tongs
Celsius thermometer
Forceps
Ice water
Paper towels

Assignment 2

1. ***Complete item 2a on the laboratory report***.
2. Fill a beaker two thirds full with water and heat it on a hot plate to about 50°C. Use beaker tongs to place the beaker on a pad of paper towels.
3. Place an equal amount of ice water (but no ice) in another beaker. Record the temperature of the water in each beaker.
4. Keeping both beakers motionless so that the water is also motionless, add a crystal of potassium permanganate to each. Record the time. Observe the rate of diffusion by the potassium permanganate molecules during a 15-min period. Is the rate identical in each beaker?
5. ***Complete item 2 on the laboratory report***.

Diffusion and Molecular Weight

In this experiment, you will test the hypothesis: Molecular weight has no effect on the rate of diffusion. You are to do so by determining the rate of diffusion of two substances that have different molecular weights: methylene blue (mol. wt. 320) and potassium permanganate (mol. wt. 158). Both substances are soluble in water and readily diffuse through an agar gel that is about 98% water.

Materials

Forceps
Petri dish containing agar gel
Crystals of:
    methylene blue
    potassium permanganate

Assignment 3

1. Place equal-sized crystals of potassium permanganate and methylene blue about 5 cm apart on an agar plate. Gently press the crystals into the agar with forceps to assure good contact. Record the time.
2. After 1 hr, determine the distance each substance has diffused by measuring the diameter of each colored circle on the agar plate.
3. ***Complete item 3 on the laboratory report***.

Diffusion, Molecular Size, and Membrane Permeability

At times, materials seem to diffuse through cell membranes if their molecules are sufficiently small while larger molecules are prevented from passing through. This selectivity based on molecular size can be observed in an experiment in which a cellulose membrane simulates a cell membrane. The cellulose membrane has many microscopic pores scattered over its surface, and molecules smaller than the pores are able to pass through the membrane.

You will use the iodine test for starch and Benedict's test for reducing sugars to determine the results of the experiment.

Starch Test

1. Place one full dropper of the substance to be tested in a clean test tube.
2. Add 3 drops of iodine solution to the test tube, and shake to mix. A gray to blue-black coloration indicates the presence of starch in that order of increasing concentration.

Benedict's Test

1. Place one full dropper of the substance to be tested in a clean test tube.
2. Add 5 drops of Benedict's solution to the test tube, and mix by shaking.
3. Heat to near boiling for 2 to 3 min in a boiling water bath. A water bath is set up by placing the test tube in a beaker, which is about half full of water, and by heating the beaker on a hot plate. In this way the test tube is heated by the water in the beaker

rather than directly by the hot plate. A light green, yellow, orange, or brick-red coloration of the liquid in the test tube indicates the presence of reducing sugars in that order of increasing concentration.

Materials

Hot plate
Beakers, 250 ml
Cellulose tubing, 1-in. width
Rubber bands

Test tubes, 24 × 200 mm
Test tubes, 14 × 150 mm
Test-tube brush
Test-tube holders
Test-tube rack
Dropping bottles of:
   Benedict's solution
   iodine (I_2 + KI) solution
Squeeze bottles of:
   glucose, 20%
   soluble starch, 0.1%

---

1. Fill a large test tube two thirds full with water and add 4 droppers of iodine solution. Place test tube in a beaker.

2. Soak a 20-cm length of cellulose tubing in water. Then tie a knot in one end to form a sac.

3. Fill the sac half full with starch solution and add 5 droppers of glucose solution.

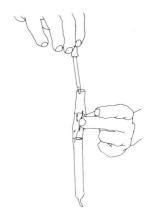

4. Hold sac closed and rinse outside of sac under the tap.

5. Insert the sac into the test tube.

6. Bend top of sac over lip of test tube and secure it with a rubber band. Let stand for 20 min.

Figure 5.2. Preparation of an experiment to study the effect of molecular size on the diffusion of molecules through a cellulose membrane

Assignment 4

1. Set up the experiment as shown in Figure 5.2. *Be certain* to rinse the outside of the sac before inserting it into the test tube.
2. Place the test tube in a beaker or rack for 20 min.
3. Examine your test tube and record any color change that has occurred.
4. Test the solution outside the sac to see if glucose diffused from the sac. Place one dropper of the solution in a clean test tube and perform Benedict's test.
5. ***Complete item 4 on the laboratory report.***
6. Clean the test tubes thoroughly using a test-tube brush.

Selective Permeability of Living Membranes

Living membranes function in a more complex fashion than the cellulose membrane that you have observed. A living membrane may prevent the passage of small diffusable molecules while allowing the passage of large molecules. You will test the hypothesis: Anthocyanin is not retained within living beet root cells.

Materials

Hot plate
Beakers, 250 ml
Dissecting instruments
Glass marking pen
Test tubes, 14 × 150 mm
Isopropyl alcohol
Beet roots, fresh

Assignment 5

1. Use a scalpel or sharp knife to cut a portion of a beet root into narrow pieces that will fit into a test tube. Pieces about 0.5 × 1.0 cm are fine.
2. Place the pieces in a beaker, and rinse them with two washings of tap water to remove the anthocyanin released by cutting the cells. **Anthocyanin** is the pigment in the central vacuoles of beet root cells that gives the color to beet roots.
3. Place equal numbers of beet pieces into three numbered test tubes. Add sufficient

water to cover the pieces in tubes 1 and 2, and add a similar amount of alcohol to tube 3.
4. Place tubes 1 and 3 in a test-tube rack. Heat the water in tube 2 to boiling for 3 min in a water bath, and then place tube 2 in the test-tube rack. After 30 min, note the color of the liquid in each tube. Alcohol is a poison that kills the cells but leaves the membranes intact, while boiling kills the cells and ruptures the cell membranes.
5. ***Complete item 5 on the laboratory report.***

OSMOSIS

Water is essential for life. It is the **solvent** of living systems and is the aqueous medium in which the chemical reactions of life occur. Substances dissolved in a solvent are called **solutes**. Water is the most abundant substance in cells, and it contains both organic and inorganic solutes. Water is a small molecule that diffuses freely in and out of cells in accordance with the concentration of water inside and outside the cells. **Osmosis** is simply the diffusion of water through a semipermeable or selectively permeable membrane. Like all substances, water diffuses down a concentration gradient. It moves from an area of higher concentration of water to an area of lower concentration of water. Remember that the concentration of water increases as the concentration of solutes decreases. Thus, a 5% salt solution contains 95% water while a 10% salt solution contains 90% water.

Osmosis and Concentration Gradients

In this experiment, you will test the hypothesis: The magnitude of the concentration gradient does not effect the rate of osmosis. Your instructor has set up two osmometers like those in Figure 5.3. The equal-sized cellulose sacs are attached to rubber stoppers into which glass tubing has been inserted. The sacs have been filled with 10% and 20% sucrose, respectively, and immersed in tap water. The molecules of sucrose are too large to pass through the pores of the membrane; however, water will diffuse into the sacs, increasing the volume of the fluid,

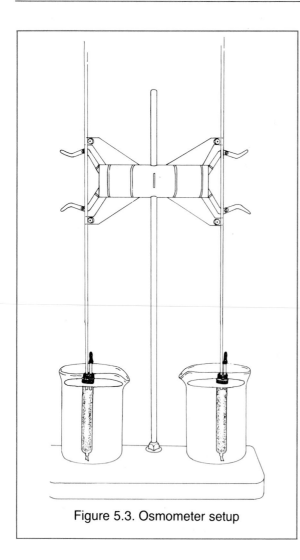

Figure 5.3. Osmometer setup

which generates pressure, forcing the fluid up the glass tubes.

When unequal solute concentrations in aqueous solutions are on each side of a semipermeable membrane, water always moves from the **hypotonic solution** (lower solute concentration) into the **hypertonic solution** (higher solute concentration), that is, down the concentration gradient of water. If the concentrations of solutes on each side of the membrane are equal, the solutions are said to be **isotonic solutions**. They are at **osmotic equilibrium**, and no net movement of water occurs through the membrane.

Assignment 6

1. Use a meterstick to determine the distance the fluid rises in each tube during a 30-min period.
2. *Complete item 6 on the laboratory report*.

Osmosis and Living Cells

In this section, you will observe the movement of water into and out of living cells that are exposed to hypertonic and hypotonic solutions.

Materials

Beakers, 250 ml
Microscope slides and cover glasses
Scalpel or sharp knife
10% salt solution in:
    dropping bottles
    flasks
Elodea shoots

Assignment 7

This experiment uses the crispness of celery sticks as an indicator of water content in the cells. The more water that is in the cells, the greater the crispness.

1. Place about 200 ml of distilled water in one beaker and the same amount of salt solution in another beaker.
2. Cut four celery sticks (about 0.5 × 7 cm), and immerse two sticks in each beaker. Record the time.
3. Examine the crispness of the celery sticks after 1 hr.
4. *Complete items 7a and 7b on the laboratory report*.
5. Mount an *Elodea* leaf in tap water. Note the size of the central vacuoles and the location of the protoplasm in the cells.
6. *In the space for item 7c on the laboratory report, draw an optical midsection view of a normal Elodea cell*. Show the distribution of the protoplasm and the central vacuole.
7. Place 1 drop of 10% salt solution at the edge of the cover glass, and draw it under the cover glass as shown in Figure 5.4. Repeat the process. Now the leaf is mounted in salt solution. This experiment allows

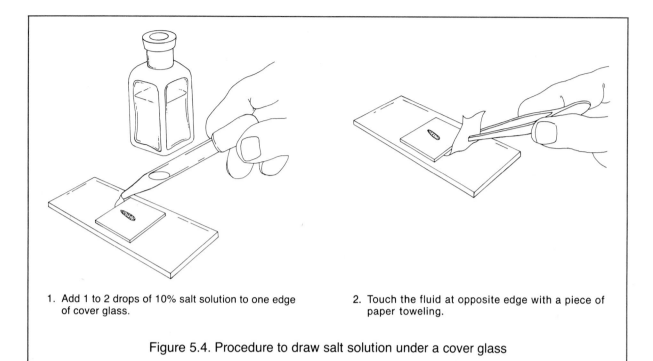

1. Add 1 to 2 drops of 10% salt solution to one edge of cover glass.

2. Touch the fluid at opposite edge with a piece of paper toweling.

Figure 5.4. Procedure to draw salt solution under a cover glass

you to observe at the cellular level the effect of water moving into or out of plant cells. What do you think will happen?

8. Examine the cells with your microscope. Observe the protoplasm as it shrinks. The shrinking of protoplasm caused by the loss of water is called **plasmolysis**.

9. ***Make a drawing of a cell mounted in salt solution in the space for item 7c on the laboratory report***.

10. Remove the leaf from the salt solution, rinse it, and remount it on a clean slide in tap water. Quickly observe it with your microscope. What happens?

11. ***Complete the laboratory report***.

12. Clean and dry the objectives and stage of your microscope to remove any salt water that may be present.

PART II

DIVERSITY
OF ORGANISMS

6

MONERANS AND PROTISTS

This is the first of six exercises designed to acquaint you with the variety of organisms in the biotic world. The emphasis in these exercises is on the distinguishing features of the major organismic groups, their evolutionary relationships, and their adaptations.

Biologists use a five-kingdom system for classifying organisms according to their similarities and evolutionary relationships. The five kingdoms are **Monera, Protista, Fungi, Plantae**, and **Animalia**. Their relationships are shown in Figure 6.1. See Appendix D for a classification of organisms.

Several **taxonomic categories** are used in classifying organisms, and these are shown in Table 6.1 in the classification of the human species. The kingdom category contains the largest number of different species (kinds) of organisms, and the species category contains the fewest—only one. The kingdom contains related phyla, the phylum contains related classes, and so forth.

The genus and species names are combined to form the **scientific (binomial) name** of an organism, in this case *Homo sapiens*. Note that the species name is not capitalized. Botanists use the term "division" in place of "phylum," which is used by zoologists.

KINGDOM MONERA

Monerans are the simplest organisms. They are either **unicellular** (composed of a single cell) or **colonial** (composed of a group of independently functioning cells). All cells are **prokaryotic** (lack membrane-bound organelles). A rigid, nonliving cell wall is formed external to the cell membrane.

Bacteria (Division Schizophyta)

Bacteria are the smallest organisms with cell lengths of 1 to 10 μm, and they occur in all habitats supporting life. Bacteria exhibit three characteristic shapes: rodlike (bacilli), spherical (cocci), and spiral (spirilla). See Figure 6.2 and Plates 1.1 to 1.3. A slimy capsule may be present external to the cell wall. Some species possess **flagella** that enable them to swim about. Bacteria are identified not only by shape but also by their physiology and biochemistry.

Almost all bacteria are **heterotrophs** and must obtain nutrients from outside sources. Most of these bacteria are **saprotrophs** and are useful in decaying nonliving organic matter to form inorganic compounds that can be used by

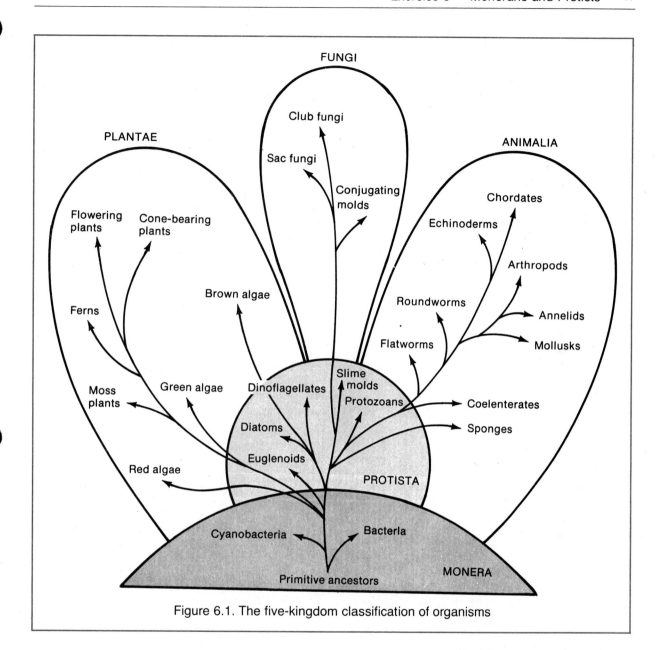

Figure 6.1. The five-kingdom classification of organisms

other organisms. A few heterotrophic bacteria are **parasites** and cause diseases.

Some bacteria are **autotrophs** capable of producing their own organic nutrients. **Chemosynthetic bacteria** are able to oxidize inorganic compounds, such as ammonia or sulfur, and capture the released energy. **Photosynthetic bacteria** possess a light-capturing pigment, **bacteriochlorophyll**, which enables them to use light as their energy source for the synthesis of organic compounds.

Bacteria are used to produce many products useful to humans, such as alcohols and other

TABLE 6.1
Classification of the Human Species,
Homo sapiens

Taxonomic Category	Classification of Humans
Kingdom	Animalia
Phylum	Chordata
Class	Mammalia
Order	Primates
Family	Hominidae
Genus	*Homo*
Species	*sapiens*

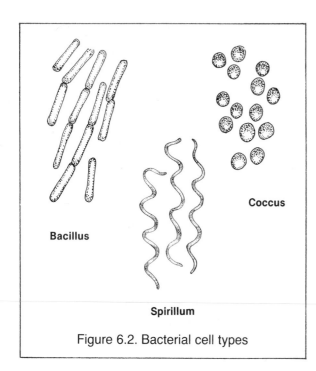

Bacillus

Coccus

Spirillum

Figure 6.2. Bacterial cell types

chemicals. Genetic engineering has enabled the use of bacteria to produce biomedical chemicals, such as insulin and interferon, that are useful in the treatment of human diseases.

Blue-Green Bacteria (Division Cyanobacteria)

Formerly called blue-green algae, all species of cyanobacteria are photosynthetic autotrophs that occur in freshwater, marine, and moist soil habitats. Hot springs and glaciers are often colored by their presence. They possess **chlorophyll a** and a variety of other pigments that give them their coloration, but they lack plastids. The pigments are concentrated in the **chromoplasm**, the outer portion of the cell. Most forms are blue-green, but many other colors occur, such as yellow, orange, red, and purple. The cells of cyanobacteria are a bit larger than bacteria and are usually covered with a gelatinous coat. See Plate 1.4.

Cyanobacteria are often responsible for the "algal blooms" that occur in polluted lakes. Some species can fix (utilize) atmospheric nitrogen in the production of proteins.

Prochlorophytes (Division Chloroxybacteria)

These primitive marine organisms, which were discovered living symbiotically in tunicates, are of evolutionary interest because they contain **chlorophyll a** and **b**. Some biologists believe that these organisms may be the forerunners of the chloroplasts found in green algae and higher plants.

Materials

Antibiotic discs
Medicine droppers
Metric ruler
Microscope slides and cover glasses
Broth cultures of motile bacteria:
    Pseudomonas or *Bacillus cereus*
Cultures of cyanobacteria:
    Gloeocapsa
    Oscillatoria
Pour plates of bacteria:
    Escherichia coli B
    Staphyloccus epidermidis
Prepared slides of:
    bacterial types
    Clostridium botulinum
    Staphylococcus aureus
    Treponema pallidum

Assignment 1

1. Examine a prepared slide of bacterial types. There are three smears of stained bacteria on the slide, and each consists of a different morphological type. Examine each smear with reduced light at 400×. Bacteria probably are smaller than you expect, so focus carefully.
2. Examine the prepared slides of bacteria set up under oil-immersion objectives of demonstration microscopes. Note their small size and shape at 1000×.
   a. *Staphylococcus aureus* causes pimples and boils when it gains access into the pores of the skin.
   b. *Treponema pallidum* causes syphilis.
   c. *Clostridium botulinum* produces a potent poison that causes botulism poisoning when it is eaten in contaminated foods.

3. Examine the living flagellated bacteria set up in a hanging drop slide under a demonstration microscope. You will not be able to see the flagella, but you can distinguish true motility from Brownian movement.
4. Examine the demonstration agar plates of *Escherichia coli* B and *Staphylococcus epidermidis* on which antibiotic discs were placed prior to incubation. The antibiotic diffuses into the agar and kills susceptible bacteria where its concentration is lethal. The size of the clear no-growth area around each antibiotic disc indicates the effectiveness of the antibiotic against the bacterial species. Measure and record the diameter of the no-growth areas around each disc for each species, but *do not open the plates*.
5. **Complete item 1a to 1f of Laboratory Report 6 that begins on page 373.**
6. Make wet-mount slides of *Gloeocapsa* and *Oscillatoria* and observe them at 400×. Note the size, color, shape, and arrangement of the cells. Locate the chromoplasm and the gelatinous sheath, if present. Note the absence of a nucleus and chloroplasts.
7. **Complete item 1g and 1h on the laboratory report.**

KINGDOM PROTISTA

Protists are a heterogeneous group of unicellular or colonial organisms that exhibit animal-like, plantlike, or funguslike characteristics. The major groups probably are not closely related but are products of evolutionary lines that diverged millions of years ago. Protists and all higher organisms are composed of **eukaryotic cells** that contain membrane-bound organelles.

Biologists are not in agreement regarding the organismic groups that compose the Protista. Some biologists include slime molds and multicellular (green, brown, and red) algae; others include one or the other, or neither. In this book, slime molds are included as protists, but multicellular algae are placed in the plant kingdom.

Protozoans: Animal-like Protists

These animal-like protists lack a cell wall and are usually motile. Most are **holotrophs** that engulf food into vacuoles, where it is digested. Some absorb nutrients through the cell membrane, and a few are parasitic. Protozoans occur in most habitats where water is available. Water tends to diffuse into fresh-water forms, and some protozoans possess **contractile vacuoles** that repeatedly collect and pump out the excess water to maintain their water balance. Three groups of protozoans are shown in Figure 6.3.

Flagellated Protozoans (Phylum Mastigophora)

These primitive protozoans possess one or more **flagella** that provide a means of movement. Food may be engulfed and digested in vacuoles, or nutrients may be absorbed through the cell membrane.

Some zooflagellates have established symbiotic relationships with other organisms. *Trypanosoma brucei* (formerly *T. gambiense*) is a parasitic form that causes African sleeping sickness. It lives in the blood and nervous system of its vertebrate host and is transmitted by the bite of tsetse flies. *Trichonympha collaris* is a mutualistic symbiont that lives in the gut of termites. It digests the cellulose (wood) to produce simple carbohydrates that can be digested or utilized by the termite. See Plates 2.1 to 2.3.

Amoeboid Protozoans (Phylum Sarcodina)

These protists move by means of **pseudopodia**, flowing extensions of the cell. Prey organisms are engulfed and digested in food vacuoles. Some forms secrete a shell for protection. Calcareous shells of foraminiferans and siliceous shells of radiolarians are abundant in ocean sediments. *Entamoeba histolytica* is a parasitic form causing amoebic dysentery in humans. See Plates 2.4, 2.5, 3.1 and 3.2.

Ciliated Protozoans (Phylum Ciliophora)

Ciliates are the most advanced and complex of the protozoans. They are characterized by the presence of a **macronucleus**, one or more **micronuclei**, and movement by means of numer-

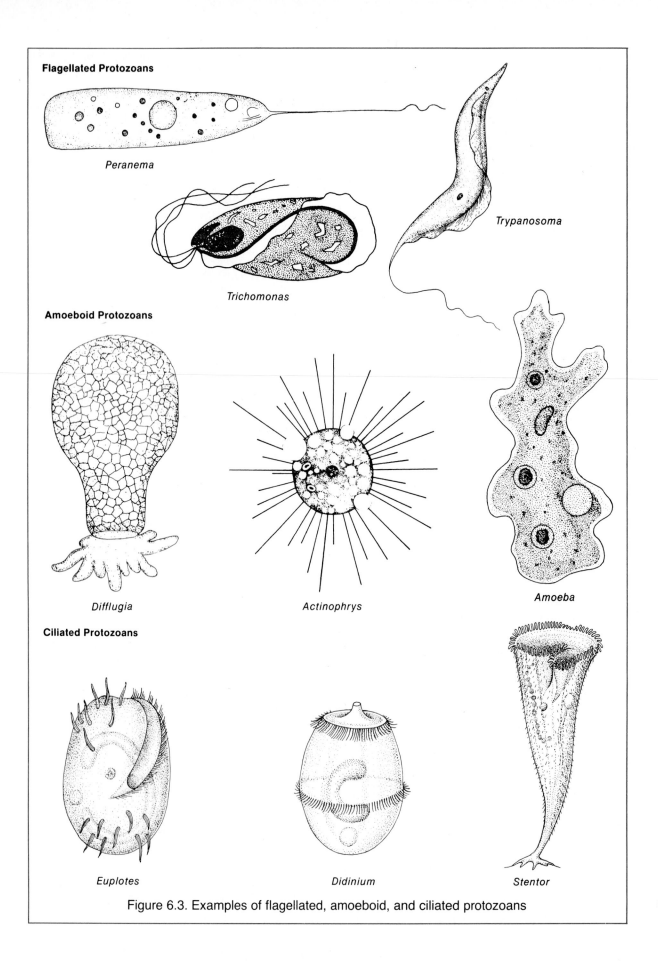

Flagellated Protozoans

Peranema

Trichomonas

Trypanosoma

Amoeboid Protozoans

Difflugia

Actinophrys

Amoeba

Ciliated Protozoans

Euplotes

Didinium

Stentor

Figure 6.3. Examples of flagellated, amoeboid, and ciliated protozoans

ous **cilia**, hairlike processes that cover the cell. A flexible outer covering, the **pellicle**, is located exterior to the cell membrane. *Paramecium* is a common example that also possesses **trichocysts**, tiny dartlike weapons of offense and defense located just under the cell surface. Food organisms are swept down the **oral groove** by cilia and into food vacuoles, where digestion occurs. See Figures 6.3 and 6.4 and Plates 3.3 and 3.4.

Sporozoans (Phylum Sporozoa)

All species lack motility and are internal parasites of animals. *Plasmodium vivax,* a pathogen causing malaria, is a typical example. It is transmitted by the bite of *Anopheles* mosquitoes and infests red blood cells and liver cells of human hosts. Malaria has probably caused the death of more humans than any other disease. See Plate 3.5.

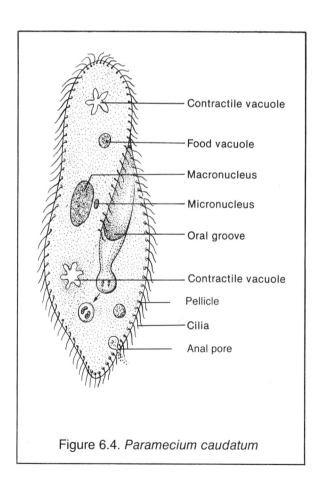

Figure 6.4. *Paramecium caudatum*

Materials

Colored pencils
Medicine droppers
Microscope slides and cover glasses
Protoslo
Toothpicks
Cultures of protozoa:
   Paramecium
   Pelomyxa
   Peranema
Pond water
Termites, living
Yeast cells stained with congo red
Prepared slides of protozoa:
   foraminifera shells
   Plasmodium vivax
   radiolaria shells
   Trypanosoma brucei

Assignment 2

1. Place a drop of the *Peranema* culture on a slide and examine the specimens at 100×. Note the use of the flagellum in movement. Use a toothpick to mix some Protoslo with the drop on your slide, add a cover glass, and examine cellular detail at 400×.
2. Examine a prepared slide of *Trypanosoma brucei* in human blood.
3. Remove the abdomen from a live termite and crush it in a drop of water on a clean slide. Discard the abdomen, add a cover glass, and locate the large multiflagellated *Trichonympha collaris*. What other types of organisms are present?
4. ***Complete item 2a on the laboratory report.***
5. Place a drop from the *bottom* of the *Pelomyxa* culture on a slide and observe a specimen at 100× without a cover glass. This is a large, multinucleate amoeboid protozoan. Compare it with the common *Amoeba* in Figure 3.9 that you observed in Exercise 3. Note the flowing motion, food vacuoles, nuclei, and contractile vacuoles.
6. ***Complete item 2b on the laboratory report.***
7. Examine the prepared slides of foraminifera and radiolaria shells and note the differences in their structures.

8. ***Complete items 2b and 2c on the laboratory report***.
9. Use colored pencils to color-code labeled structures in Figure 6.4.
10. Place a drop from the *Paramecium* culture on a slide and observe it at 40× or 100× without a cover glass. Note the movement of *Paramecium*. What happens when it bumps into an object?
11. Add a drop of Protoslo to your slide, and use a toothpick to mix it with the drop of *Paramecium* culture. Add a cover glass and observe. Locate an entrapped specimen and examine it carefully at 100× and 400×. Locate the structures shown in Figure 6.4. The food vacuoles are easily seen since the *Paramecium* has been feeding on dead yeast cells stained with congo red. As digestion proceeds, the color of the yeast changes from bright red to purple and finally to blue. Is there any pattern to the location of food vacuoles and the progression of digestion?
12. ***Complete items 2d–2i on the laboratory report***.
13. Examine the prepared slide of *Plasmodium vivax* in red blood cells that is set up under a demonstration microscope.

Plantlike Protists

The plantlike protists, sometimes called unicellular algae, possess chloroplasts containing chlorophyll, and most of them have a cell wall. You will study representatives of three major groups.

Euglenoids (Division Euglenophyta)

These unicellular protists possess both plantlike and animal-like characteristics. They have **chlorophyll a** and **b** in chloroplasts, a flagellum for movement, and an "eyespot" (stigma) that detects light intensity. A cell wall is absent. See Figure 6.5 and Plates 4.1 and 4.2.

Dinoflagellates (Division Pyrrophyta)

Most of these unicellular forms are marine and photosynthetic with **chlorophyll a** and **c** in their chloroplasts. Some are holotrophic, and a few can live as either holotrophs or photosynthetic autotrophs, depending on the circum-

stances. Most have a cell wall of cellulose, and all forms have two flagella. One lies in a groove around the equator of the cell, and the other hangs free from the end of the cell. When nutrients are abundant, certain marine species reproduce in such enormous numbers that the water turns a reddish color, a condition known as the red tide. Toxins produced by these forms may cause massive fish kills and make shellfish unfit for human consumption. See Plates 4.3 and 4.4.

Diatoms (Division Chrysophyta)

These protists are unicellular and have a cell wall of silica, a natural glass. The cell wall consists of two halves that fit together like the top and bottom of a box. **Chlorophyll a** and **c** are found in their chloroplasts. When diatoms die, the siliceous walls settle to the bottom. In some areas, they have formed massive deposits of diatomaceous earth that are mined and used commercially in a variety of products such as fine abrasive cleaners, toothpaste, filters, and insulation. Much of the atmospheric oxygen has been produced by photosynthesis in marine diatoms. See Plate 4.5.

Materials

Colored pencils
Construction paper, black
Medicine droppers
Microscope slides and cover glasses
Petri dish
Protoslo
Toothpicks
Cultures of:
   Euglena
   pond water
Diatomaceous earth
Prepared slides of:
   diatoms
   Gonyaulax.

Assignment 3

1. Use colored pencils to color-code chloroplasts and nucleus of *Euglena* in Figure 6.5.
2. Make a slide of a drop from the *Euglena* culture and examine it at 100× and 400×. Note the size, color, motility, and

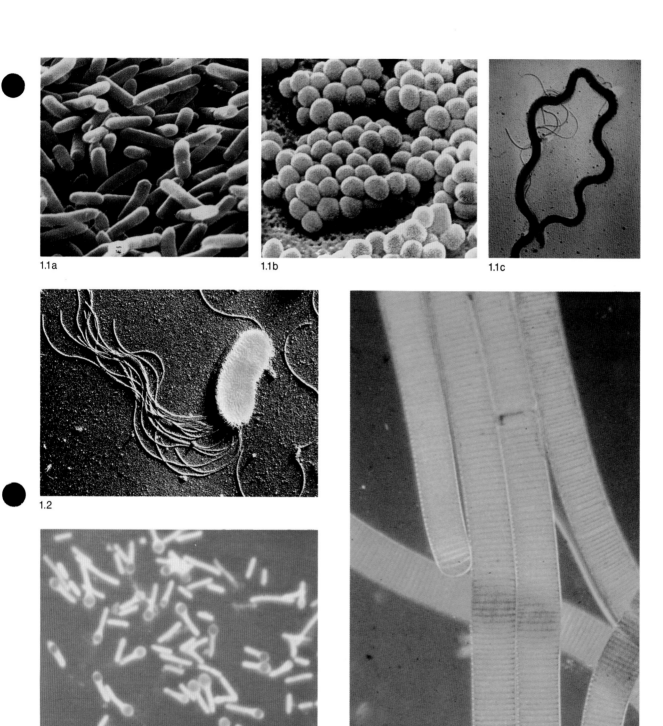

1.1a

1.1b

1.1c

1.2

1.3

1.4

Plate 1.1 The three basic shapes of bacterial cells. (a) A false-color scanning electron micrograph (SEM) of the rod-shaped cells of *Pseudomonas aeruginosa*, a chlorine-resistant form that sometimes causes skin infections among swimmers and users of hot tubs. 20,000X. (b) A false-color SEM of the spherical cells of *Staphylococcus aureus*, a pathogen causing boils and abscesses. 3,000X. (c) The spiral-shaped cell of *Treponema pallidum* which causes syphilis. 60,000X.

Plate 1.2 A false-color transmission electron micrograph (TEM) of *Pseudomonas fluorescens*, a motile soil bacterium with flagella. 10,000X.

Plate 1.3 *Clostridium tetani* causes tetanus (lockjaw). The enlarged end of some cells contains a spore that is resistant to unfavorable environmental conditions. 800X.

Plate 1.4 Filaments of *Oscillatoria*, a photosynthetic cyanobacterium, are composed of thin, circular cells joined together like stacks of coins. 100X.

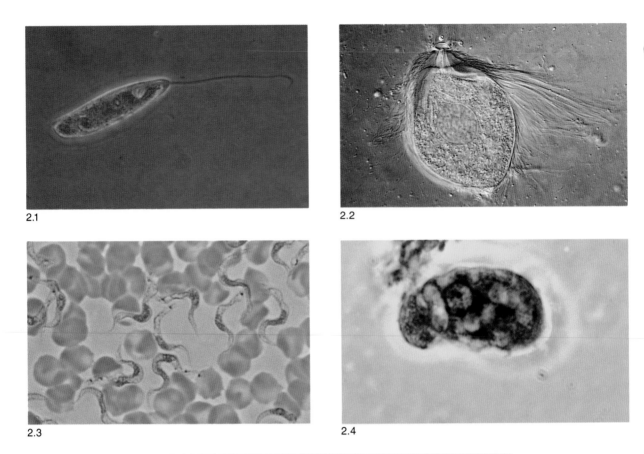

2.1

2.2

2.3

2.4

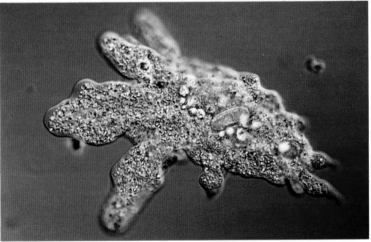

2.5

Plate 2.1 *Peranema trichophorum*, a freshwater flagellated protozoan, is propelled by a thick anterior flagellum held rigid and straight except for its tip. It engulfs prey through an opening near the base of the flagellum. A thin trailing flagellum is also present but is not visible here. 150X.

Plate 2.2 *Trichonympha* is a complex multiflagellated protozoan. It digests wood in the gut of termites. 135X.

Plate 2.3 *Trypanosoma brucei*, shown among human red blood cells, causes African sleeping sickness. It is transmitted by the bite of tsetse flies. 320X.

Plate 2.4 *Entamoeba histolytica* invades the lining of the intestine causing amoebic dysentery in humans. 600X.

Plate 2.5 *Amoeba proteus*, a common freshwater amoeboid protozoan. Food is digested within food vacuoles. Excess water is collected in contractile vacuoles and pumped out of the cell. 160X.

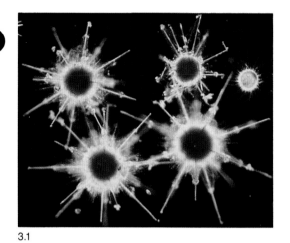

3.1

3.2

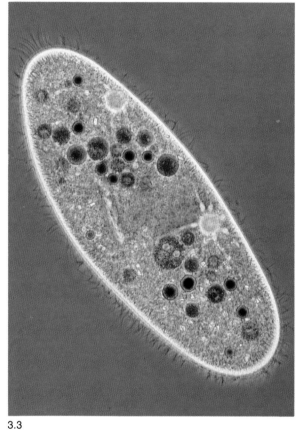

3.3

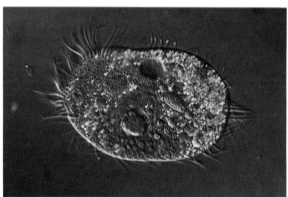

3.4

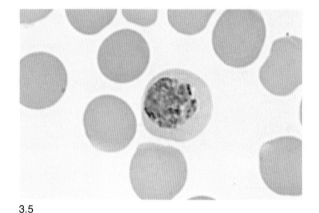

3.5

Plate 3.1 Living radiolarians. Radiolarians are marine amoeboid protozoans that secrete a siliceous shell. Slender pseudopodia extend through small openings in the shell.

Plate 3.2 A living foraminiferan. Foraminiferans are marine amoeboid protozoans that secrete a calcareous shell. Thin, ray-like pseudopodia extend through tiny openings in the shell.

Plate 3.3 *Paramecium* is a freshwater ciliate. Beating cilia function like little oars that move the cell through the water. Captured food organisms are digested in food vacuoles. Excess water is collected in contractile vacuoles and pumped out of the cell.

Plate 3.4 *Euplotes* is an advanced ciliate. Tufts of cilia unite to form cirri that function almost like legs as the ciliate "walks" over the bottom of a pond. Note the food vacuoles and a contractile vacuole that appears like a cavity in this view.

Plate 3.5 *Plasmodium vivax*, a sporozoan, in a human red blood cell. It causes malaria and is transmitted by the bite of female anopheline mosquitoes.

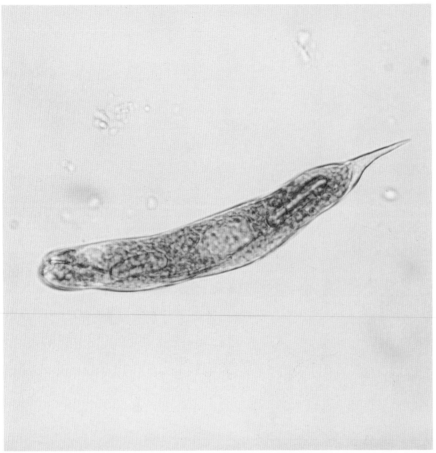

4.1

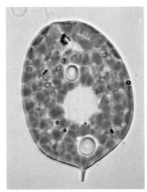

4.2

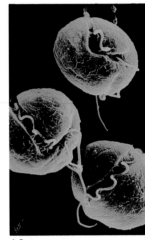

4.3

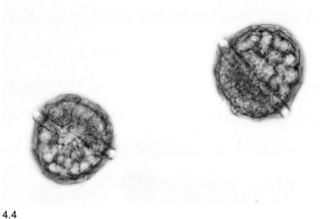

4.4

4.5

Plate 4.1 *Euglena* is a photosynthetic freshwater euglenoid protist. The eye spot detects light intensity. Food reserves are stored as paramylon, a complex carbohydrate. The flagellum is recurved along the underside of the cell in this photomicrograph.

Plate 4.2 *Phacus* is a freshwater euglenoid. Note the flagellum, eye spot, and chloroplasts.

Plate 4.3 Like all dinoflagellates, *Gymnodinium*, a marine form, possesses two flagella. One lies in an equatorial groove in the cell wall, and the other hangs from the end of the cell. (SEM 10,000X.)

Plate 4.4 *Peridinium* is a freshwater dinoflagellate common in acid-polluted lakes. Here, chloroplasts and equatorial groove are visible, but the flagella are not.

Plate 4.5 The siliceous walls of diatoms are formed of two parts that fit together like the top and bottom of a pillbox as in this SEM of *Cyclotella meneghinians*. 875X.

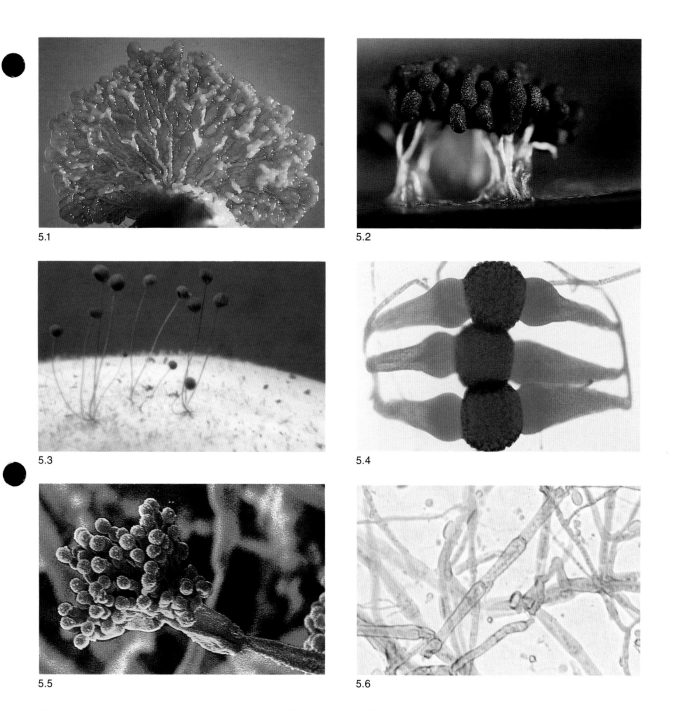

5.1

5.2

5.3

5.4

5.5

5.6

Plate 5.1 The plasmodium of the slime mold *Physarum* lives like a giant, multinucleate amoeba as it creeps along absorbing nutrients and engulfing food organisms.

Plate 5.2 When either food or moisture becomes inadequate, a *Physarum* plasmodium transforms into spore-forming sporangia supported by short stalks.

Plate 5.3 In *Rhizopus stolonifer*, black bread mold, sporangiophore hyphae support sporangia that form asexual spores. Each sporangiophore is attached to the substrate by rhizoid hyphae.

Plate 5.4 The fusion of (+) and (−) gametes in *Rhizopus stolonifer* results in the formation of a zygospore sandwiched between a pair of gametangia. Each zygospore later produces a spore-forming sporangium.

Plate 5.5 This false-color SEM of the asexual conidiospores of *Penicillium* shows their spherical shape and attachment to a sporangiophore. 5,750X.

Plate 5.6 The hyphae of *Nectria* are composed of separate cells joined end-to-end. Such hyphae are said to be septate because cross walls separate the cells.

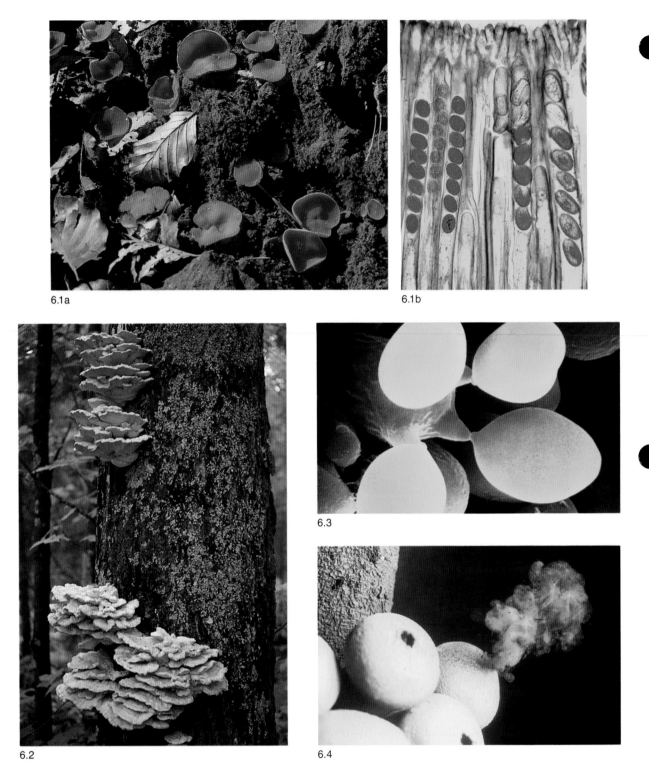

6.1a

6.1b

6.2

6.3

6.4

Plate 6.1   (a) Ascocarps (fruiting bodies) of *Peziza aurantia*, a sac fungus. (b) Ascospores are formed in the sac-like asci after a complex sexual process.

Plate 6.2   Basidiocarps (fruiting bodies) of shelf fungi, a club fungus, contain spore-forming basidia. The basidiospores are released through tiny pores on the lower surface of the basidiocarps.

Plate 6.3   Each basidium of a club fungus produces four basidiospores as shown in this SEM. 1,100X.

Plate 6.4   A discharge of millions of basidiospores from a puffball, a club fungus.

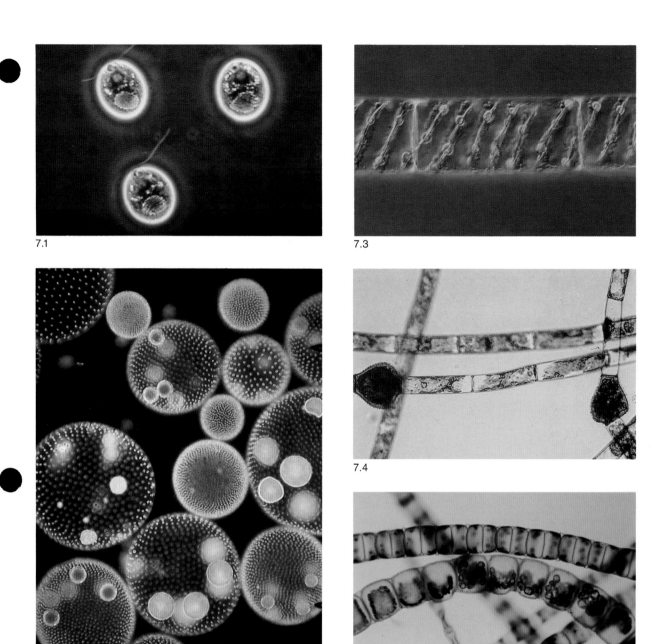

7.1

7.3

7.4

7.2

7.5

Plate 7.1 *Chlamydomonas* is a tiny unicellular green alga that moves by means of two flagella. Only the basal portion of the cup-shaped chloroplast is visible here. An eye spot, not visible here, enables detection of light intensity.

Plate 7.2 *Volvox* colonies may consist of up to 50,000 *Chlamydomonas*-like cells. *Volvox* may reproduce asexually by forming daughter colonies within the sphere. Some forms also have a few cells specialized for sexual reproduction.

Plate 7.3 The spiral chloroplast in a cell of *Spirogyra*, a filamentous green alga. The enlarged regions are pyrenoids, sites of starch storage.

Plate 7.4 Cellular specialization for sexual reproduction is evident in *Oedogonium*, a filamentous green alga. The enlarged, dark cells are oogonia which contain an oospore formed after union of sperm and egg cells.

Plate 7.5 Cellular specialization for reproduction also occurs in *Ulothrix*, another filamentous green alga. The cells in the lower filament are sporangia containing asexual zoospores.

8.1

8.2

8.3

8.4

8.5

Plate 8.1 *Ulva* or sea lettuce is a marine green alga that exhibits alternation of generations. Note the holdfast.

Plate 8.2 Brown marine algae are characterized by a robust body that can withstand strong wave action. Note the holdfast, stipe, and divided blade in *Durvillaea*, a brown alga from the Australian coast.

Plate 8.3 A forest of giant kelp, *Macrocystis*, along the California coast provides an important habitat for marine animals. The long stipes may be over 100 ft. in length.

Plate 8.4 The delicate, feathery body structure of red algae sharply contrasts with the robust brown algae. Red algae, like *Callithamnion*, live at ocean depths where wave action is minimal.

Plate 8.5 Gas-filled floats of *Macrocystis* lift the blades and stipe toward the water surface where more sunlight is available.

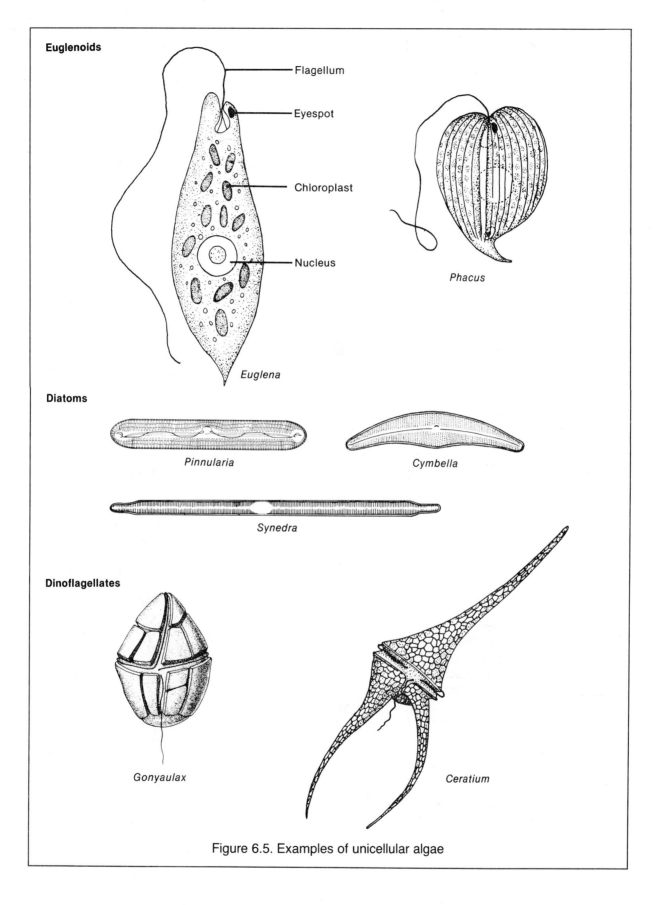

Euglenoids

Flagellum

Eyespot

Chloroplast

Nucleus

Euglena

Phacus

Diatoms

Pinnularia

Cymbella

Synedra

Dinoflagellates

Gonyaulax

Ceratium

Figure 6.5. Examples of unicellular algae

the absence of a cell wall. Is more than one method of motility evident? Which end is anterior?

3. Observe the distribution of *Euglena* in the partially shaded petri dish on the demonstration table. *Do not move the dish.* Remove the cover to make your observation, and immediately return the cover to its *exact prior position.*

4. **Complete item 3a to 3c on the laboratory report**.

5. Observe a prepared slide of *Gonyaulax*, a dinoflagellate that causes the red tide along the Pacific Coast. Note the positions of the flagella and the cell wall.

6. Make a slide of a drop from the mixed dinoflagellate culture and examine it with your microscope. Note the cellulose cell wall and the spinning movement of the organisms caused by the positions of the flagella.

7. **Complete item 3d on the laboratory report**.

8. Examine a prepared slide of diatoms. Note the symmetry and pattern of the cell walls.

9. Examine the sample of diatomaceous earth, noting its fine texture. Make a wet-mount slide of a small amount of the dustlike particles, and observe the sample with a microscope. Does your sample contain more than one species?

10. **Complete item 3e on the laboratory report**.

Slime Molds: Funguslike Protists

The feeding stages of slime molds resemble amoeboid protozoans, but they reproduce by forming sporangia that produce spores. This latter funguslike characteristic explains why they previously were classified as fungi.

Cellular Slime Molds (Division Acrasiomycota)

Members of this group live as single-celled amoeboid organisms engulfing bacteria in leaf litter and soil. Poor environmental conditions cause the cells to congregate in what is known as the "swarming stage" that results in the formation of a sporangium containing spores. *Dictyostelium* is an example of this group.

Plasmodial Slime Molds (Division Myxomycota)

Feeding stages of plasmodial slime molds consist of strands of protoplasm streaming along in amoeboid fashion and engulfing bacteria on leaf litter, rotting wood, and the like. The large plasmodium is multinucleate. Unfavorable conditions stimulate the plasmodium to migrate and ultimately to form a sporangium with spores. *Physarum* is an example of a plasmodial slime mold. See Plates 5.1 and 5.2.

Materials

Cultures of slime molds:
  Dictyostelium
  Physarum

Assignment 4

1. Examine the macroscopic and microscopic demonstrations of cellular and plasmodial slime molds. Note the organization of the nonreproductive stages and the sporangia formed during the reproductive stages.

2. **Complete item 4 on the laboratory report**.

Assignment 5

1. Make and examine slides of pond water, and observe the monerans and protists that are present. Use reduced illumination for best viewing. **Draw several observed specimens in item 5a on the laboratory report**. Make your drawings large enough to show the major features of the organisms.

2. Review what you have learned in the laboratory session, and **complete item 5b on the laboratory report**.

3. Your instructor has set up several "unknown specimens" for you to identify under numbered demonstration microscopes. **Match the specimens with the correct responses in item 5c on the laboratory report by placing the numbers of the microscopes in the spaces provided**.

7

FUNGI

OBJECTIVES

On completion of the laboratory session, you should be able to:
1. Describe the distinguishing characteristics of the organismic groups studied.
2. Identify representatives of fungi, lichens, and algae.
3. Define all terms in bold print.

The **kingdom Fungi** contains a large and diverse group of heterotrophic organisms that occur in freshwater, marine, and terrestrial habitats. Terrestrial fungi reproduce by **spores**, dormant reproductive cells that are dispersed by wind and that germinate to form a new fungus when conditions are favorable. Most fungi are multicellular; only a few are unicellular. Fungi are either **saprotrophs** or **parasites**. Most species are saprotrophs and play a beneficial role in decomposing nonliving organic matter, but some saprotrophs cause serious damage to stored food products. Rusts and mildews are important plant parasites. Ringworm and athlete's foot are common human ailments caused by fungi.

The vegetative (nonreproductive) body of a multicellular fungus is called a **mycelium**, and it is composed of threadlike filaments, the **hyphae**. Hyphae are formed of cells joined end to end. The cells of hyphae may be separated by cell walls (septate hyphae), or the cell walls may be incomplete or lacking (non-septate hyphae). The cell walls are formed of chitin. Nutrients are obtained by hyphae secreting digestive enzymes into the surrounding substrate, which is digested extracellularly. The resulting nutrients are then absorbed into the hyphae.

Spores of true fungi are formed either from terminal cells of reproductive hyphae or in **sporangia**, enlarged structures at the ends of specialized hyphae. Some fungi form **fruiting bodies** that contain the spore-forming hyphae.

CONJUGATING MOLDS (DIVISION ZYGOMYCOTA)

The common black bread mold, *Rhizopus stolonifer,* is an example of this group that is characterized by the three types of hyphae shown in Figure 7.1. **Stolon hyphae** spread over the surface as the mycelium grows. **Rhizoid hyphae** penetrate the substrate to digest it and to provide anchorage. **Sporangiophores** are upright hyphae that form a **sporangium** at their tips. Asexual spores are formed within the sporangia. See Plates 5.3 and 5.4. The life cycle of *Rhizopus* is studied in Exercise 28.

SAC FUNGI (DIVISION ASCOMYCOTA)

Yeasts, mildews, molds, and cup fungi belong to this group. The division name derives from the fact that the **ascospores**, which are formed after a complex sexual process, occur in tiny

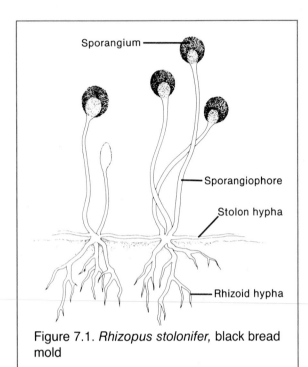

Figure 7.1. *Rhizopus stolonifer,* black bread mold

sacs (**asci**). Most forms, especially mildews and molds, produce asexual spores, **conidiospores**, at the end of upright hyphae. See Plates 5.5, 6.1 and 6.2.

Many of the Ascomycota are important parasites of plants and animals, but some have commercial value. Some members of the genus *Penicillium* are used to either produce antibiotics or to flavor cheeses. The yeast, *Saccharomyces cerevisiae,* is used to produce ethyl alcohol in the brewing and wine-making industries and to release bubbles of carbon dioxide in the baking industry.

CLUB FUNGI (DIVISION BASIDIOMYCOTA)

The familiar mushrooms, puffballs, and shelf fungi belong to this group. Also included are rusts and smuts, serious plant parasites that cause enormous losses in wheat, corn, and other cereal crops each year. **Basidiospores** are formed after a sexual process by club-shaped **basidia**, terminal cells of reproductive hyphae, which are usually located in a fruiting body.

You will study the mushroom as an example of club fungi. The hyphae of the mycelium grow

through the soil where they absorb nutrients derived from extracellular digestion of nonliving organic matter. After fusion of uninucleate hyphae in a sexual process, a fruiting body is formed by the binucleate hyphae. This fruiting body is recognized as a mushroom. The basidiospores are formed by **basidia**, specialized cells on the gills of the fruiting body. The life cycle of a mushroom will be studied in Exercise 28. See Plates 6.3 to 6.5.

Materials

Colored pencils
Dissecting instruments
Medicine droppers
Microscopes slides and cover glasses
Water-detergent solution in dropping bottle
Mushrooms, fresh
Penicillium sp. and *Rhizopus stolonifer*
   growing on:
     plate of Sabouraud's agar
     germinating spores on slide coated with
       Sabouraud's agar
     citrus fruit and moist bread, respectively
Peziza sp. cups
Yeast culture in 5% glucose
Representative fungi
Prepared slides of:
   mushroom gills with basidia
Peziza sp. cup, l.s.

Assignment 1

1. ***Complete item 1 on Laboratory Report 7 that begins on page 377***.

Assignment 2

1. Examine the black bread mold growing on agar with a dissecting microscope. Spores were placed in the center of the agar plate to start the colony. Note the pattern of growth.
2. Use forceps to remove a small portion of the fungus growing on bread and mount it in a drop of water-detergent solution on a clean slide. Tease away the bread as necessary and observe at 40× and 100× without a cover glass. Locate the different types of hyphae and sporangia.
3. Examine hyphae growing on an agar-coated microscope slide at 400×. Are they septate? (See Plate 5.6.) Locate a growing

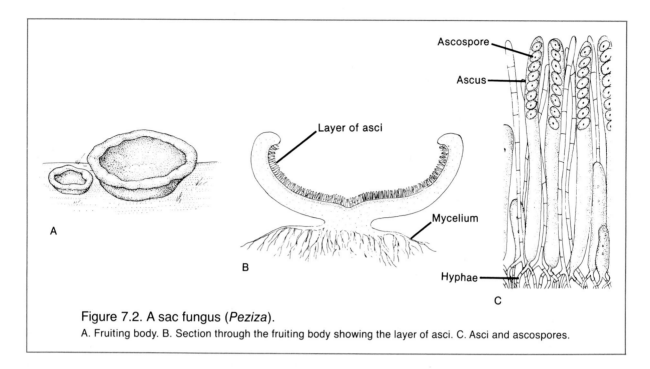

Figure 7.2. A sac fungus (*Peziza*).
A. Fruiting body. B. Section through the fruiting body showing the layer of asci. C. Asci and ascospores.

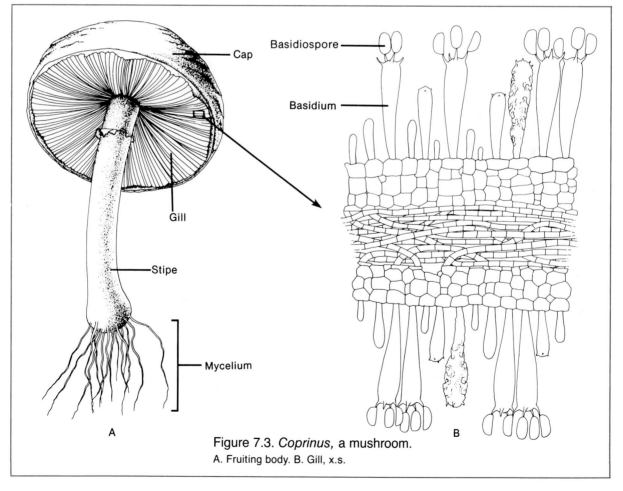

Figure 7.3. *Coprinus*, a mushroom.
A. Fruiting body. B. Gill, x.s.

tip and observe it for a short time. If you are patient, you can see it grow.

4. ***Complete item 2 on the laboratory report***.

Assignment 3

1. Examine a colony of *Penicillium* with a dissecting microscope. Recognizing that the colony grows outward, locate the older and new hyphae. Note the color of immature and mature conidiospores.

2. Use the tip of a scalpel to remove a few hyphae with mature conidiospores and make a water-mount slide of them. Examine the hyphae at 400× to note the pattern of conidiospore formation.

3. Examine hyphae of *Penicillium* growing on an agar-coated slide at 400×. Are they septate?

4. ***Complete item 3a on the laboratory report***.

5. Make a slide of the yeast culture, add a cover glass, and observe at 400×. Yeasts are unicellular sac fungi that reproduce asexually by budding. ***Draw a few cells in item 3b on the laboratory report***.

6. Examine the fruiting body of a cup fungus *Peziza* and compare it with Figure 7.2. Use colored pencils to color-code labeled structures. The sacs (asci) containing the ascospores line the cup. Asci and ascospores are formed by binucleate hyphae following the fusion of uninucleate hyphae in a sexual process.

7. Examine a prepared slide of *Peziza* cup, l.s. Observe the asci, ascospores, and the two types of hyphae forming the cup. Note the number of nuclei in the hyphal cells. How many nuclei are in the cells of hyphae that form asci? Are the hyphae septate?

8. ***Complete item 3 on the laboratory report***.

Assignment 4

1. Examine the fruiting body of a mushroom. Locate the parts shown in Figure 7.3. Use colored pencils to color-code labeled structures.

2. Use a scalpel to remove a very thin longitudinal section of mushroom stipe and make a water-mount slide of it. Examine your slide to see if the hyphae are septate.

3. Examine a prepared slide of mushroom gill. Locate the hyphae, basidia, and basidiospores. How many nuclei are in the hyphal cells?

4. Examine other examples of club fungi.

5. ***Complete item 4 on the laboratory report***.

LICHENS

A lichen is a **symbiotic relationship** of a fungus and an alga. The "organism" resulting from this association has unique characteristics that are different from those of either the alga or fungus composing it. Lichens are usually classified according to the fungus species involved, however. Most of the fungi found in lichens are members of the sac fungi, and the algae are green algae.

The fungus of a lichen provides the algal cells with support and protection from high light intensities and desiccation. The alga provides the fungus with organic nutrients produced by photosynthesis. This association allows a lichen to live in habitats that are unfavorable to either fungus or alga if living alone. Most, but not all, biologists consider this symbiotic relationship to be **mutualism**, where both members benefit from the association. Lichens serve as important soil builders by eroding rock surfaces and paving the way for more advanced forms of life.

Lichens have three types of body forms. **Crustose lichens** consist of a thin crust that is tightly attached to rocks. **Foliose lichens** are a bit thicker than crustose lichens but have a flattened body form. They are the most common types of lichens and usually occur on rocks. **Fruticose lichens** have an erect, often branched, body form and may occur on trees and soil as well as on rocks.

Materials

Representative lichens
Prepared slides of lichen, x.s.

Assignment 5

1. Examine the representative lichens and note the type of body form of each one.
2. Examine a prepared slide of lichen, x.s., and locate the fungal hyphae and the algal cells. Is there a pattern to the location of the algal cells?
3. ***Complete item 5 on the laboratory report.***

Assignment 6

1. Review your understanding of fungi and the distinguishing characteristics of each group.
2. Your instructor has set up several "unknown specimens" for you to examine and identify as to the fungus group to which they belong.
3. ***Complete item 6 on the laboratory report.***

8

ALGAE

OBJECTIVES

On completion of the laboratory session, you should be able to:
1. Describe the characteristics of the algal groups studied.
2. Identify representatives of the three lines of algae.
3. Identify and state the functions of the parts of the algae studied.
4. Define all terms in bold print.

Some biologists place green, brown, and red algae in the kingdom Protista because of the simpler forms of these groups, especially the unicellular and colonial members of the green algae. Other biologists consider them to be plants, due to the complex life cycles of advanced forms and the affinity of complex green algae with land plants. In this book, the green, brown, and red algae join the multicellular land plants in the **kingdom Plantae**.

Plants are photosynthetic **autotrophs**. They have cells with walls of cellulose, pigments contained in **plastids**, and a large **central vacuole** in mature cells. Plants also have an **alternation of generations** in their life cycles. See Figure 8.1. In such a life cycle, there are two adult generations instead of only one. The **gametophyte** produces gametes (sperm and eggs) that unite in fertilization to yield a **zygote**. The zygote grows to become a mature **sporophyte** that produces spores. Upon germination, the spores develop into the gametophytes of the next generation.

Algae do not possess vascular tissue or true roots, stems, or leaves, although they may have structures that resemble them. There are three evolutionary lines of complex algae, as evidenced by the pigments in their chloroplasts. See Table 8.1.

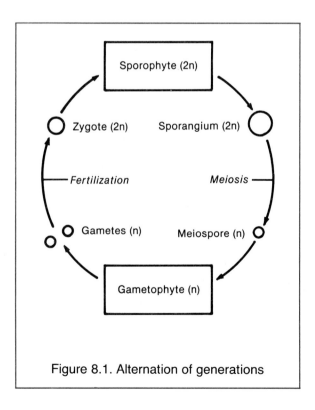

Figure 8.1. Alternation of generations

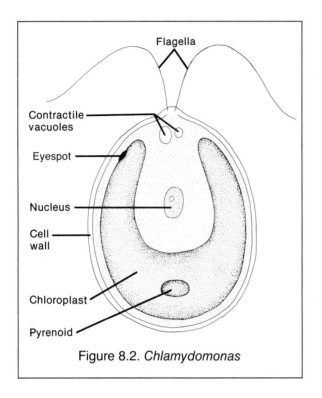

Figure 8.2. *Chlamydomonas*

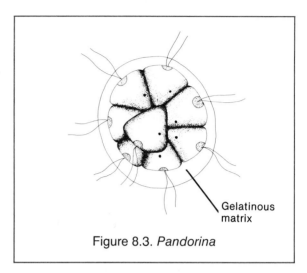

Figure 8.3. *Pandorina*

GREEN ALGAE (DIVISION CHLOROPHYTA)

Green algae may be unicellular, colonial, or multicellular. Most species occur in freshwater or marine habitats, but a few occur in moist areas on land. The presence of (1) **chlorophylls a** and **b** in chloroplasts, (2) cellulose cell walls, (3) starch as the nutrient storage form, and (4) whiplash flagella on motile cells suggests that green algae are ancestral to higher plants. See Figures 8.2–8.4, and Plates 7.1 to 7.5 and 8.1.

TABLE 8.1
Evolutionary Lines of Algae

Evolutionary Line	Characteristic Pigments
Green	Chlorophylls a and b Carotenoids
Brown	Chlorophylls a and c Fucoxanthin
Red	Chlorophylls a and d Phycobilins

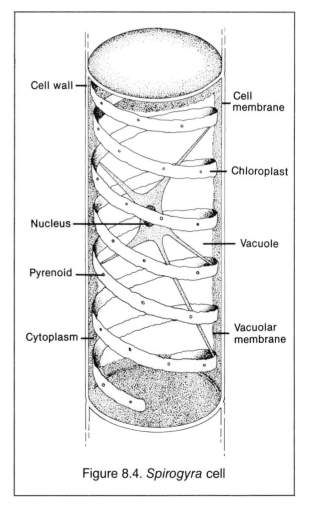

Figure 8.4. *Spirogyra* cell

BROWN ALGAE
(DIVISION PHAEOPHYTA)

These multicellular algae are almost exclusively marine and are often called seaweeds because of their abundance along rocky seacoasts. Their brownish color results from the presence of the pigment **fucoxanthin** in addition to **chlorphylls a** and **c** in the chloroplasts. The typical structure of a brown alga includes a **holdfast**, which anchors the alga to a rock, a **stipe** (stalk), and **blades**, which are leaflike structures. See Figure 8.5. All parts of the plant carry on photosynthesis, but the blades are the most important photosynthetic organs. Blades of large brown algae often have air-filled bladders associated with them to hold them near the surface of the water. Flagellated cells have one whiplash and one tinsel flagellum. See Plates 8.2 to 8.4.

RED ALGAE
(DIVISION RHODOPHYTA)

Most red algae are marine and occur at greater depths than brown algae. They have a delicate body structure, probably because they are not subjected to strong wave action. Like brown algae, they are attached to rocks by holdfasts. See Figure 8.6 and Plate 8.5. The presence of red pigments, **phycobilins**, in addition to **chlorophylls a** and **d**, allows them to absorb the deeper-penetrating wavelengths of light for photosynthesis. Flagellated cells are absent.

Materials

Colored pencils
Construction paper, black
Dissecting instruments
Medicine droppers
Microscope slides and cover glasses
Petri dish
Cultures of:
    Chlamydomonas (or *Carteria*)
    Pandorina
    Oedogonium
    Spirogyra
    Volvox
Representative green, brown, and
    red algae

Assignment 1

1. Use colored pencils to color code labeled structures in the figures.
2. ***Complete item 1 on Laboratory Report 8 that begins on page 381.***

Assignment 2

1. Make a slide of a drop from the *Chlamydomonas* culture and examine it at 400×. Note the color, movement, and size of the organisms. Compare your specimens with Figure 8.2 and Plate 7.1. Note the "eyespot" (stigma), which is sensitive to light intensity, and the two flagella, which enable motility.
2. Observe the distribution of *Chlamydomonas* in a partially shaded petri dish. Remove the cover to make your observation; then replace it exactly as you found it. How do you explain the distribution?
3. Make and observe slides of the *Pandorina* and *Volvox* cultures. See Figure 8.3 and Plate 7.2. Note the similarity of the individual cells of these colonial algae to *Chlamydomonas*. Colonial forms are believed to have originated from unicellular organisms that remained attached after cell division.
4. Make and examine a slide of *Spirogyra*, a filamentous green alga. Note the spiral chloroplast in the cells. Compare your specimen with Figure 8.4 and Plate 7.3. Pyrenoids are sites of starch accumulation.
5. Make and examine a slide of *Oedogonium*, a filamentous green alga, and note the different types of cells in the filament. This represents an early stage leading to the division of labor among various types of cells found in multicellular forms. See Plate 7.4.
6. ***Complete item 2 on the laboratory report.***

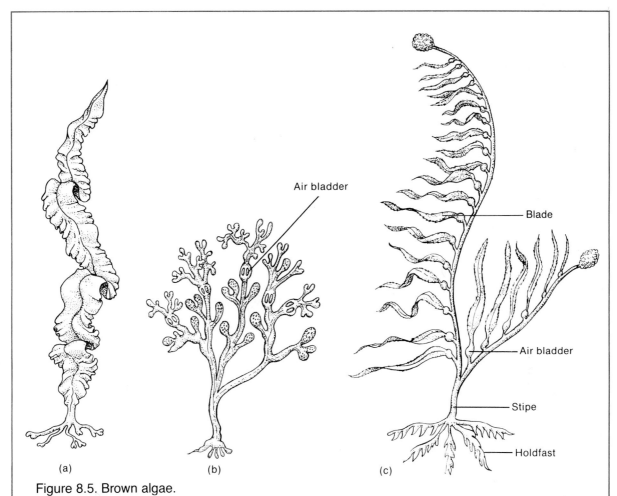

Figure 8.5. Brown algae.
The illustrations are not drawn to scale. *Laminaria* (a) and *Fucus* (b) are much smaller than *Macrocystis* (c), a giant kelp.

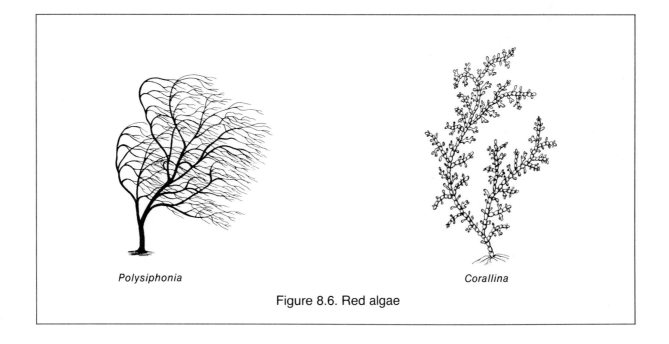

Polysiphonia

Corallina

Figure 8.6. Red algae

Assignment 3

1. Examine the specimens of brown and red algae, and locate the parts shown in Figures 8.5 and 8.6.
2. ***Complete item 3 on the laboratory report.***

Assignment 4

1. Examine the "unknown" algae.
2. ***Complete the laboratory report.***

TERRESTRIAL PLANTS

9

OBJECTIVES

On completion of the laboratory session, you should be able to:
1. Describe the characteristics of the plant groups studied.
2. Identify representatives of moss plants and vascular plants.
3. Identify and state the functions of the parts of flowering plants.
4. Define all terms in bold print.

The two major lineages of land plants are bryophytes and vascular plants. Nearly all are photosynthetic autotrophs that have **chlorophylls a** and **b** and **carotenoids** in their **chloroplasts**. Cell walls are composed of cellulose. Motile cells possess whiplash flagella. The life cycle consists of an **alteration of generations**, which originated in advanced algae. See Figure 8.1. Note the *two* adults for each plant species: a **gametophyte** and a **sporophyte**. Modification of this life cycle has played a central role in the colonization of land by plants.

MOSS PLANTS (Division Bryophyta)

Moss plants, liverworts, and hornworts compose the bryophytes. They occur in habitats that are moist for at least part of the year because surface water is required for sperm to swim to the eggs. Most are not well adapted for terrestrial life because they lack vascular tissues (xylem and phloem) that conduct water, minerals, and nutrients. They also lack true leaves, stems, and roots.

Mosses are the most common bryophytes. The basic structure of a moss gametophyte and sporophyte is shown in Figure 9.1. The gametophyte is the **dominant generation**, which means that it is larger and lives longer than the sporophyte. The gametophyte possesses **leaflike** and **stemlike structures** and **rhizoids**, filaments that anchor it to the soil. Gamete-forming organs are located at the top of the gametophyte. After the egg has been fertilized in the female reproductive organ, the resulting zygote grows to become the sporophyte that remains attached to the gametophyte and, in fact, is a partial parasite.

The sporophyte consists of a **seta** (stalk) and **sporangium** (capsule). When mature, spores are released from the capsule and distributed by wind. The sporophyte dies and later falls off the gametophyte.

Materials

Colored pencils
Representative mosses and liverworts, living
Moss gametophytes with sporophytes, living or preserved
Prepared slides of:
   moss "stem," x.s.
   moss "leaf," x.s.

Assignment 1

1. Color-code labeled structures in Figure 9.1.

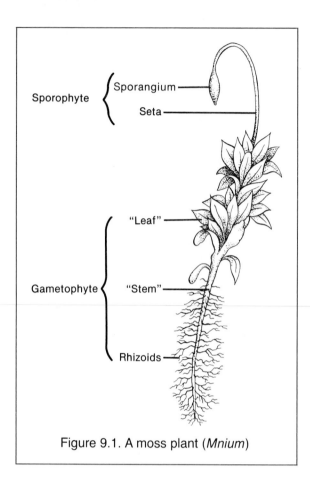

Figure 9.1. A moss plant (*Mnium*)

2. Examine the living moss and liverwort gametophytes. Although their appearance is quite different, both are anchored to the soil by rhizoids. Note their similarities and differences.

3. Examine a moss gametophyte with sporophyte attached. Use a dissecting microscope for your observations, if desirable. Locate the parts labeled in Figure 9.1.

4. Examine a prepared slide of moss "stem," x.s., and note that it is composed of only a few types of tissues. The central column of cells aids water movement, but it is not composed of vascular tissue.

5. Examine a prepared slide of moss "leaf," x.s. Observe that it also is composed of only a few types of tissues. Vascular tissue is absent.

6. ***Complete item 1 on Laboratory Report 9 that begins on page 383***.

VASCULAR PLANTS

Vascular plants are more advanced than moss plants in at least three important ways: (1) **vascular tissue** is present, (2) the sporophyte is the dominant generation, and (3) true **roots, stems**, and **leaves** are found in the sporophyte.

Vascular tissue enables efficient transport of materials within the plant. Without this adaptation, all plants would not be much bigger than mosses and could not exist in drier habitats. There are two vascular tissues: xylem and phloem. **Xylem** transports water and minerals upward from the roots to the shoot system of the plant. **Phloem** carries organic nutrients upward or downward within the plant.

The sporophyte is the dominant generation and is the adult form that is recognizable as a plant. The gametophyte is small and usually microscopic in size. Reproduction in land plants will be studied in Exercise 28.

In this exercise, you first will study the basic characteristics of the major plant groups and then examine the structure of flowering plants.

Ferns (Division Pterophyta)

Fern sporophytes typically have an underground stem, a **rhizome**, which is anchored by roots. The large **leaves** are attached to the stem and consist of many leaflets joined to the midrib. On the undersurface of the leaflets of certain leaves are small brown spots, the **sori**. These consist of numerous sporangia that form the spores. The vascular tissue in ferns is not as well developed as it is in more advanced vascular plants, but it still makes ferns better adapted to terrestrial life than mosses.

Fern gametophytes are small, flat plants that are anchored to the soil by rhizoids. The gamete-forming organs are located on the lower surface of the gametophytes. Like sporophytes, gametophytes are photosynthetic autotrophs, but they are shortlived. Surface water is required for the sperm to swim to the eggs as in mosses, so ferns usually are restricted to habitats that are moist for a part of the year.

Seed Plants

The most successful vascular plants produce both pollen and seeds. **Pollen** is male gametophytes that produce sperm nuclei. **Pollination** is the transfer of pollen from the pollen-producing portion to the egg-producing portion of sporophytes. Pollination occurs by either wind or insects and eliminates the need for water in sperm transport. A **seed** contains an **embryo sporophyte** and stored nutrients for the growth of the young sporophyte. Both are surrounded by a protective **seed coat** that enables survival of the young sporophyte during periods of unfavorable conditions.

A sporophyte consists of a **shoot system**, composed of stem and leaves, and a **root system** that grows underground in most cases.

Seed plants may be categorized into two groups according to the location of their seeds. Plants whose seeds are borne exposed on a leaf or scale are called **gymnosperms**. Plants whose seeds are enclosed within a fruit are called **angiosperms**.

Cone-Bearing Plants (Division Coniferophyta)

Conifers are the best known and largest group of gymnosperms. The cones contain the reproductive organs, and two types of cones are formed. **Male** (pollen) **cones** are small with paper-thin scales. **Female** (seed) **cones** are large with woody scales. Pollen is transferred from male cones to female cones by wind. Seeds are borne exposed on the upper surface of the scales of mature female cones. The leaves are either needlelike or scalelike. Conifers may attain considerable size and may live in rather dry habitats because their vascular tissue is well developed, their leaves restrict water loss, and water is not required for sperm transport.

Flowering Plants (Division Anthophyta)

Flowering plants are the most advanced plants. Their success in colonizing the land is due to well-developed vascular tissues and **flowers** (reproductive organs) that greatly enhance reproductive success. Most flowers attract insects that bring about pollination, a process leading to the fertilization of egg cells by sperm nuclei. Seeds are enclosed within **fruits** that facilitate dispersal.

Flowering plants are subdivided into two classes: **monocotyledonous** (monocots) **plants** and **dicotyledenous** (dicots) **plants**. Grasses,

palms, and lilies are examples of monocots. Daisies, beans, and oaks are examples of dicots. Dicots may be either **herbaceous**, usually annuals or biennials, or **woody**, usually perennials. Figure 9.2 shows the distinguishing characteristics of each class.

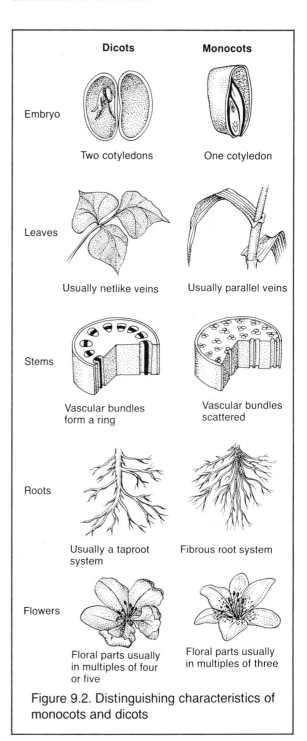

Figure 9.2. Distinguishing characteristics of monocots and dicots

External Structure of Flowering Plants

Figure 9.3 shows the basic parts of a flowering plant. The shoot system consists of a **stem** that supports the **leaves, flowers**, and **fruits**. Leaves branch from the stem at sites called **nodes**. A section of stem between nodes is an **internode**. Leaves are the primary photosynthetic organs of the plant, and they exhibit two types of venation: net venation and parallel venation. Leaves with **net venation** have a central vascular bundle (vein) called a midrib from which smaller lateral veins branch. Such leaves consist of a thin, expanded portion called a **blade** and a leaf stalk called a **petiole**. Leaves with **parallel venation** lack a midrib and have veins running parallel to each other for the length of the blade. Such leaves usually lack a petiole.

The root system is located in the ground, and it may be even more highly branched than the shoot system. Roots not only anchor the plant, but they absorb water and nutrients as well.

Figure 9.4 shows the basic structure of a flower. The colorful **petals** are used to attract insects for pollination. The **sepals** provide protection during the bud stage and are often green in color. The **stamens** are the male portion of the flower. A stamen consists of a thin **filament** that supports a pollen-producing **anther**. The **pistil** is the female portion. It consists of three parts. The upper tip, the **stigma**, receives pollen in pollination, and the slender **style** connects the stigma to the ovary. The enlarged basal portion, the **ovary**, contains egg cells in **ovules** (female gametophytes). After pollination, **pollen grains** (male gametophytes) grow long tubes that pass down the style to the ovary and carry sperm nuclei that fertilize the egg cells in the ovules. After fertilization, the ovules grow to become seeds. The **receptacle** is the base of the flower to which the ovary, sepals, and petals are attached.

After fertilization, the seeds and ovary grow. A fully developed (ripened) ovary is a **fruit** containing mature seeds. Apples, tomatoes, pea pods, and coconuts are examples of fruits. Fruits may be fleshy or dry. Some dry fruits split open to release the seeds. Reproduction in flowering plants will be studied in more detail Exercise 28.

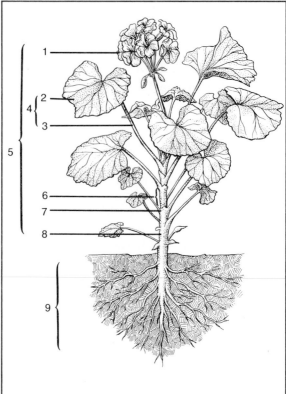

Figure 9.3. External structure of a flowering plant.

Label the parts of the plant by placing the number of each structure in the space by the correct label.

_____Blade	_____Leaf	_____Root system
_____Flower	_____Node	_____Shoot system
_____Internode	_____Petiole	_____Stem

Materials

Colored pencils
Razor blades, single edge
Red food coloring solution
Ferns:
   sporophytes, living
   prepared slides of gametophytes
Conifers:
   scalelike and needlelike leaves
   female and male cones
Flowering plants:
   representative monocots and dicots
   Zea (corn) seedlings

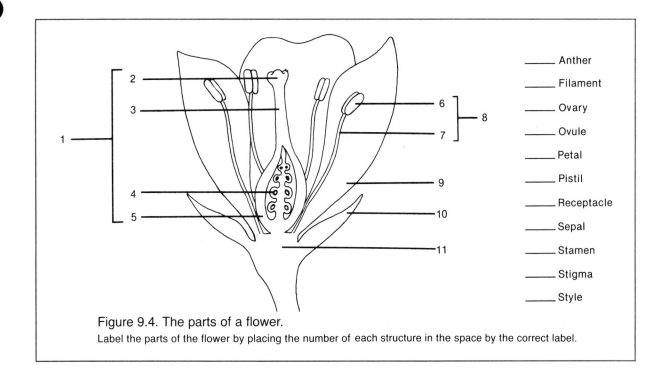

Figure 9.4. The parts of a flower.

Label the parts of the flower by placing the number of each structure in the space by the correct label.

_____ Anther
_____ Filament
_____ Ovary
_____ Ovule
_____ Petal
_____ Pistil
_____ Receptacle
_____ Sepal
_____ Stamen
_____ Stigma
_____ Style

Coleus plants, small
Gladiolus flowers
fruits, various types

Assignment 2

1. ***Complete items 2a and 2b on the laboratory report.***
2. Examine the living fern sporophytes that have been washed to expose the rhizomes and roots. Note the relationships of the rhizome, leaves, and roots.
3. Observe that a fern leaf is large and consists of numerous leaflets that branch from the midrib. The midrib and its branches contain vascular tissue that is continuous with vascular tissue of the rhizome and roots. Are sori present on the undersurface of some leaflets?
4. Observe the prepared slide of a fern gametophyte. Note the rhizoids on its undersurface that anchor it to the soil. Male (near the rhizoids) and female (near the notch) reproductive organs are also visible on the undersurface of the gametophyte.

5. ***Complete item 2 on the laboratory report.***

Assignment 3

1. Observe the twigs of conifers with scalelike and needlelike leaves. Note how the scalelike leaves are arranged on a twig and how needlelike leaves occur either singly or in groups of a definite number.
2. Examine the female and male cones. Compare their size, weight, and hardness. Are seeds still present on the scales of the female cone?
3. ***Complete item 3 on the laboratory report.***

Assignment 4

1. Label Figures 9.3 and 9.4. Color-code the roots, stem, and leaves in Figure 9.3 and the sepals, petals, pistil, and stamens in Figure 9.4. ***Complete items 4a–4c on the laboratory report.***
2. Obtain and examine a small *Coleus* plant. Compare it with Figures 9.2 and 9.3. Note the venation of the leaves.

3. Stems of several *Coleus* plants have been severed and placed in a water-soluble dye. As the dye is transported up the stem, it stains the xylem, allowing easy identification of vascular tissue. Use a razor blade to cut a thin cross section from one of these *Coleus* stems. Place the section on a clean slide and examine it with a dissecting microscope. Determine the distribution of the vascular tissue by comparing your observations with Figure 9.2.

4. ***Complete items 4d–4f on the laboratory report***.

5. Obtain a corn seedling. Gently wash the soil from the roots. Note the arrangement of the roots and the venation of the leaves. Compare it with Figures 9.2 and 9.3.

6. Use a razor blade to cut a thin cross section from a corn stem that was placed in water-soluble dye. Examine the section with a dissecting microscope and determine the distribution of the vascular tissue. Compare it with Figure 9.2.

7. ***Complete items 4g–4k on the laboratory report***.

8. Obtain and examine a *Gladiolus* flower. Locate the parts as shown in Figure 9.4. Examine an anther with a dissecting microscope and observe the pollen grains. Use a razor blade to make a cross section of the ovary, and examine it with a dissecting microscope to see how the ovary is divided into chambers and to locate the ovules.

9. ***Complete items 4l–4n on the laboratory report***.

10. Examine the fruits provided. Some fruits have been sectioned to expose the seeds. What evidence indicates that fruits are ripened ovaries? What flower parts are still attached, if any?

11. Examine the representative monocots and dicots provided. Compare each plant with the characteristics shown in Figure 9.2, and learn to recognize the distinctive characteristics of monocots and dicots.

12. ***Complete item 4 on the laboratory report***.

Assignment 5

Your instructor has set up several "unknown" plants for you to categorize as moss plants, ferns, conifers, monocots, and dicots. Examine these plants, and ***complete item 5 on the laboratory report***.

ANATOMY OF FLOWERING PLANTS

Keep in mind the differences between monocots and dicots as you study the basic anatomy of flowering plants.

Roots

Roots perform three important functions: (1) anchorage and support, (2) absorption and transport of water and minerals, and (3) storage and transport of organic nutrients.

The Root Tip

The basic structure of a root tip is shown in Figure 9.5. Cells formed in the **region of cell division** become either part of the root proper

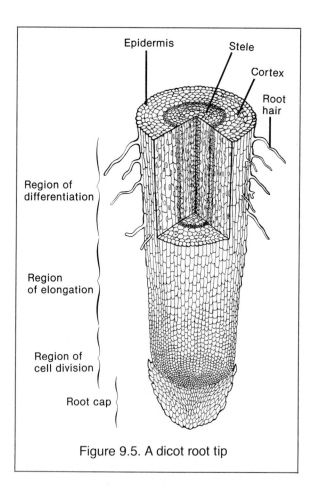

Figure 9.5. A dicot root tip

or form the root cap. The **root cap** is composed of rather large cells that protect the region of cell division. It also provides a sort of lubrication as its cells are eroded by the root growing through the soil. Cells of the root are enlarged in the **region of elongation**, and this accounts for the greatest increase in the linear growth of the root. As the cells become older, they develop their specialized characteristics in the **region of differentiation** where the **primary root tissues** are formed. Note that the cells of the root are arranged in vertical rows. The row in which a cell is located determines the type of cell it will become. For example, cells in the outer rows become epidermal cells, and those in the innermost rows become xylem cells. **Root hairs** are extensions of epidermal cells in the region of differentiation. They greatly increase the surface area of the root tip and are the primary sites of water and mineral absorption.

Growth in length of both roots and stems results from the formation of new cells in the region of cell division and their subsequent enlargement in the region of elongation. Thus, growth of both roots and stems occurs at their tips, and this growth is continuous throughout the life of the plant.

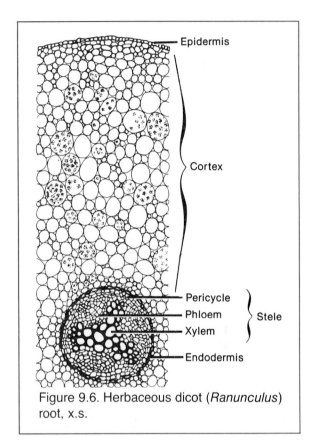

Figure 9.6. Herbaceous dicot (*Ranunculus*) root, x.s.

Root Tissues

Figure 9.6 shows the tissues found in a root of a herbaceous dicot (*Ranunculus*) as viewed in cross section. Note the three major divisions: epidermis, cortex, and stele.

The **epidermis** is the outermost layer of cells that provides protection for the underlying tissues and reduces water loss.

The **cortex** composes the bulk of the root and consists mostly of large, thin-walled cells used for food storage. The **endodermis**, the innermost layer of the cortex, is composed of thickwalled, water-impermeable cells and a few water-permeable cells with thinner walls. This ring of cells controls the movement of water and minerals into and out of the xylem.

The **stele** is that portion of the root within the endodermis. It is sometimes called the central cylinder. The outer layer of the stele is composed of thin-walled cells, the **pericycle**, from which branch roots originate. The thick-walled cells in the center of the stele are the **vessel cells** of the **xylem**. Between the rays of the xylem are

the **sieve tubes** and **companion cells** of the **phloem**.

Root Types

There are three types of roots. **Tap root systems** have a single dominant root from which branch roots arise. **Fibrous root systems** consist of a number of similar-sized roots that branch repeatedly. Fibrous roots are characteristic of monocots. Both types of root systems may be shallow or deep, depending on the species of plant, but tap roots are capable of the deepest penetration. **Adventitious roots** are unique in that, unlike other roots, they do not grow from the primary root of the embryo. They originate from stems or leaves and are typically fibrous in nature.

Materials

Colored pencils
Dissecting instruments
Microscope slides and cover glasses

Dropping bottles of methylene blue, 0.01%
Examples of adventitious, fibrous, and tap
  roots
Germinated grass seed
Prepared slides of:
  Allium (onion) root tip, l.s.
  Ranunculus (buttercup) root, x.s.

Assignment 6

1. Color-code the root cap and regions of cell division, elongation, and differentiation in Figure 9.5 and the xylem and phloem in Figure 9.6.
2. Obtain a germinated grass seed. Place it in a drop of water on a microscope slide and examine it at 40×. Note the root hairs. Locate the oldest and youngest root hairs. Use a scalpel to cut the seed from the root, add a drop of methylene blue, and add a cover glass. Examine an epidermal cell and its root hair at 100× and 400×. Try to locate the cell nucleus.
3. Examine a prepared slide of *Allium* (onion) root tip. Locate the root cap, region of cell division, and region of elongation. The region of differentiation is not on your slide. Note the arrangement and size of the cells.
4. Examine a prepared slide of *Ranunculus* root, x.s., and locate the tissues shown in Figure 9.6. Note the starch granules in the cortical cells and the differences between xylem and phloem.
5. Examine the examples of root types and note their characteristics.
6. ***Complete item 6 on the laboratory report.***

Stems

The stem serves as a connecting link between the roots and the leaves and reproductive organs. It also may serve as a site for food storage. The arrangement of vascular tissue in stems varies among the subgroups of vascular plants.

Monocot Stems

Figure 9.7 illustrates the cross-sectional structure of part of a corn stem. Note the *scattered* **vascular bundles** surrounded by large thin-walled cells, a characteristic of monocots.

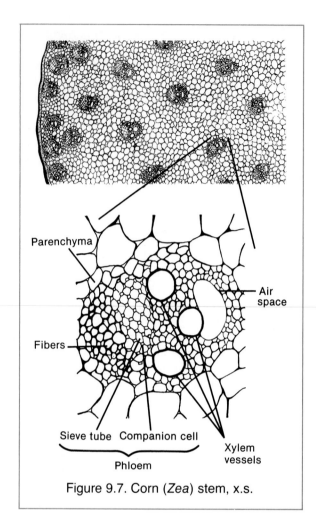

Parenchyma

Air space

Fibers

Sieve tube Companion cell

Xylem vessels

Phloem

Figure 9.7. Corn (*Zea*) stem, x.s.

Each vascular bundle has thick-walled, fibrous cells around the edges surrounding the large xylem vessels and the smaller sieve tubes and companion cells of phloem. Most of the support for the stem is provided by the xylem and fibrous cells.

Herbaceous Dicot Stems

Figure 9.8 shows the structure of a portion of an alfalfa stem in cross section. Note that the vascular bundles are arranged in a *broken ring* just interior to the **cortex**. This pattern is characteristic of herbaceous dicots. Each vascular bundle is composed of four tissues (from outside): **phloem, vascular cambium**, and **xylem**; the central portion of the stem, the **pith**, is composed of large, thin-walled cells.

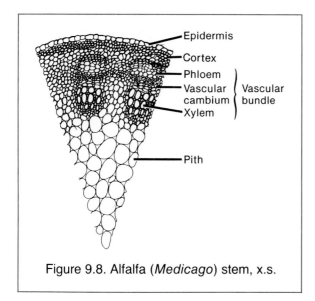

Figure 9.8. Alfalfa (*Medicago*) stem, x.s.

Woody Dicot Stems

The structure of a young woody stem in cross section is shown, in part, in Figure 9.9. Note the characteristic *continuous ring* of primary vascular tissue and the arrangement of the tissues within the stem. The **wood** of woody plants is actually xylem tissue.

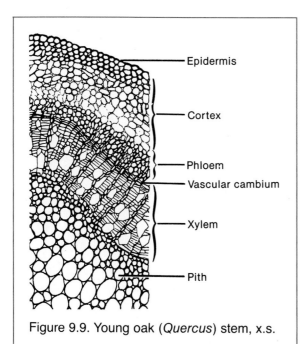

Figure 9.9. Young oak (*Quercus*) stem, x.s.

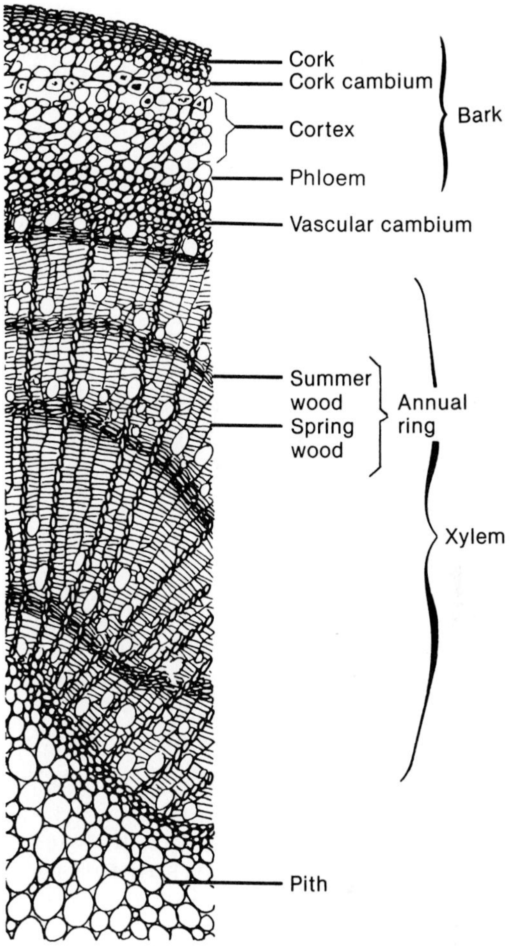

Figure 9.10. Older *Quercus* stem, x.s.

In each growing season, the vascular cambium forms new (secondary) xylem and phloem. Each growing season is identifiable by an **annual ring** of xylem, which is composed of large-celled **spring wood** (lighter color) and small-celled **summer wood** (darker color). No cells are formed in fall or winter in temperate climates. The stem grows in diameter by the formation and enlargement of new xylem and phloem. Since the growth is actually from the inside of the stem and since the epidermis and cortex cannot grow, they tend to fracture and slough off. The **cork cambium** forms in the cortex, and it produces cork cells that assume the function of protecting the underlying tissues and preventing water loss. At this stage, the stem is subdivided into (1) bark, (2) vascular cambium, and (3) wood. The **bark** is everything exterior to the vascular cambium: phloem, cortex, cork cambium, and cork. Figure 9.10 illustrates the structure of an older woody stem.

Materials

Colored pencils
Tree stem, x.s.
Prepared slides of:
  Quercus (oak) stem, 1 yr, x.s.
  Quercus (oak) stem, 4 yr, x.s.
  Medicago (alfalfa) stem, x.s.
  Zea (corn) stem, x.s.

Assignment 7

1. Color-code the xylem and phloem in Figures 9.7 to 9.9 and the spring wood, summerwood, vascular cambium, phloem, cork, and cork cambium in Figure 9.10.
2. Examine a prepared slide of *Zea* (corn) stem and locate the parts shown in Figure 9.7.
3. Examine a prepared slide of *Medicago* (alfalfa) stem and locate the structures shown in Figure 9.8.
4. Examine prepared slides of young and four-year-old *Quercus* (oak) stems. Locate the structures shown in Figures 9.9 and 9.10.
5. ***Complete item 7 on the laboratory report***.

Leaves

Leaves are the principal organs of organic nutrient production by photosynthesis, gaseous exchange with the atmosphere, and water loss by evaporation. See Figure 9.11. The major portion of a leaf is the broad, thin **blade** that is usually attached to a stem by a small stalk, the **petiole**. Some leaves, especially those of monocots, lack a petiole. The blade is supported by numerous **veins** that may be arranged in a parallel pattern (monocots) or in a net-veined pattern (dicots). In net-veined leaves, the large central vein from which others branch is called the **midrib**. Veins are composed of vascular tissue plus supporting fibers.

Leaf Tissues

The tissues of a leaf are shown in Figure 9.12. The outer surface of the epidermis is coated with a waxy **cuticle** that retards water loss by evaporation. Epidermal cells do not contain chloroplasts except for the **guard cells** that control the size of the **stomata**, tiny openings in the epi-

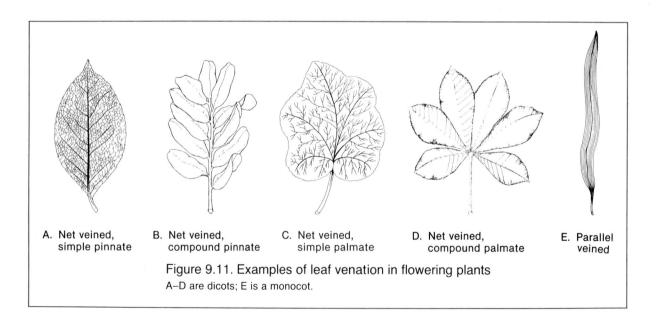

A. Net veined, simple pinnate   B. Net veined, compound pinnate   C. Net veined, simple palmate   D. Net veined, compound palmate   E. Parallel veined

Figure 9.11. Examples of leaf venation in flowering plants
A–D are dicots; E is a monocot.

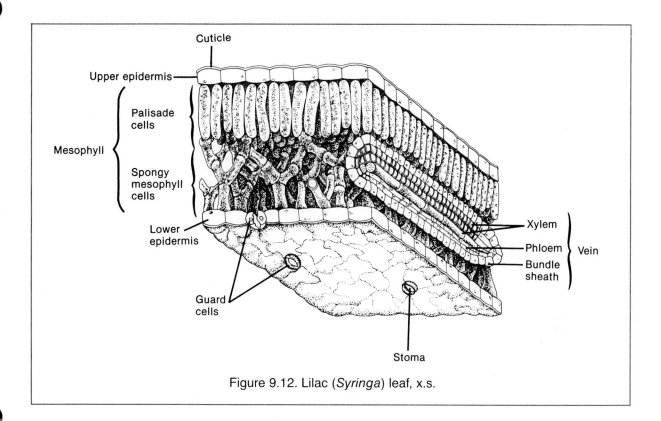

Figure 9.12. Lilac (*Syringa*) leaf, x.s.

dermis that enhance the exchange of oxygen and carbon dioxide with the surrounding air. The **mesophyll** is the primary photosynthetic tissue, especially the **palisade layer**. The **spongy mesophyll** contains numerous spaces between the cells to aid the movement of gases within the leaf.

Materials

Colored pencils
Examples of leaf types
Prepared slides of:
 Syringa (lilac) leaf, x.s.

Assignment 8

1. Color-code the epidermis, guard cells, mesophyll, xylem, and phloem in Figure 9.12.
2. Examine the different types of leaves available. Note the type of venation and compare them to Figure 9.11.
3. Examine a prepared slide of *Syringa* leaf and locate the tissues shown in Figure 9.12.
4. ***Complete the laboratory report***.

10

SIMPLE ANIMALS

Animals are members of the **kingdom Animalia** and are classified according to their similarities and evolutionary relationships. All animals are characterized by (1) a **multicellular body** formed of different types of **eukaryotic cells** that lack plastids and cell walls, (2) **heterotrophic nutrition**, and (3) movement by the shortening of **contractile fibers**. Animals reproduce sexually by forming gametes: large nonmotile eggs and small motile sperms. Some reproduce asexually as well. In this exercise, you will study the principal distinguishing characteristics of the major groups of "simple animals."

Biologists use several major criteria, and many lesser ones, to classify animals. The major criteria that will be used in your study are noted in this section. You will develop a better understanding of these criteria and their significance as you study the various animal groups. Figure 10.1 shows how certain major criteria are used to establish the presumed evolutionary relationships among the principal animal phyla.

MAJOR CRITERIA FOR CLASSIFYING ANIMALS

Symmetry

Symmetry refers to the proportion of body parts on each side of a median plane. Animals may exhibit **radial symmetry, bilateral symmetry**, or none. Radially symmetrical animals may be divided into two similar halves by *any* plane passing through the longitudinal axis. Bilaterally symmetrical animals may be divided into two similar halves by only *one* plane, a **midsaggital plane** passing through the longitudinal axis of the body.

Level of Organization

An animal may exhibit a **tissue, organ**, or **organ system** level of body organization.

Embryonic Tissues (Germ Layers)

Animals with an organ level of organization, or higher, possess all three embryonic tissues: **ectoderm, endoderm**, and **mesoderm**. Those with a tissue level of organization possess only ectoderm and endoderm.

Body Plan

Animals with a tissue level of organization, or higher, exhibit one of two types of body plans. A

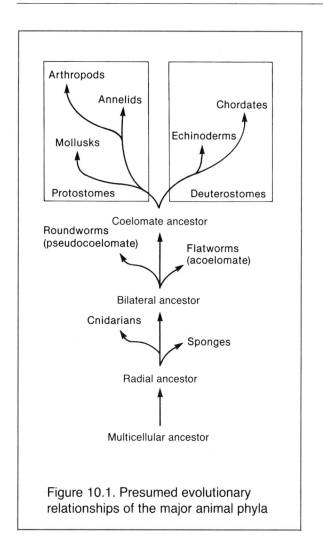

Figure 10.1. Presumed evolutionary relationships of the major animal phyla

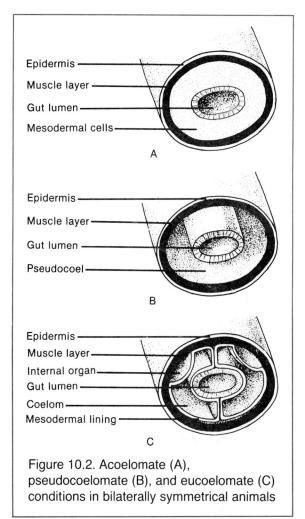

Figure 10.2. Acoelomate (A), pseudocoelomate (B), and eucoelomate (C) conditions in bilaterally symmetrical animals

saclike body plan is characterized by a single opening, the mouth, into a gastrovascular cavity. A **tube-within-a-tube body plan** has a mouth at one end of a tubelike digestive tract and an anus at the other.

Body Cavity (Coelom)

Animals with a tube-within-a-tube body plan have a fluid-filled space between the digestive tract and the body wall that is called a **coelom**. If the coelom is completely lined with mesoderm, it is a **true coelom**. If not, it is a **false coelom**. See Figure 10.2.

Segmentation

Segmentation is the serial repetition of body parts along the longitudinal axis.

Fate of the Blastopore

The first opening that is formed in an early embryo is called the **blastopore**. In some coelomate animals (protostomes), the blastopore becomes the mouth, while in others (deuterostomes), the blastopore becomes the anus.

SPONGES (Phylum Porifera)

Sponges are simple, primitive animals that probably evolved from flagellated protozoans. Most forms are marine; only a few live in fresh water. The radially symmetrical or asymmetrical adults live attached to rocks or other objects, but the larvae are ciliated and motile. Figure 10.3 shows the basic structure of a simple

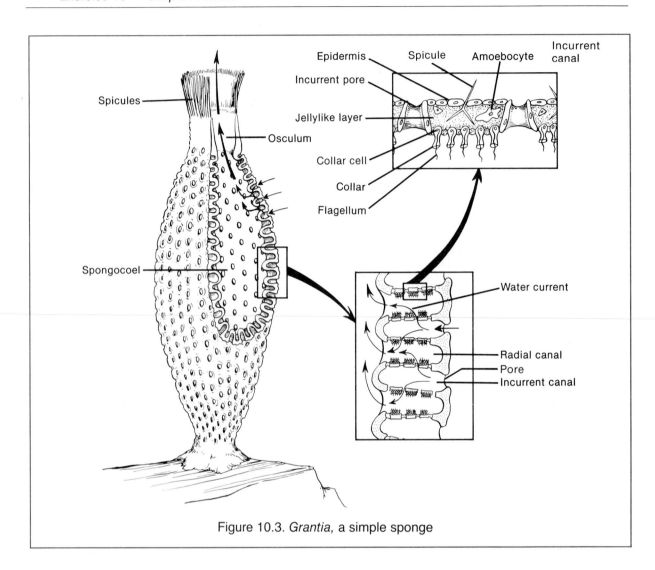

Figure 10.3. *Grantia,* a simple sponge

TABLE 10.1
Distinctive Characteristics of Sponges, Coelenterates, Flatworms, and Roundworms

	Sponges	*Coelenterates*	*Flatworms*	*Roundworms*
Symmetry	Asymmetrical or radial symmetry	Radial symmetry	Bilateral symmetry	Bilateral symmetry
Level of organization	Cellular-tissue level	Tissue level	Organ level	Organ system level
Germ layers	None	Ectoderm and endoderm	Ectoderm, mesoderm, and endoderm	Ectoderm, mesoderm, and endoderm
Coelom	N/A	N/A	Acoelomate	Pseudocoelomate
Body plan	N/A	Saclike	Saclike	Tube within a tube

N/A = Not applicable.

sponge. The cells are so loosely arranged into tissues that the level of organization is intermediate between distinct cellular and tissue levels, a **cellular-tissue level of organization**.

Sponges are classified according to the type of **spicules**, skeletal elements, that they possess. Chalk sponges have calcium carbonate spicules, glass sponges have siliceous spicules, and fibrous sponges have a proteinaceous skeleton.

The body wall of a sponge is perforated by numerous pores. The beating flagella of the **choanocytes** (collar cells) maintain a constant flow of water through the **incurrent pores**, into the **radial canals**, then into the **spongocoel** (central cavity), and out the **osculum**. Tiny food particles in the water are engulfed by the amoeboid action of the collar cells and digested in food vacuoles. Nutrients are distributed to other cells by diffusion, a process that is aided by the wandering **amoebocytes**.

Materials

Colored pencils
Grantia, preserved
Skeletons of representative sponges
Prepared slides of *Grantia,* l.s.

Assignment 1

1. Color-code the collar cells, epidermis, amoebocytes, and spicules in Figure 10.3.
2. Examine whole and longitudinally sectioned *Grantia* with a dissecting microscope. Note the incurrent pores, radial canals, spongocoel, and osculum.
3. Examine a prepared slide of *Grantia,* l.s., set up under a demonstration microscope. Observe the collar cells and spicules.
4. Examine the chalk, silica, and protein skeletons of representative sponges.
5. ***Complete item 1 on Laboratory Report 10 that begins on page 389***.

COELENTERATES
(Phylum Coelenterata)

Coelenterates (cnidarians) include jellyfish, sea anemones, corals, and hydroids. Most are marine, but a few occur in fresh water. They may live singly or in colonies and are characterized by (1) a saclike body plan with a single opening, the mouth, into a **gastrovascular cavity**, (2) a tissue level of organization evidenced by a body wall of two embryonic tissues, the ectoderm (epidermis) and endoderm (gastrodermis); (3) radial symmetry; and (4) **sting-**

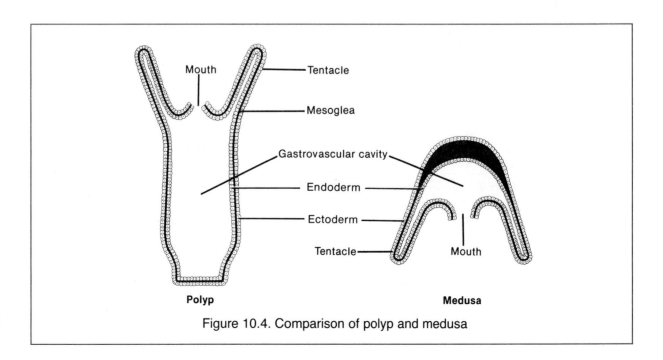

Figure 10.4. Comparison of polyp and medusa

ing cells (cnidocytes) that eject dartlike weapons of offense and defense (nematocysts).

Two morphologic types of adults are recognized. See Figure 10.4. **Polyps** have tubular bodies with a mouth and tentacles at one end and usually are attached to the substrate by the other end. **Medusae** are free-swimming jellyfish that have umbrella- or bell-shaped bodies with a mouth located in the center of the concave side and tentacles hanging from the edge of the bell. Some species exhibit *both* adult forms in an **alternation of generations** (metagenesis). In such a life cycle, medusae, which reproduce sexually by gametes, alternate with polyps, which reproduce asexually by **budding**. See Figure 10.5.

Figure 10.6 shows the structure of *Hydra*, a common freshwater coelenterate that exhibits the basic characteristics of the phylum. Study it carefully. Locate the coelenterate features described previously, and note the various types of cells in the two layers of the body wall. What is located between the tissue layers?

Coelenterates are capable of simple movements enabled by the contraction of **myofibrils** located in cells of both tissue layers. Contraction of the longitudinally oriented myofibrils in the epidermal cells shortens the body, while contraction of the circularly oriented myofibrils in the

gastrodermis extends the body. A simple **nerve net** located in the **mesoglea**, a noncellular layer between the two tissue layers, coordinates body functions.

Small prey organisms are paralyzed by poison injected by darts from the stinging cells and are engulfed into the gastrovascular cavity where digestive enzymes initiate extracellular digestion. Subsequently, minute particles of partially digested food are engulfed by the gastrodermal cells to complete digestion within their food vacuoles. The distribution of nutrients to other cells is by diffusion.

Materials

Beaker, 1000 ml
Colored pencils
Medicine droppers
Syracuse dish
Pond water
Daphnia, small, living
Elodea shoots
Hydra, living
Representative coelenterates
Prepared slides of:
    Hydra, l.s.
    Obelia polyps, w.m.
    Obelia medusae, w.m.

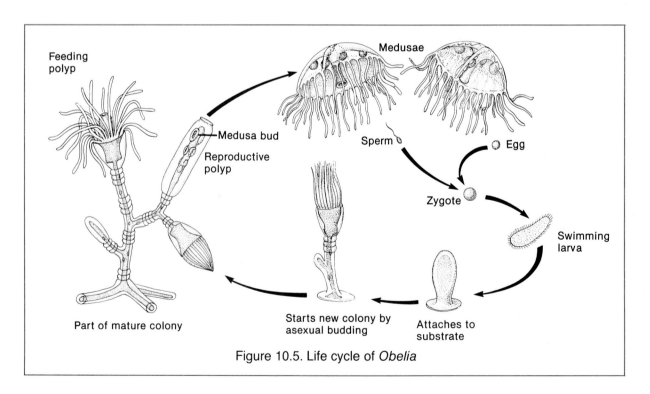

Figure 10.5. Life cycle of *Obelia*

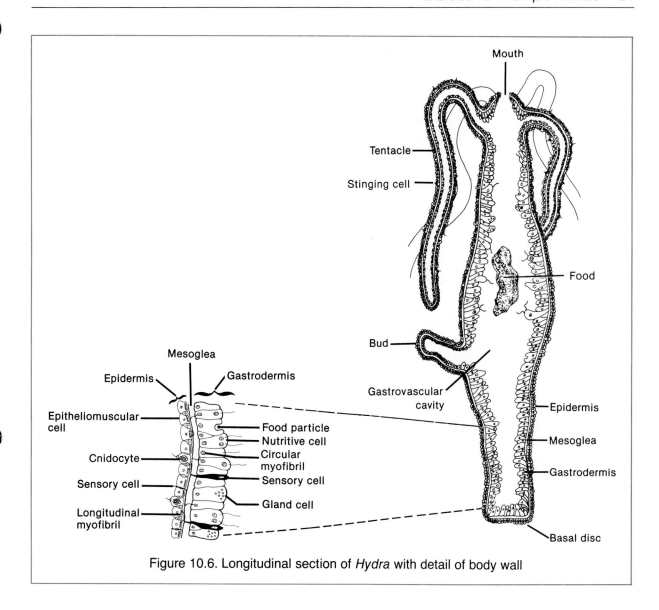

Figure 10.6. Longitudinal section of *Hydra* with detail of body wall

Assignment 2

1. Color-code the ectoderm blue and the endoderm yellow in Figures 10.4 and 10.6 (detail of body wall only).
2. Examine the representative coelenterates. Learn their names and characteristics. Identify the polyps and medusae. Note the calcium carbonate skeletons of the corals.
3. Examine prepared slides of *Obelia* polyps and medusae. Compare your observations with Figure 10.5.
4. Watch the living *Hydra* in the large beaker or small acquarium capture and engulf *Daphnia*. Look for *Hydra* buds growing out

of the side of the adult specimens. Note how the *Hydra* attach themselves by the **basal disc** and usually hang downward with their tentacles extended. What is their feeding strategy?
5. Place a *Hydra* in a Syracuse dish containing water from the culture jar. Then place the dish on the *black surface* of the stage of a dissecting microscope for observation. Keep the lamp some distance away so that the heat from the bulb does not affect the specimen. Observe how the *Hydra* extends itself when not disturbed and contracts when the dish is tapped. Which movements are more rapid, extension or retraction? Lo-

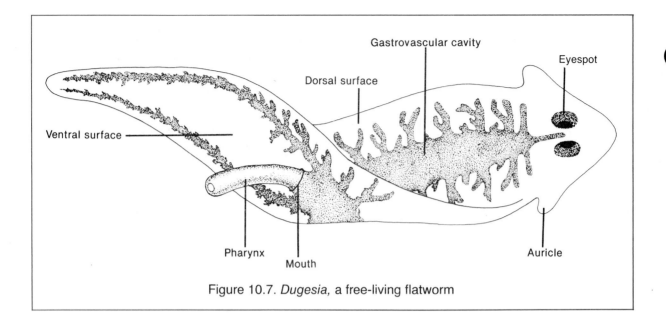

Figure 10.7. *Dugesia,* a free-living flatworm

cate the stinging cells that appear as bumps on the tentacles. Keep the lamp turned off except when observing.

6. Examine a prepared slide of *Hydra,* l.s., and locate the parts shown in Figure 10.6.
7. ***Complete item 2 on the laboratory report***.

FLATWORMS
(Phylum Platyhelminthes)

Flatworms are distinguished by (1) a saclike body plan modified into an elongate, bilaterally symmetrical, and dorsoventrally flattened body; (2) the presence of all three embryonic tissues; (3) an organ level of organization; and (4) the absence of a coelom. The embryonic tissues develop into the tissues and organs of the adult. **Ectoderm** forms the epidermis and nervous tissue, while **endoderm** forms the inner lining of the gastrovascular cavity. **Mesoderm** forms muscles, reproductive organs, excretory organs, and the parenchyma cells that fill the space around the gastrovascular cavity. Animals taxonomically higher than flatworms are composed of all three embryonic tissues. Flatworms may be either free living or parasitic. All flatworms are **hermaphroditic** since each adult possesses both male and female reproductive organs.

Free-Living Flatworms

A common planarian, *Dugesia,* is typical of nonparasitic forms. See Figure 10.7. The mouth and extensible **pharynx** are located at the middle of the ventral surface. Food is passed by the pharynx into the branched gastrovascular cavity where digestion begins extracellularly and is completed intracellularly, as in coelenterates. Nutrients are distributed to other cells by diffusion.

Figure 10.8 shows the major organs of *Dugesia.* The concentration of neural structures at the anterior end is called **cephalization**, a condition that begins here and reaches its peak in humans. In *Dugesia,* the **auricles** contain touch and chemical receptors, and the **eyespots** are photoreceptors. The **cerebral ganglia** consist of concentrations of nerve cells that coordinate body movements via two widely separated **ventral nerve cords** and transverse connecting nerves. Waste products of metabolism diffuse into the excretory ducts and are flushed from the body through **excretory pores** by the current created by the beating flagella of **flame cells**.

Parasitic Flatworms

Tapeworms are parasitic flatworms that infest humans and other meat-eating animals. Adults live attached to the inner wall of the small intestine by suckers or hooks of the **scolex** and absorb nutrients through the body wall. The body consists of a series of **proglottids**, "reproductive factories" that are constantly formed by the scolex and broken off at the posterior end when they are full of fertile eggs. A gastrovascular cavity is absent, and nutrients are ab-

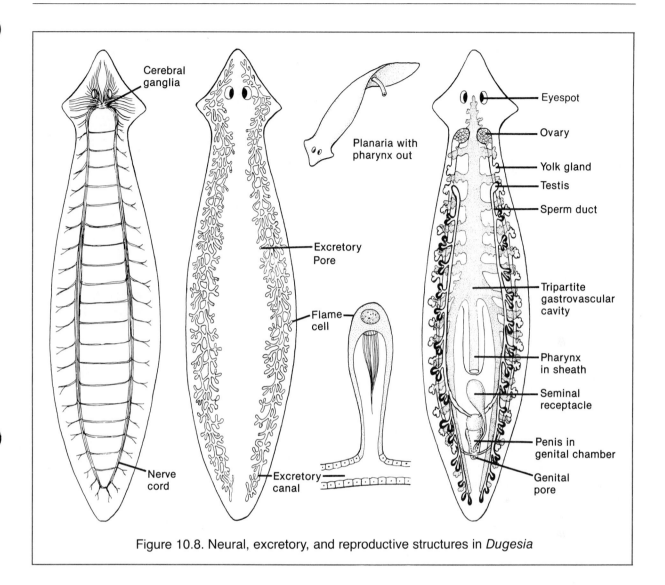

Figure 10.8. Neural, excretory, and reproductive structures in *Dugesia*

sorbed directly through the body wall. Study the life cycle of the beef tapeworm (*Taenia*) shown in Figure 10.9.

Flukes are parasitic flatworms that infest the lungs, liver, and blood of vertebrates. The adult human liver fluke (*Clonorchis*) lives in the bile duct of the human host. It is common in the Orient where the eating of raw fish is practiced. The larval forms are encysted in the muscles of certain fish. The human blood fluke (*Schistosoma*) has become a prominent parasite of humans in the Nile Valley of Egypt since the completion of the Aswan Dam. The dam increased irrigation and allowed the snails serving as the intermediate host of the blood fluke to flourish.

Materials

Construction paper, black
Finger bowl
Medicine droppers
Petri dish
Stirring rod
Watch glass
Model of *Dugesia*
Dugesia, living
Tapeworms, preserved
Prepared slides of:
    Clonorchis (human liver fluke), w.m.
    Dugesia, w.m. and x.s.
    Schistosoma (human blood fluke), w.m.

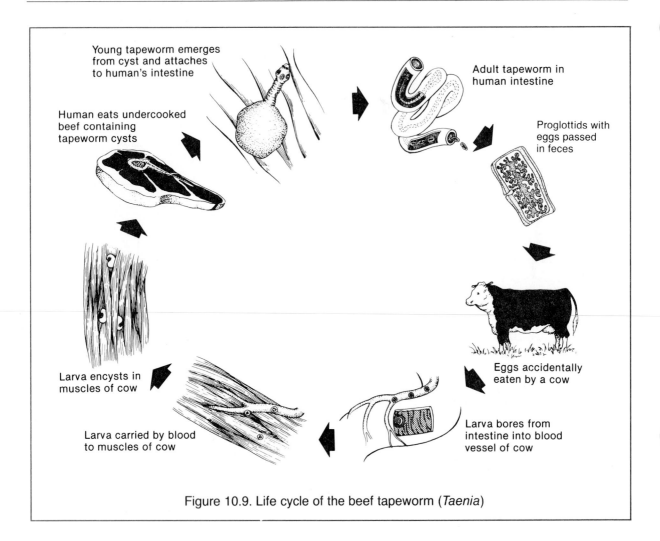

Young tapeworm emerges from cyst and attaches to human's intestine

Human eats undercooked beef containing tapeworm cysts

Adult tapeworm in human intestine

Proglottids with eggs passed in feces

Eggs accidentally eaten by a cow

Larva encysts in muscles of cow

Larva carried by blood to muscles of cow

Larva bores from intestine into blood vessel of cow

Figure 10.9. Life cycle of the beef tapeworm (*Taenia*)

tapeworm scolex, mature and gravid proglottids
tapeworm larvae in muscle

Assignment 3

1. Examine the model of *Dugesia* and locate the structures shown in Figures 10.7 and 10.8.
2. **Complete items 3a–3d on the laboratory report**.
3. Place a living planarian, *Dugesia,* in a watch glass with enough water from the culture jar so that it can move about. Observe it with dissecting microscope over the white surface of the stage. Locate the external structures and note the manner of movement. Is there a pattern to the movement? What enables the movement?
4. Examine a prepared slide of *Dugesia,* x.s. Note the gastrovascular cavity, body wall, and the cilia on the ventral surface. What is the function of the cilia?
5. Remove the black paper that partially shades the dish of flatworms, and note their distribution. Replace the paper in its exact prior position.
6. On the demonstration table is a dish containing about 25 planarians. Note their position and direction of movement. Swirl the water by stirring it clockwise with a glass rod. As the water continues to swirl, note the direction of movement of the planarians.
7. **Complete items 3e–3f on the laboratory report**.
8. Examine the preserved tapeworms and

estimate their lengths. Locate the proglottids and scolex.

9. Examine a prepared slide of tapeworm scolex and mature proglottids. Note how the scolex attaches to the intestine wall and locate the fertile eggs in the mature proglottids.
10. Examine a prepared slide of tapeworm larvae encysted in beef muscle. Note the inverted scolex.
11. Examine the demonstration slides of flukes. Note the name and characteristics of each.
12. ***Complete items 3g–3j on the laboratory report***.

ROUNDWORMS (Phylum Nematoda)

Roundworms are characterized by (1) a cylindrical, bilaterally symmetrical body that is tapered at each end, (2) a tube-within-a-tube body plan, (3) an organ system level of organization, and (4) a false coelom. Most roundworms are free living, but some are important parasites of animals and plants.

Ascaris is a large roundworm parasite that lives in the small intestine of humans and swine and feeds on partially digested food. Females can be distinguished from males because they are larger and do not have a curved posterior end.

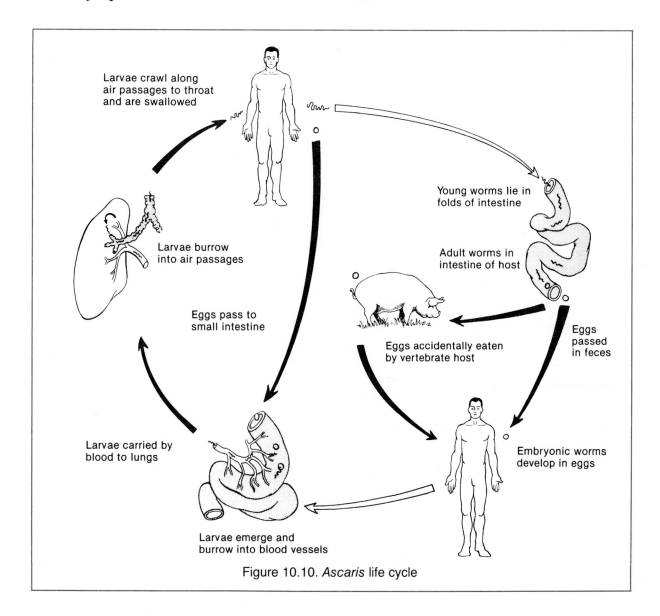

Larvae crawl along air passages to throat and are swallowed

Larvae burrow into air passages

Eggs pass to small intestine

Larvae carried by blood to lungs

Larvae emerge and burrow into blood vessels

Young worms lie in folds of intestine

Adult worms in intestine of host

Eggs accidentally eaten by vertebrate host

Eggs passed in feces

Embryonic worms develop in eggs

Figure 10.10. *Ascaris* life cycle

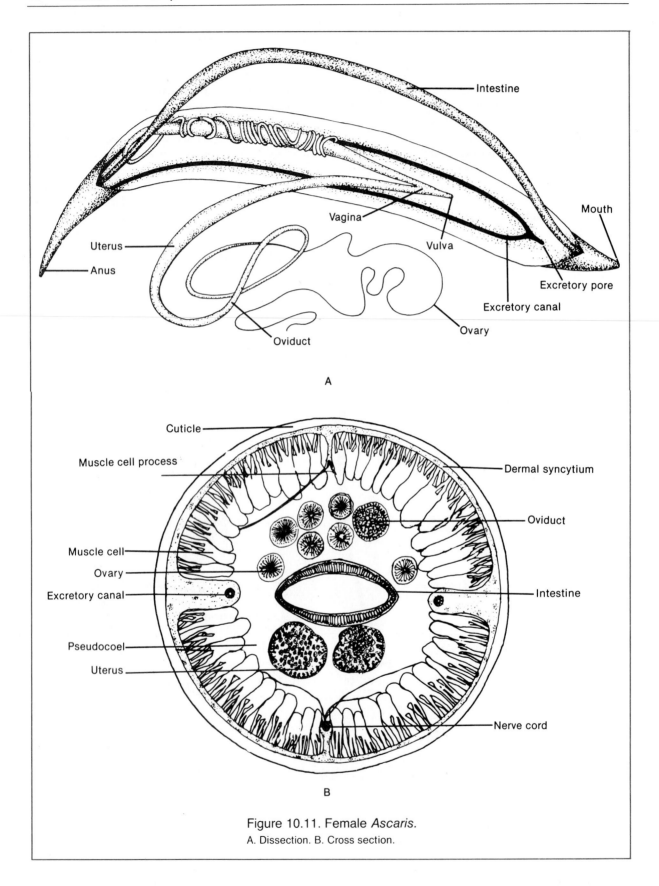

A

B

Figure 10.11. Female *Ascaris.*
A. Dissection. B. Cross section.

The structure of a female *Ascaris* is shown in dissection and in cross section in Figure 10.11. The body wall consists of longitudinal muscles, epidermis, and a cuticle that is secreted by the epidermis. The **cuticle** protects the worm from the digestive action of enzymes in the host's small intestine. The space between the intestine and the body wall is a **pseudocoelom** (false coelom) since it is *incompletely* lined with mesodermal tissue. The female reproductive system and the ribbonlike intestine occupy most of the pseudocoelom. Note the locations of the nerve cords and the excretory canals.

Materials

Colored pencils
Dissecting instruments
Dissecting pan, wax bottomed
Dissecting pins
Microscope slides
   and cover glasses
Protoslo
Toothpicks
Ascaris, preserved
Turbatrix culture
Prepared slides of:
   Ascaris female, x.s.
   Ascaris eggs
   Ascaris larvae
   Trichinella spiralis, encysted larvae
   Wuchereria bancrofti, w.m.

Ascaris Dissection

A dissection of a female *Ascaris* will enable you to observe the basic roundworm characteristics.

1. Pin each end of a preserved female *Ascaris* to the bottom of a dissecting pan.
2. Being careful not to damage the internal organs, slit the body wall from one end to the other with the tips of scissors. Pin out the body wall to expose the internal organs. Add water to the pan to cover the specimen so that it does not dry out.
3. Use a probe to carefully separate the parts of the coiled female reproductive organs. Locate the structures shown in Figure 10.11. The excretory pore is difficult to find. Note the ribbonlike intestine. Organisms with a tube-within-a-tube organization have a **complete digestive tract** with a mouth at one end and an anus at the other.

Assignment 4

1. On Figure 10.11, color-code the major structures.
2. Examine a prepared slide of a female *Ascaris,* x.s., and locate the structures labeled in Figure 10.11.
3. Study the *Ascaris* life cycle in Figure 10.10. Examine the prepared slides of *Ascaris* eggs and larvae.
4. Examine the prepared slides of these roundworms:
   a. *Trichinella spiralis* larvae encysted in muscles of the host. Adults live in the small intestine. Humans are infected by eating encysted larvae in undercooked pork.
   b. *Wuchereria bancrofti* are tiny roundworms that plug lymphatic vessels and cause **elephantiasis.** They are transmitted by the bite of certain mosquitoes.
5. Make and examine a slide of the *Turbatrix* (vinegar eels) culture. Note the motility of these nonparasitic roundworms. Mix a bit of Protoslo with the culture fluid on your slide. Add a cover glass and observe a nonmoving specimen.
6. ***Complete items 4 and 5 on the laboratory report.***

11

PROTOSTOMATE ANIMALS

In this exercise, you will study mollusks, segmented worms, and arthropods. They compose the **protostomate animals**—the blastopore of the early embryo forms the mouth. All of these animals exhibit (1) an **organ system level of organization**, (2) all three **embryonic tissues**, (3) **bilateral symmetry**, (4) a **tube-within-a-tube body plan**, and (5) a **true coelom**. Because the coelom is formed by the splitting of the mesoderm, these groups also are called **schizocoelomate animals**.

MOLLUSKS (Phylum Mollusca)

The molluscan body plan is characterized by (1) a **ventral muscular foot**, (2) a **dorsal visceral mass**, (3) a **mantle** that surrounds the visceral mass and often secretes a **shell**, and (4) a **radula**, a filelike mouthpart used to scrape off bits of food. This plan has been greatly modified among the different groups of mollusks, however, See Table 11.1. Mollusks are primarily marine organisms, but many occur in fresh water and some are terrestrial.

Clams (Class Pelecypoda)

The basic molluscan structure as found in the freshwater clam is shown in Figure 11.1. The body is flattened laterally and enclosed within the two valves (halves) of the shell that are hinged together at the dorsal surface. Thus, this class of mollusks is commonly called bivalves. The valves may be tightly closed by strong **adductor muscles**. A digging foot is used to bury the clam in mud or sand, often with only the tips of the siphons protruding. Water is brought into the **mantle cavity** via the **incurrent siphon** and exits via the **excurrent siphon**. Gas exchange occurs between the water in the mantle cavity and blood in the **gills**. Tiny food particles in the water are trapped in mucus coating the gills, and both mucus and food are carried by cilia to the **labial palps** and on into the mouth. Digestion occurs extracellularly in the digestive tract, and nutrients are absorbed into the blood. Blood carries nutrients and oxygen to body cells and carries metabolic wastes and carbon dioxide from the cells to excretory organs. An **open circulatory system** is present. Blood is carried from the heart in vessels that empty the blood into sinuses. The blood moves slowly through the sinuses and back into the heart. Nitrogenous metabolic wastes are removed by a "kidney."

TABLE 11.1
Distinguishing Characteristics of Molluscan Classes

Class	Characteristics
Monoplacophora (*Neopilina*)	Primitive; remnants of segmentation; single shell; dorsoventrally flattened foot; radula; marine
Amphineura (chitons)	Shell of eight overlapping plates; dorsoventrally flattened foot; radula; marine
Scaphapoda (tooth shells)	Conical shell open at each end; digging foot; radula; marine
Gastropoda (snails and slugs)	Shell absent or single, often coiled; dorsoventrally flattened foot; radula; marine, freshwater, and terrestrial
Pelecypoda (clams, oysters)	Shell of two valves, hinged dorsally; laterally flattened foot; no radula; filter-feeder; marine or freshwater
Cephalopoda (octopi, squids)	Single shell or none; foot modified to form tentacles; image-producing eyes; horny beak and radula; predaceous; water-jet propulsion; marine

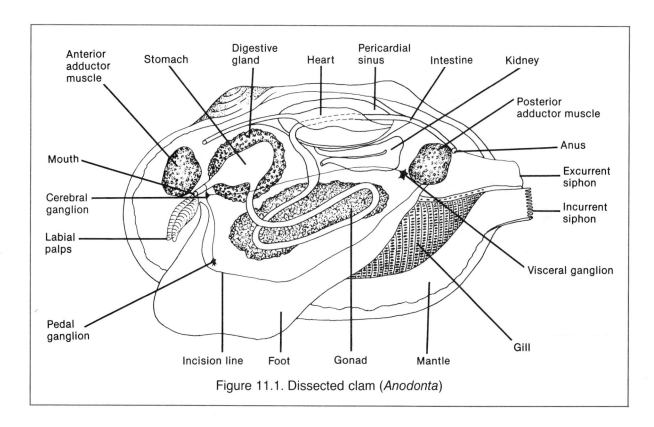

Figure 11.1. Dissected clam (*Anodonta*)

Octopi and Squids (Class Cephalopoda)

Squid and octopi represent the peak of molluscan evolution. In contrast to the filter-feeding clam, cephalopods are predaceous. See Figure 11.2. The foot is modified to form "arms" that have suction cups and are used to capture prey. Specialized arms are called tentacles. Food is eaten with the aid of a **horny beak**. Water enters the mantle cavity via a **siphon** and bathes the gills, enabling gas exchange to occur. Water

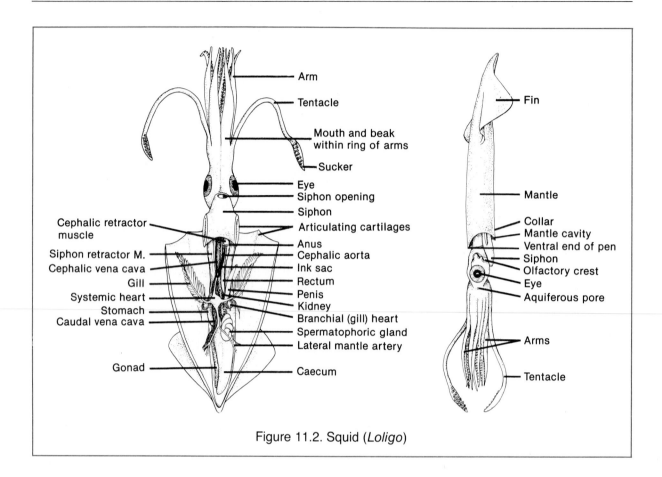

Figure 11.2. Squid (*Loligo*)

also can be forcefully ejected through the siphon, which provides a water-jet method of propulsion. Usually squid swim backward, but they also can swim forward depending on the direction in which the tip of the siphon is pointing. Ink may be ejected to form a smoke screen when cephalopods are trying to escape from enemies. Octopi lack a shell. In squids, the shell is reduced to a thin, transparent plate called a **pen** that is covered by the mantle and serves to stiffen and support the body. The large **image-forming eyes** superficially resemble vertebrate eyes but are formed quite differently.

Materials

Colored pencils
Clam, dissected demonstration
Representative mollusks, living or preserved
Squids, frozen or preserved
Model of dissected clam
Prepared slides of radula

Assignment 1

1. ***Complete item 1 on Laboratory Report 11 that begins on page 393.***
2. Color-code the major structures in Figures 11.1 and 11.2.

Assignment 2

1. Examine the representative mollusks. Identify the molluscan class to which each belongs. Note the distinguishing traits of the representatives of each group.
2. Examine a prepared slide of a radula set up under a demonstration microscope. Note the filelike teeth.
3. Use Table 11.1 to determine the class of each of the "unknown" mollusks.
4. ***Complete items 2a–2d on the laboratory report.***
5. Examine the demonstration dissected clam. Compare it with the model of a dissected

TABLE 11.2
Distinguishing Characteristics of Annelid Classes

Class	Characteristics
Polychaeta	Head with simple eyes and tentacles; segments with lateral extensions (parapodia) and many bristles (setae); sexes usually separate; predaceous; marine
Oligochaeta	No head; segments without extensions and with few, small bristles; hermaphroditic; terrestrial or freshwater
Hirudinea (leeches)	No head; segments with superficial rings and without lateral extensions or bristles; anterior and posterior suckers; hermaphroditic; parasitic, feeding on blood; terrestrial or freshwater

clam and Figure 11.1. Locate the principal parts.

6. On a preserved or frozen squid, locate the external features shown in Figure 11.2. If time permits, dissect it to observe the internal parts.

7. ***Complete item 2 on the laboratory report.***

SEGMENTED WORMS
(Phylum Annelida)

The major characteristic of these worms is a **segmented body** formed of repeating units called **somites**. Most of the somites are basically similar, including the internal arrangement of muscles, nerves, blood vessels, and excretory ducts. Most annelids are marine, but many occur in freshwater and terrestrial habitats. Earthworms, sandworms, and leeches represent the three major classes of segmented worms. See Table 11.2.

Earthworms

The earthworm exhibits the basic characteristics of segmented worms. Note the external features shown in Figure 11.3. The **prostomium** is a small projection over the mouth. The

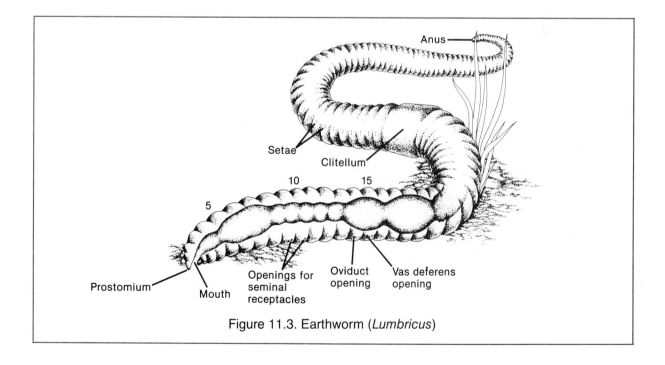

Figure 11.3. Earthworm (*Lumbricus*)

clitellum appears as a smooth band around the worm. It forms a ring of mucus around copulating worms that later becomes the egg case for the fertile eggs. Earthworms are **hermaphroditic** and exchange sperm during copulation.

The internal structure of an earthworm is shown in Figures 11.4 and 11.5. Note the **septa** that separate the segments. Parts of the digestive tract are specialized. The muscular **pharynx** aids in the ingestion of food, which passes through the **esophagus** to the **crop** and then enters the muscular **gizzard**. The gizzard grinds the food into smaller pieces before it enters the intestine where extracellular digestion occurs. Nutrients are absorbed into the blood and carried to all parts of the body. Annelids, unlike most mollusks, have a **closed circulatory system**, which means that blood is always contained in blood vessels. Contractile segmental vessels ("hearts"), located around the esophagus, pump the blood throughout the body.

Paired **ganglia** form the "brain" that lies dorsal to the pharynx. It is connected to the **ventral nerve cord** that runs the length of the body along the ventral midline. A **segmental ganglion** and **lateral nerves** occur in each somite. Metabolic wastes are removed from the blood and coelomic fluid by a pair of **nephridia** in each segment. **Seminal vesicles** contain sperm formed by the two pairs of **testes** within them. The **seminal receptacles** receive sperm during copulation and later release sperm to fertilize

the eggs that are formed in a pair of **ovaries** and released through the **oviducts**.

Earthworm structure is shown in cross section in Figure 11.6. The body wall consists of **circular** and **longitudinal muscle layers**, **epidermis**, and a **cuticle** formed by the epidermal cells. The dorsal part of the intestine wall, the **typhlosole**, folds inward and increases the surface area of the intestinal lining. Four pairs of **setae** are on each body segment. (What is their function?) Note the blood vessels, ventral nerve cord, and nephridium.

Materials

Colored pencils
Dissecting instruments
Dissecting pan, wax bottomed
Dissecting pins
Model of dissected earthworm
Earthworms, preserved
Representative annelids
Prepared slides of earthworm, x.s.

Assignment 3

1. Color-code the parts of the digestive tract and the nephridia in Figures 11.4 and 11.5.
2. Examine the representative segmented worms, and note their similarities and differences. Compare the characteristics of the specimens with those in Table 1.2. Note the adaptations of each specimen.

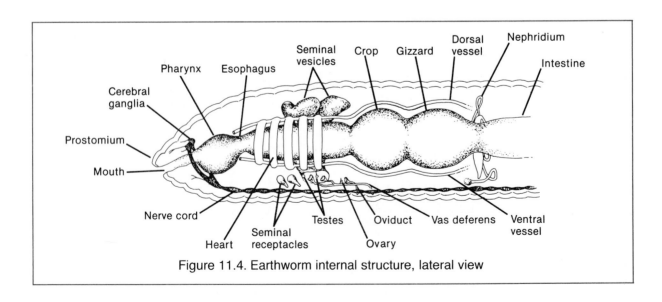

Figure 11.4. Earthworm internal structure, lateral view

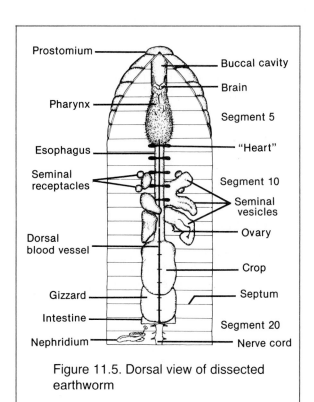

Figure 11.5. Dorsal view of dissected earthworm

3. Use Table 11.2 to determine the class of each of the "unknown" annelids.
4. ***Complete items 3a–3c on the laboratory report*.**
5. Locate the parts labeled in Figures 11.3, 11.4, and 11.5 on the model of the earthworm.
6. Examine a prepared slide of earthworm, x:s., and locate the structures shown in Figure 11.6.
7. ***Complete items 3d1 and 3d2 on the laboratory report*.**

Earthworm Dissection

1. Obtain a preserved earthworm and locate the external features as shown in Figure 11.3.
2. Pin the earthworm, *dorsal side up,* to the wax bottom of a dissecting pan by placing a pin through the prostomium and another pin through the body posterior to the clitellum. Pin the worm about 5 to 7 cm from one edge of the pan to facilitate observation of the structures with a dissecting

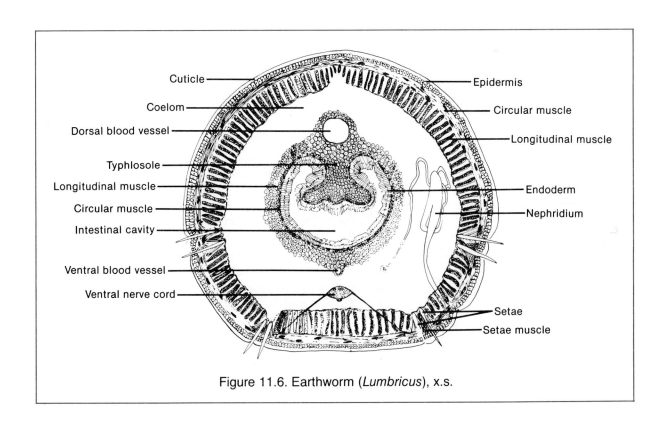

Figure 11.6. Earthworm (*Lumbricus*), x.s.

microscope. *Note*: The dorsal surface is convex and usually dark in color.

3. Use a sharp scalpel to cut *only* through the body wall, along the *dorsal midline* from the clitellum to the prostomium. Be careful to avoid cutting the underlying organs.

4. Use a dissecting needle to break the septa as you spread out the body wall and pin it to the pan. This will expose the internal organs. Add water to cover the worm to prevent it from drying out.

5. Locate the structures in Figures 11.4 and 11.5. Carefully dissect away surrounding tissue to locate the brain. Use a dissecting microscope to observe the smaller structures.

6. Remove the crop, gizzard, and part of the intestine to expose the underlying nerve cord with its segmented ganglia. Cut open the crop and gizzard and compare the thickness of their walls.

7. *Complete item 3d on the laboratory report.*

ARTHROPODS (Phylum Arthropoda)

Arthropods have more species than any other phylum of animals. They occur in almost every habitat on earth and usually are considered to be the most successful group of animals because of their vast numbers. Major characteristics of arthropods are (1) a segmented body divided into either **cephalothorax** and **abdomen**, or **head**, **thorax**, and **abdomen**; (2) **jointed appendages**; (3) an **exoskeleton** of chitin; (4) **simple** or **compound eyes**; and (5) an **open circulatory system**. Table 11.3 lists the characteristics of the major classes.

The Crayfish

The freshwater crayfish is a common crustacean that exhibits the basic characteristics of arthropods. Examine its external structure in Figure 11.7. Most of the exoskeleton is hardened by the accumulation of mineral salts. It provides protection for the soft underlying body parts, and it is flexible at the joints for easy movement of the appendages. The exoskeleton is shed periodically by molting to allow for growth of the crayfish.

The segmented body is divided into a cephalothorax and abdomen. Each body segment has a pair of jointed appendages, and most have been modified for special functions. **Mouthparts** are modified for feeding, and the two pairs of **antennae** contain numerous sensory receptors. The first three pairs of **walking legs** have pincers for grasping, and the first pair of these, the **chelipeds**, are used for offensive and defensive purposes. The **swimmerets** are relatively unmodified appendages on the abdominal segments. In females, they serve as sites of attachment for fertile eggs, and in males, the first two pairs are modified for sperm transport. The

TABLE 11.3
Distinguishing Characteristics of Arthropod Classes

Class	Characteristics
Arachnida (spiders, scorpions)	Cephalothorax and abdomen; four pairs of legs; simple eyes; no antennae; mostly terrestrial
Crustacea (crabs, shrimp)	Cephalothorax and abdomen; compound eyes; two pairs of antennae; five pairs of legs; mostly marine or freshwater
Chilopoda (centipedes)	Elongate, dorsoventrally flattened body; each body segment with a pair of legs; one pair of antennae; simple eyes; terrestrial
Diplopoda (millipedes)	Elongate, dorsally convexed body; segments fused in pairs giving *appearance* of two pairs of legs per segment; one pair of antennae; simple eyes; terrestrial
Insecta (insects)	Head, thorax, and abdomen; three pairs of legs on thorax; wings, if present, on thorax; simple and compound eyes; one pair of antennae; freshwater or terrestrial

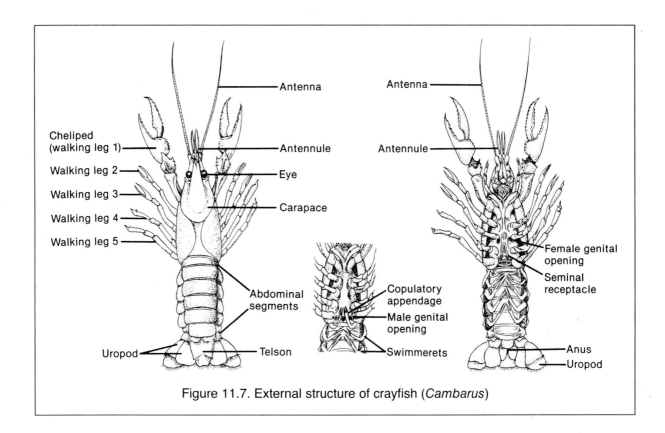

Figure 11.7. External structure of crayfish (*Cambarus*)

uropods are the last pair of appendages. Along with the **telson**, the uropods form a "paddle" used in swimming. The **compound eyes** are formed of many individual photosensory units and occur in only certain groups of arthropods.

The internal structure of the crayfish is shown in Figure 11.8. The **heart** consists of a single chamber and lies just under the **carapace** at the dorsal midline in the posterior part of the cephalothorax. **Arteries** carry blood anteriorly, posteriorly, and ventrally into the coelom, which spreads the blood so that it is near the body tissues for the exchange of materials. Blood returns from the coelom into the **pericardial sinus** surrounding the heart and then flows into the heart through tiny, valved openings called **ostia**. Just ventral to the heart lie a pair of elongated **gonads** (testes or ovaries) whose ducts lead to the genital openings. The large digestive glands surround the gonads and stomach and fill much of the cephalothorax.

The digestive system consists of the mouth, esophagus, stomach, digestive glands, intestine,

and anus. Food is crushed by **mandibles** and passed into the anterior **cardiac chamber** of the stomach. The stomach contains a **gastric mill** that further grinds the food into particles small enough to pass through a "strainer" formed by bristles into the posterior **pyloric chamber**. The finest particles enter ducts of the digestive glands where digestion is completed and nutrients are absorbed into the blood. Larger particles pass into the intestine and are removed from the body via the anus.

Metabolic wastes are removed from the blood by the **green glands** (flesh colored in preserved crayfish) and excreted via ducts that open at the base of the antennae.

The nervous system includes a pair of **supraesophageal ganglia** ("brain"), a ventral nerve cord, a series of segmented ganglia, and lateral nerves in each segment. Some of the segmental ganglia have fused to form larger ganglia in the cephalothorax. Crayfish have a variety of sensory receptors, especially on the anterior appendages.

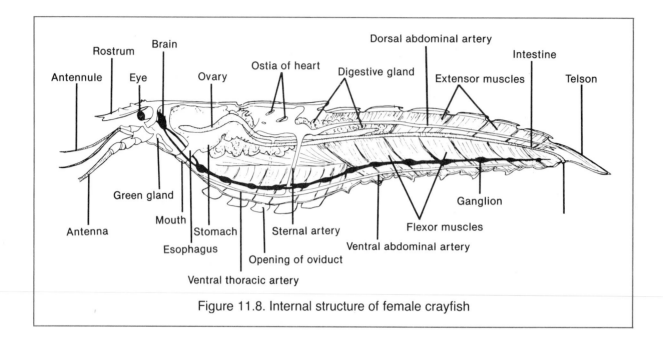

Figure 11.8. Internal structure of female crayfish

Insects (Class Insecta)

Insects are the largest group of arthropods. They are extremely abundant in terrestrial habitats, and many occur in fresh water. They do not occur in marine habitats, however. Figure 11.9 shows the structure of a grasshopper, a common insect. Note how the body is modified from the more generalized structure of the crayfish. The body is divided into head, thorax, and abdomen. Only three pairs of legs are present, and they are attached to the thorax, which also supports two pairs of **wings**. Abdominal appendages are absent. Both simple and compound eyes are present, but only one pair of antennae occurs. The **tympanum** is a sound receptor. The **spiracles** are openings into the **tracheal system**, a network of tubules that carry air directly to the body tissues.

Internally, the **gastric caeca** are glands that aid in digestion. Metabolic wastes are removed by **Malpighian tubules** that empty into the intestine. The dorsal blood vessel contains a series of "hearts" that pump blood anteriorly where it empties into the coelom and then slowly returns to enter the dorsal vessel.

Most insects undergo some form of **metamorphosis**, a hormonally controlled process that changes the body form of an insect at certain stages of the life cycle. Four stages occur in **complete metamorphosis**. The **egg** hatches to yield a wormlike **larva** that grows through several molts and then turns into a **pupa**, a nonfeeding stage. The **adult** emerges from the pupa. Incomplete or **gradual metamorphosis** has only three stages because it lacks a pupal stage. The egg hatches into a **nymph** (terrestrial form) or a **naiad** (aquatic form) that grows through several molts. The adult emerges at the final molt.

Materials

Colored pencils
Dissecting instruments
Dissecting pans, wax bottomed
Dissecting pins
Examples of insect metamorphosis
Preserved dissection specimens:
  crayfish (*Cambarus*)
  grasshoppers (*Romalea*)
Representative arthropods

Assignment 4

1. Color-code the major structures in Figures 11.7 to 11.9.
2. Examine the representative arthropods. Compare the characteristics of the specimens with those listed in Table 11.3. Record the names and characteristics of each

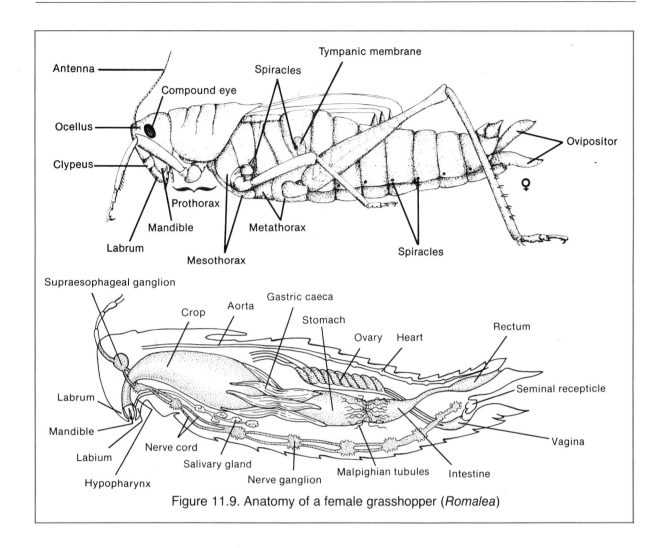

Figure 11.9. Anatomy of a female grasshopper (*Romalea*)

specimen. Note the adaptations of each based on the arthropod body plan.

3. Use Table 11.3 to determine the class of each of the "unknown" arthropods.

4. Examine the insect metamorphosis display. Note the components of each type.

5. ***Complete items 4a–4e on the laboratory report***.

Crayfish Dissection

1. Obtain a preserved crayfish and locate the external parts as shown in Figure 11.7. Correlate structure with function. Locate the flexible and rigid parts of the exoskeleton. Note the similarities and differences among the walking legs. Observe an eye and some of the appendages under a dissecting microscope. ***Draw the appearance of the eye surface in item 4g5***.

2. The primary incisions to be made in the dissection of a crayfish are shown in Figure 11.10. Make the first incision on the left side by inserting the tips of scissors under the posterior margin of the carapace and cutting forward to the cervical groove and downward to the ventral edge of the carapace. This removes the left side of the carapace and exposes the gills.

3. Examine the gills. To what are they attached? Note the thin transparent body wall medial to the gills.

4. Make the second incision anteriorly, nearly to the eye, and then downward to remove

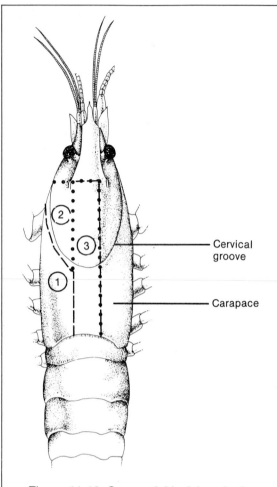

Cervical groove

Carapace

Figure 11.10. Sequential incisions in the crayfish dissection

gans, and gently remove them after observation to expose the underlying structures. Proceed in this manner until all of the organs have been identified and removed. After removing the stomach, cut it open lengthwise and wash out the contents to expose the gastric mill and the bristles forming the "strainer."

7. After the internal organs have been removed, you should be able to observe the ganglia along the ventral nerve cord. Try to locate the "brain" dorsal to the esophagus.

8. Make a transverse cut through the abdomen and examine the cut surface. Locate the muscles, intestine, and nerve cord.

9. ***Complete item 4f on the laboratory report***.

Grasshopper Dissection

1. Obtain a preserved grasshopper and locate the external parts shown in Figure 11.9. Note that both legs and wings arise from the thorax. Locate the spiracles on the abdominal segments. Observe the compound and simple eyes with a compound microscope. ***Draw the appearance of the compound eye in item 4g5***.

2. To observe the internal structures, make an incision to the right of the midline with your scissors. Start at the posterior end and cut anteriorly. Pin out the body wall to expose the internal organs. When you do this, you will notice the tracheal trunks extending from each spiracle into the body tissues.

3. The dorsal vessel is difficult to locate unless you have made a careful incision. Start with the most dorsal organs and remove them gently after observation to expose the underlying structures. After the digestive tract has been removed, you will see the large thoracic ganglia and the ventral nerve cord.

4. ***Complete item 4g on the laboratory report***.

the anterior lateral portion of the carapace. This will expose the heavy mandibular muscle and lateral wall of the stomach.

5. Make the third incision as shown, and carefully remove the median strip of carapace by gently separating it from the underlying tissues. This will expose the heart and other dorsally located structures. Remove the heart and locate the ostia.

6. Locate the internal structures shown in Figure 11.8. Start with the most dorsal or-

12

DEUTEROSTOMATE ANIMALS

OBJECTIVES

On completion of the laboratory session, you should be able to:
1. Describe the distinguishing characteristics of the animal groups studied and the evolutionary advances of each.
2. Identify representatives of the animal groups studied.
3. Identify and state the function of the structural components of the animals studied.
4. Define all terms in bold print.

Echinoderms and chordates are **deuterostomate animals** because the blastopore of the embryo becomes the anus and the mouth is formed from the second opening to appear. These groups also are called **enterocoelomate animals** because the coelom is formed from mesodermal pouches that "bud off" the embryonic gut, the enteron. Thus, the embryonic development suggests that echinoderms are more closely related to chordates than to other invertebrates.

ECHINODERMS (Phylum Echinodermata)

These "spiny-skinned" animals occur only in marine habitats. Their name is derived from the spines that project from the calcareous plates forming the **endoskeleton** just under the thin epidermis. A major distinctive characteristic is the presence of a **water vascular system** that provides a means of locomotion via the associated **tube feet** in most forms.

The **radial symmetry** that is characteristic of adult echinoderms is secondarily acquired since their larvae are bilaterally symmetrical. Adults have no anterior or posterior ends, only **oral** (surface with the mouth) and **aboral** surfaces. A pentamerous (five-part) organization is common in echinoderms. Tiny **dermal gills** project between the spines on the aboral surface. They are involved in gas exchange, as are the tube feet. Microscopic pincerlike structures, **pedicillariae**, protect the gills and keep the aboral surface free of debris. Table 12.1 describes the characteristics of the classes of echinoderms.

Sea Stars (Class Asteroidea)

Sea stars are perhaps the most familiar echinoderms. See Figure 12.1. The radially symmetrical body consists of five **arms** radiating from a **central disc**. The mouth is located in the center of the disc on the **oral surface**. The anus is located on the **aboral surface**, along with the **madreporite** that filters seawater as it enters the water vascular system.

The water vascular system consists of interconnecting tubes filled with seawater. Water passes from the madreporite through a **stone canal** to the **ring canal** that encircles the mouth. A **radial canal** extends from the ring canal into each arm, where it is concealed by the ambulacral bridge. The oral surface of each arm has an **ambulacral groove** that is bordered with rows of **tube feet**, the locomotor organs.

TABLE 12.1
Classes of Echinoderms

Class	*Characteristics*
Asteroidea (sea stars)	Star-shaped forms with broad-based arms; ambulacral grooves with tube feet that are used for locomotion
Ophiuroidea (brittle stars)	Star-shaped forms with narrow-based arms that are used for locomotion; ambulacral grooves covered with ossicles or absent
Echinoidea (sea urchins)	Globular forms without arms; movable spines and tube feet for locomotion; mouth with five teeth
Holothuroidea (sea cucumbers)	Cucumber-shaped forms without arms or spines; tentacles around mouth
Crinoidea (sea lilies)	Body attached by stalk from aboral side; five arms with ciliated ambulacral grooves and tentaclelike tube feet for food collection; spines absent

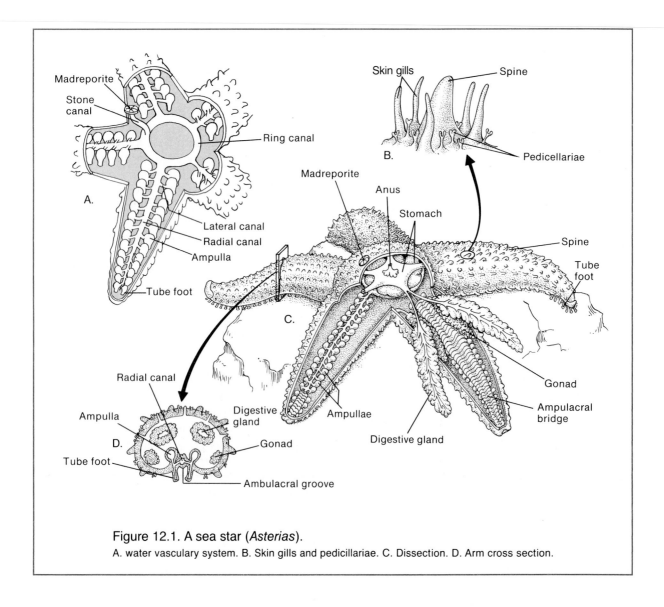

Figure 12.1. A sea star (*Asterias*).
A. water vascular system. B. Skin gills and pedicillariae. C. Dissection. D. Arm cross section.

The tube feet are connected to a **radial canal** by short **transverse canals**. Contraction or relaxation of the **ampulla** causes a tube foot to be extended or retracted, respectively, due to changes in the water pressure within the system. A suction cup at the end of the foot enables firm attachment to the substrate.

The central disc is largely filled with the digestive tract. The mouth opens into the **cardiac stomach** that leads to the **pyloric stomach** from which a pair of **digestive glands** extends into each arm. A very short intestine leads to the anus. A pair of **gonads** also extends from the central disc into each arm near the oral surface. The nervous and circulatory systems are greatly reduced and will not be considered here.

Sea stars feed on bivalves. Persistent pressure pulling against the valves of the shell weakens the adductor muscles and opens the valves slightly. The sea star then extrudes its stomach into the mantle cavity so that the prey is digested within its own shell.

Materials

Colored pencils
Dissecting instruments
Dissecting pan
Representative echinoderms
Sea stars, living in small aquarium
Sea stars, preserved

Assignment 1

1. Color-code the water vascular system and digestive organs in Figure 12.1.
2. Examine the representative echinoderms. Note their distinguishing characteristics and modifications of the basic echinoderm body plan. Identify the group to which each belongs. See Table 12.1.
3. Observe the living sea stars in the aquarium. Watch how the tube feet are used in locomotion.
4. ***Complete items 1a–1c on Laboratory Report 12 that begins on page 399.***

Sea Star Dissection

1. Obtain a preserved sea star and locate the external features shown in Figure 12.1.

2. Cut off one of the arms and examine the severed end. Locate the digestive glands, gonads, ambulacral groove, radial canal, and tube feet.
3. Use scissors to remove the aboral surface of the central disc and one of the arms. Locate the stomach which recieves secretions from the five pairs of digestive glands. Note the ampullae of the tube feet located on each side of the ambulacral bridge.
4. ***Complete item 1 on the laboratory report***.

CHORDATES (Phylum Chordata)

Chordates possess four distinguishing characteristics that are present in the embryo and may persist in the adult: (1) a **dorsal**, **tubular nerve cord**; (2) **pharyngeal gill slits**; (3) a **notochord**; and (4) a **postanal tail**. The three major groups of chordates are tunicates, lancelets, and vertebrates.

Tunicates (Subphylum Urochordata)

Figure 12.2 shows the basic structure of larval and adult tunicates (sea squirts). All chordate characteristics are present in the free-swimming larva but not in the immobile adult. Adult tunicates feed by filtering food particles from seawater passing over the gills. The food particles are passed by cilia into the intestine, where digestion occurs. Thus, pharyngeal gills serve in both feeding and gas exchange.

Lancelets (Subphylum Cephalochordata)

The lancelet amphioxus is shown in Figure 12.3. Note that all chordate characteristics are present in the adult. It is a poor swimmer and usually is buried in mud with the anterior end protruding. A true head is lacking. The **cirri** around the **buccal cavity** sweep seawater into the **pharynx**. Water passes from the pharynx through the **gills** into the **atrium** and out the **atriopore**. Food particles are trapped in mucus on the gills and moved by cilia into the **intestine**, where digestion occurs. Thus, the gills serve as both filter feeding and gas exchange organs. The **notochord**, an evolutionary forerun-

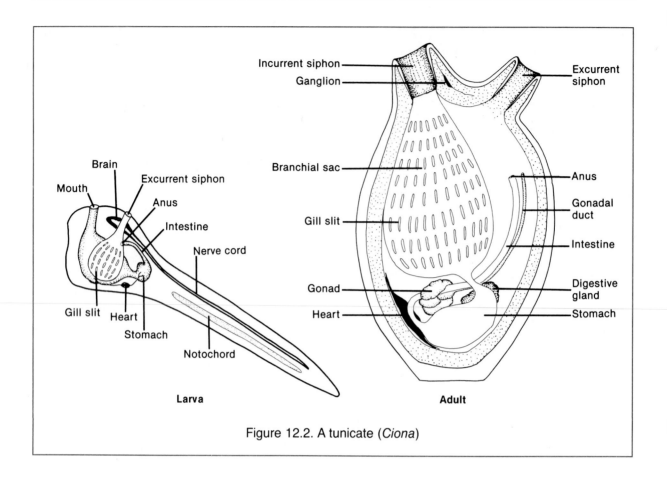

Figure 12.2. A tunicate (*Ciona*)

ner of the vertebral column, provides a flexible support for the body. Weak swimming motions are enabled by contractions of the segmentally arranged muscles.

Vertebrates (Subphylum Vertebrata)

All four chordate characteristics are usually not present in adult vertebrates, but they are present in embryonic stages. An **endoskeleton** of cartilage or bone provides structural support, including a **vertebral column** that partially or completely replaces the notochord.

The anterior part of the dorsal tubular nerve cord is modified to form a **brain** that is encased within a protective **cranium**. **Skin**, composed of dermal and epidermal layers, protects the body surface. The body is typically divided into **head**, **trunk**, and **tail**, although a tail does not persist in all adult vertebrates. Most forms have **pec-**

toral and **pelvic appendages**. Parts of the digestive tract are modified for special functions. Metabolic wastes are removed from the blood by a pair of **kidneys**, and gas exchange occurs by either **pharyngeal gills** or **lungs**. The **closed circulatory system** consists of a ventral heart that is composed of two to four chambers, arteries, capillaries, and veins. Table 12.2 lists the major characteristics of the vertebrate classes.

Materials

Colored pencils
Syracuse dishes
Model of amphioxus
Amphioxus adults, preserved
Tunicate adults, preserved
Prepared slides of:
    amphioxus, w.m. and x.s.
    tunicate larva, w.m.

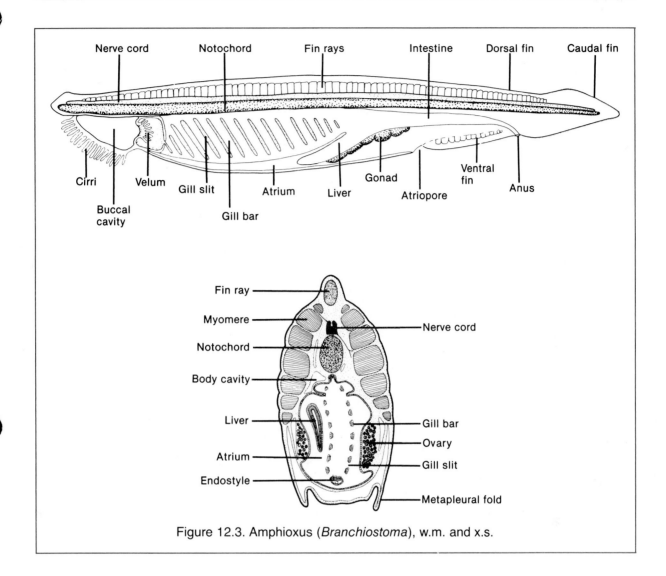

Figure 12.3. Amphioxus (*Branchiostoma*), w.m. and x.s.

Assignment 2

1. Color-code the nerve cord, notochord, gill slits, and digestive tract in Figures 12.2 and 12.3.
2. Examine the adult tunicates and the prepared slide of tunicate larva under a demonstration microscope. Locate the basic chordate features in the larval tunicate.
3. Examine the amphioxus model and locate the chordate characteristics. Compare the preserved amphioxus under a demonstration dissecting microscope with the model. Locate the **dorsal**, **caudal**, and **ventral fins**, the two **metapleural fin folds**, and the segmentally arranged muscles.

4. Examine prepared slides of amphioxus, w.m. and x.s., and locate the structures labeled in Figure 12.3. Note the positions of the nerve cord, notochord, and pharynx.
5. Study the characteristics of vertebrate classes in Table 12.2.
6. ***Complete item 2 on the laboratory report***.

Jawless Fish (Class Agnatha)

The first vertebrates were small jawless fish, somewhat like amphioxus, but which had a true head and a bony endoskeleton. The only living jawless fish, lampreys and hagfish, are

TABLE 12.2
Classes of Vertebrates

Class	Characteristics
Agnatha (jawless fish)	No jaws; scaleless skin; median fins only; cartilaginous endoskeleton; persistent notochord; two-chambered heart
Chondrichthyes (cartilaginous fish)	Jaws; subterminal mouth; placoid dermal scales; median and paired fins; cartilaginous endoskeleton; two-chambered heart
Osteichthyes (bony fish)	Jaws; terminal mouth; membranous median and paired fins; bony endoskeleton; lightweight dermal scales; two-chambered heart; operculum covers gills; swim bladder
Amphibia (amphibians)	Aquatic larvae with gills; air-breathing adults with lungs; scaleless skin; paired limbs; bony endoskeleton; three-chambered heart
Reptilia (reptiles)	Bony endoskeleton with rib cage; well-developed lungs; epidermal scales; claws; paired limbs in most; leathery-shelled eggs; three-chambered heart
Aves (birds)	Wings; feathers; epidermal scales; claws; endoskeleton of lightweight bones; rib cage; horny beak without teeth; hard-shelled eggs; four-chambered heart
Mammalia (mammals)	Hair; reduced or absent epidermal scales; claws or nails; mammary glands; diaphragm; bony endoskeleton with rib cage; young develop in uterus nourished via placenta; four-chambered heart

degenerate forms that have lost the bony skeleton. They live as scavengers or parasites on other fish. See Figure 12.4. Note the number of gill slits and the absence of paired fins.

Cartilaginous Fish (Class Chondrichthyes)

Cartilaginous fish constitute a side branch from jawed bony fish that appeared about 425 million years ago. Sharks and rays are the best known members of this group. See Figure 12.4. Paired **pectoral fins** and **pelvic fins** provide stabilization when swimming. The **dermal scales**, characteristic of fish, are of the placoid type, making the skin rough to the touch. The scales provide protection and yet permit freedom of movement. Note the **lateral line system** that runs along each side of the body. It consists of a fluid-filled tube containing sensory receptors for sound vibrations. The number of gill slits is less than in jawless fish, since the anterior gill arches form the jaws.

Bony Fish (Class Osteichthyes)

Most modern fish are bony fish. See Figure 12.4. Lightweight dermal scales and fins and a **swim bladder** facilitate swimming. Note the anal fin and the anterior location of the pelvic fins. The number of gills is fewer than in cartilaginous fish, and the gills are covered by a bony **operculum**. Note the lateral line system. Fish and lower animals are **poikilothermic** (body temperature fluctuates with the environmental temperature) and **ectothermic** (source of body heat is external). Fertilization and development is external in most fish.

Materials

Live fish in an aquarium
Lamprey, dogfish shark, and yellow perch: preserved, whole, and in cross section
Perch; preserved, skinned on one side
Perch skeleton
Representative fish
Prepared slides of ammocoete (lamprey) larvae, w.m.

Assignment 3

1. Observe a prepared slide of ammocoete (lamprey) larva, and note its similarity to amphioxus. Look for the basic chordate characteristics. Is a notochord present?

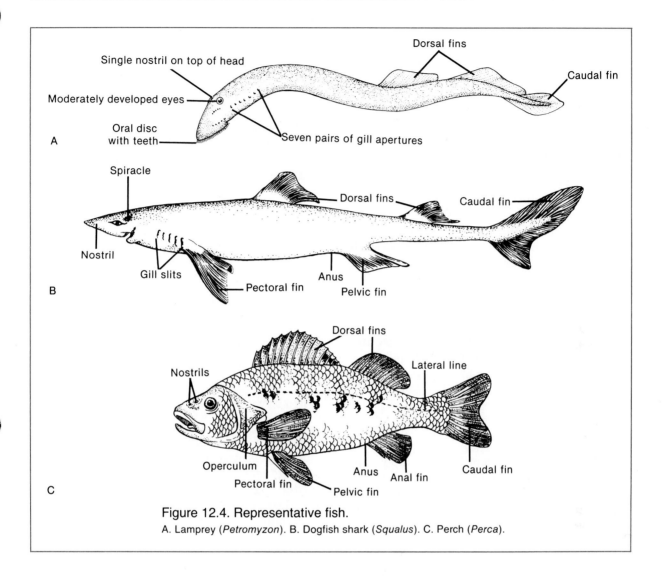

Figure 12.4. Representative fish.
A. Lamprey (*Petromyzon*). B. Dogfish shark (*Squalus*). C. Perch (*Perca*).

2. Examine the whole and sectioned lamprey, shark, and perch. Note the characteristics of each. Look for the presence of a notochord, nerve cord, and cartilaginous or bony skeleton in the cross sections. Rub your hand on the skin of each specimen posteriorly and then anteriorly. Do you detect any differences? Examine shark and perch skin with a dissecting microscope to see the differences in the dermal scales.

3. Observe the perch skeleton and a skinned perch. Is there any evidence of segmentation? Note how the cranium protects the brain and how the vertebral column protects the spinal cord but allows flexibility in lateral swimming movements.

4. Observe the live fish in the aquarium. Note how they swim and use the paired fins for stability. Observe their respiratory movements.

5. ***Complete item 3 on the laboratory report.***

Amphibians (Class Amphibia)

Amphibians are believed to have evolved from primitive lobe-finned fish. Fertilization and development are external. The aquatic larvae have functional gills and a tail for swimming. Adults are air breathing and have paired appendages. The moist skin is an auxiliary gas-exchange organ. Some adult amphibians have tails; others do not. Amphibians are poikilothermic and ectothermic.

Reptiles (Class Reptilia)

Reptiles were the first truly terrestrial vertebrates since they can reproduce without returning to water. This is possible because (1) fertilization is internal and (2) a leathery shelled, evaporation-resistant egg contains nutrients to nourish the embryo until hatching. Pharyngeal gill slits and a notochord are present only in the embryo of reptiles and higher vertebrates. In most reptiles and in all higher vertebrates, the **ribs** join ventrally to the **sternum** to form a rib cage that protects the heart and lungs.

Epidermal scales first appear in reptiles. Formed by the epidermis of the skin, these scales provide protection and retard evaporative water loss. The feet typically have five toes, and the **claws** are formed from modified epidermal scales. Reptiles have peglike teeth. Like all lower animals, reptiles are poikilothermic and ectothermic.

Birds (Class Aves)

Birds evolved from reptiles, and they still have many reptilian characteristics. Birds possess numerous adaptations for flight, including (1) **feathers**, which are derived from epidermal scales; (2) hollow, lightweight bones; (3) air sacs associated with the lungs; (4) a fused skeleton except for neck, tail, and paired appendages; and (5) pectoral appendages modified into **wings**. The jaws are modified to form a **bony beak** without teeth.

Unlike lower animals, birds are **homeothermic** (maintain a relatively constant body temperature) and **endothermic** (heat produced by cellular respiration maintains the body temperature). This enables birds to be active even when the ambient temperature is low.

Fertilization is internal, and the hard-shelled eggs contain nutrients for the developing embryo. Parents incubate the eggs until hatching and provide considerable care for the young birds until they can care for themselves.

Mammals (Class Mammalia)

Like birds, mammals evolved from reptiles. Key mammalian characteristics are the presence of **hair**, **mammary glands** (milk-producing glands), and a **diaphragm** that separates the abdominal and thoracic cavities. Mammals are homeothermic and endothermic.

After internal fertilization, most modern mammals nourish the developing embryo *in utero* via a **placenta**. A placenta brings the maternal and embryo blood vessels close together so that materials can diffuse from one to the other. **True placental mammals** have a relatively long **gestation period** so that the young mammals are rather well developed at birth. In contrast, **marsupial mammals** have a short gestation period and give birth to underdeveloped young that complete development in the pouch of the female. Only two species of **egg-laying mammals** have survived to the present: the duck-billed platypus and the spiny anteater. Both occur in Australia, along with most marsupials, and represent a primitive side branch of the main evolutionary line. They are not ancestral to higher mammals.

Materials

Colored pencils
Live salamander, lizard, and rat
Representative amphibians, reptiles, birds, and mammals
Study skins of seed-eating, insect-eating, and flesh-eating birds
Prepared slides of:
   frog embryo, neurula stage, x.s.
   chick embryo, 96 hr, w.m.

Assignment 4

1. Color-code the three embryonic tissues and their derivative's as follows in Figures 12.5 and 12.6: ectoderm—blue; mesoderm—red, endoderm—yellow.
2. Observe the prepared slide of a frog embryo set up under a demonstration microscope. Locate the notochord and nerve cord, and compare your observations with Figure 12.5.
3. Observe the demonstration slide of a chick embryo, w.m., showing pharyngeal gill slits. Compare your observations with Figure 12.6.
4. Examine the representative amphibians, reptiles, birds, and mammals, and note their distinguishing characteristics.

5. Examine a living salamander, lizard, and rat.
   a. Compare the position of the legs in each and determine whether the backbone bends laterally when the animal is walking.
   b. Place the salamander in the small aquarium and watch it swim. Is the movement of the backbone similar when swimming and walking? How does the swimming movement compare with that of fish?
   c. Compare the skin and feet. Look for epidermal scales and claws.
6. Note the adaptations of beaks and feet in the birds. How do these correlate with their different ways of life?
7. ***Complete the laboratory report.***

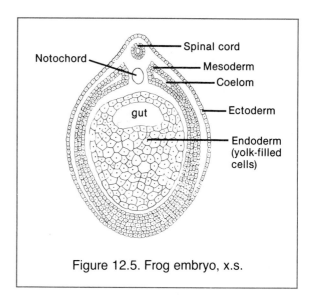

Figure 12.5. Frog embryo, x.s.

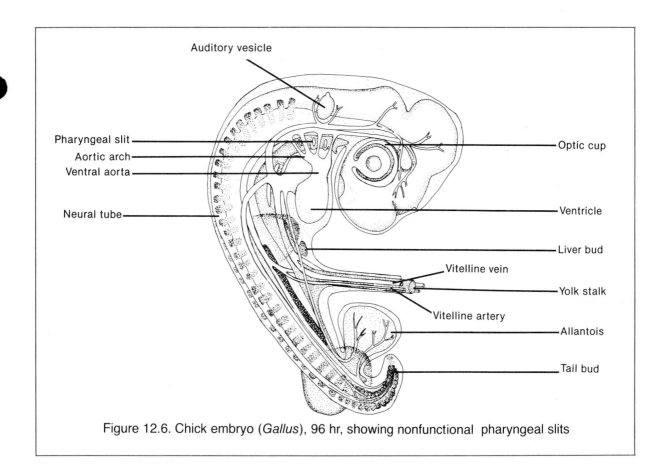

Figure 12.6. Chick embryo (*Gallus*), 96 hr, showing nonfunctional pharyngeal slits

PART III

ORGANISMIC BIOLOGY

13

DISSECTION OF THE FROG

OBJECTIVES

On completion of the laboratory session, you should be able to:
1. Identify the major external features and internal organs of the frog.
2. Describe the relative positions of the internal organs.
3. Define all terms in bold print.

A study of the frog will give you a better understanding of the body organization of vertebrates. The dissection focuses on the major organs of the body cavity. The integumentary, skeletal, muscle, nervous, and endocrine systems are not covered. The dissection will also prepare you for a practical examination on frog or vertebrate structure. Use the scalpel in your dissection *only when absolutely necessary*. Follow the directions carefully and in sequence. The terms of direction in Table 13.1 will help clarify the location of body parts as you perform the dissection. Note that members of each pair have opposite meanings.

Materials

Dissecting instruments and pins
Dissecting pan, wax bottomed
Frogs, preserved, double injected

EXTERNAL FEATURES

The frog's body is divided into a **head** and **trunk**; there is no neck. Paired appendages are attached to the trunk. The **posterior appendages** consist of a thigh, lower leg, and foot. The foot has five toes plus a rudimentary medial sixth toe. The elongation of two ankle bones has increased the length of the foot and improved leverage for hopping. The **anterior appendages** consist of an upper leg, lower leg, and foot. The foot has four toes plus a rudimentary medial fifth toe. If your specimen is a male, the medial toes of the front feet will have a swollen pad.

The bulging eyes have immovable upper and lower eyelids. However, a **nictitating membrane**, a third eyelid attached to the inner portion of the lower eyelid, can cover the eye. Observe the nictitating membrane in the live frogs. Locate a large **tympanum** (eardrum) and the paired **external nares** (nostrils).

THE MOUTH

Use heavy scissors to cut through the bones at the angle of each jaw so the mouth may be opened as in Figure 13.1. Locate the structures shown. Note that the tongue is attached at its anterior end. Feel the teeth. Locate the **internal nares** (nostrils). The narrowed region at the back of the mouth is the **pharynx** that leads to the **esophagus**, a short tube that carries food into the stomach. Just ventral to the esophageal

TABLE 13.1
Anatomical Terms of Direction

Anterior	Toward the head
Posterior	Toward the hind end
Cranial	Toward the head
Caudal	Toward the hind end
Dorsal	Toward the back
Ventral	Toward the belly
Superior	Toward the back
Inferior	Toward the belly
Lateral	Toward the sides
Medial	Toward the midline

opening is the **glottis**, a slit through which air passes into the **larynx** and on to the lungs. The **eustachian tubes**, located near the angles of the jaws, communicate with the middle ears, spaces interior to the eardrums. Male frogs have small openings ventral to the eustachian tubes that lead to vocal sacs ventral to the lower jaw. Air-filled vocal sacs increase the volume of vocalizations.

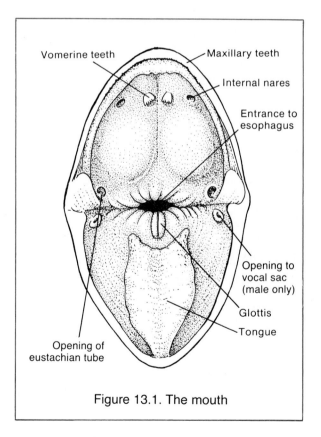

Figure 13.1. The mouth

OPENING THE BODY CAVITY

1. Pin the frog on its back to the wax bottom of a dissecting pan by placing pins through its feet.
2. Lift the skin at the ventral midline with forceps and make a small incision with the scissors. Extend the cut anteriorly to the lower jaw and posteriorly to the junction of the thighs as shown in Figure 13.2. Make incisions 2 and 3 as shown. Then use a dissecting needle and forceps to separate the skin from the underlying muscles to the lateral margins of the body. Use scissors to cut off the skin flaps to expose the muscles of the ventral body wall. The blue ventral

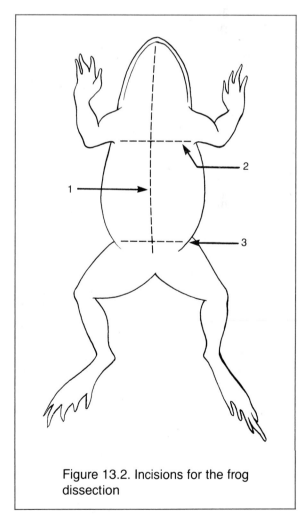

Figure 13.2. Incisions for the frog dissection

abdominal vein will be visible through the body wall.

3. Being careful to avoid damaging the underlying organs, make incision 1 through the ventral body wall just to the right of the midline to avoid cutting the **ventral abdominal vein** attached to the body wall at the midline. Free this vein from the body wall with a dissecting needle, and make transverse cuts 2 and 3 as before. Cut off the flaps of body wall.

4. Lift the muscles and bones in the "chest" region with forceps and use scissors to cut anteriorly through them at the midline. Carefully, separate the tissue from the underlying organs and avoid cutting the major blood vessels. Pin back or remove the severed tissues to expose the heart and other anteriorly located organs.

5. Note how the organs are oriented in the body cavity. The arteries are filled with red latex and veins with blue latex. The yellow, fingerlike structures are fat bodies attached to the anterior portions of the kidneys. You may need to remove most of the fat bodies to get them out of the way. As you locate the internal organs, it will be

necessary to move or remove the ventrally located organs to see the more dorsally located organs. However, save those portions attached to blood vessels if you plan to study the circulatory system.

DIGESTIVE SYSTEM

Compare your specimen with Figure 13.3. Note the large **liver**. How many lobes does it have? The liver receives nutrient-rich, venous blood from the stomach and intestines. It removes excess nutrients from the blood for processing before the blood returns to the heart. Find the **gall bladder**, a greenish sac lying between lobes of the liver. The gall bladder stores bile secreted by the liver and empties it into the small intestine via the **bile duct**. Bile aids fat digestion. Remove most of the liver to expose underlying organs, but keep those portions with blood vessels.

Food passes from the mouth into the very short **esophagus** that carries food to the **stomach**. The esophagus is anterior and dorsal to the heart and bronchi. (You can locate the esophagus after studying the circulatory system.) Food

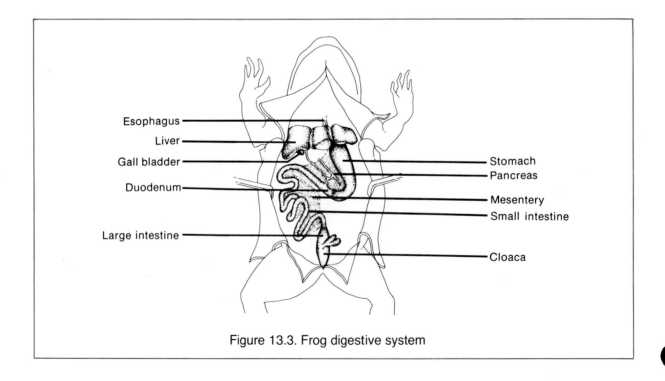

Figure 13.3. Frog digestive system

enters the stomach through the **cardiac valve**, a sphincter muscle located at the esophagus-stomach junction. Locate the stomach, an enlarged portion of the digestive tract that lies along the left side of the body cavity. After digestion has begun in the stomach, the food mass passes through the **pyloric valve**, a sphincter muscle, into the **small intestine**, where digestion of food and absorption of nutrients is completed. Locate the small intestine.

Find the **mesentery**, the transparent connective tissue supporting the digestive organs and spleen. It is formed of two layers of the **peritoneum**, a thin membrane that lines the body cavity and covers the internal organs. Locate the narrow, elongate **pancreas** in the mesentery between the stomach and the first portion of the small intestine. It secretes digestive enzymes into the small intestine. The spherical, reddish-brown (blue in injected specimens) **spleen** is also supported by mesentery and lies dorsal and posterior to the stomach. It is a circulatory organ, not a digestive organ, that removes worn-out blood cells from the blood.

The nondigestible food residue passes from the small intestine into the short **large intestine** that temporarily stores the nondigestible wastes. The large intestine extends into the pelvic region and terminates with the **anus** that opens into the **cloaca**. You will observe the cloaca when you study the urogenital system.

RESPIRATORY SYSTEM

The reddish-purple lungs are located on each side of the heart and dorsal to the liver. Air entering the glottis passes into the very short **larynx** from which two short **bronchi** branch and lead to the lungs. Inserting a probe through the glottis will help you locate a bronchus.

The lungs in a preserved frog are constricted and do not show the appearance or structure of a living lung. Amphibian lungs are essentially saclike in structure with ridges on their inner surfaces that increase the lung surface area exposed to air in the lungs. These ridges form a definite pattern, somewhat like the surface of a waffle iron. Cut open a lung and note its structure. If your instructor has set up a demonstration of an inflated lung under a dissecting microscope, lung structure will be evident.

CIRCULATORY SYSTEM

Refer to Figures 13.4, 13.5, and 13.6 as you study the circulatory system. Your dissection will be simplified if you have a good understanding of heart structure and the location and names of *both* arteries and veins *before proceeding*.

Major Arteries

Locate the heart. Carefully dissect away the overlying tissue. Find the transparent, saclike **pericardium** that envelopes the heart and remove it. Remove any excess latex that may be within the pericardial sac. Locate the single **ventricle**, the **left atrium**, and the **right atrium**. The ventricle pumps blood to the lungs and body, the left atrium receives blood from the lungs, and the right atrium receives blood from the body.

To locate the arteries described below and the veins described later, you will need to carefully dissect away surrounding connective tissue to expose them.

Locate the **conus arteriosus**, the large artery visible on the anterior surface of the heart that carries blood from the ventricle. Near the top of the heart it divides in a Y-shaped manner to form left and right branches. Each branch is a **truncus arteriosus** that, in turn, divides to form three arterial arches. (1) The anterior arch is the **carotid arch** that carries blood to the head. (2) The middle arch is the **aortic arch**. It curves posteriorly to join with the aortic arch from the other side to form the **dorsal aorta**. The **subclavian artery** branches from the aortic arch to supply the shoulder and foreleg. (3) The posterior arch is the **pulmocutaneous arch**. It branches into the **cutaneous artery** that supplies the skin and body wall and the **pulmonary artery** that supplies the lung.

The **coeliaco-mesenteric artery** is the most anterior artery to branch from the dorsal aorta. It is supported by the mesentery and supplies the stomach, small intestine, pancreas, and spleen. Posteriorly, several **urogenital arteries** branch off the aorta to supply the kidneys and gonads. Just posterior to the urogenital arteries, the dorsal aorta gives off a small branch, the **posterior mesenteric artery**, that carries blood to the large intestine. Then, the aorta

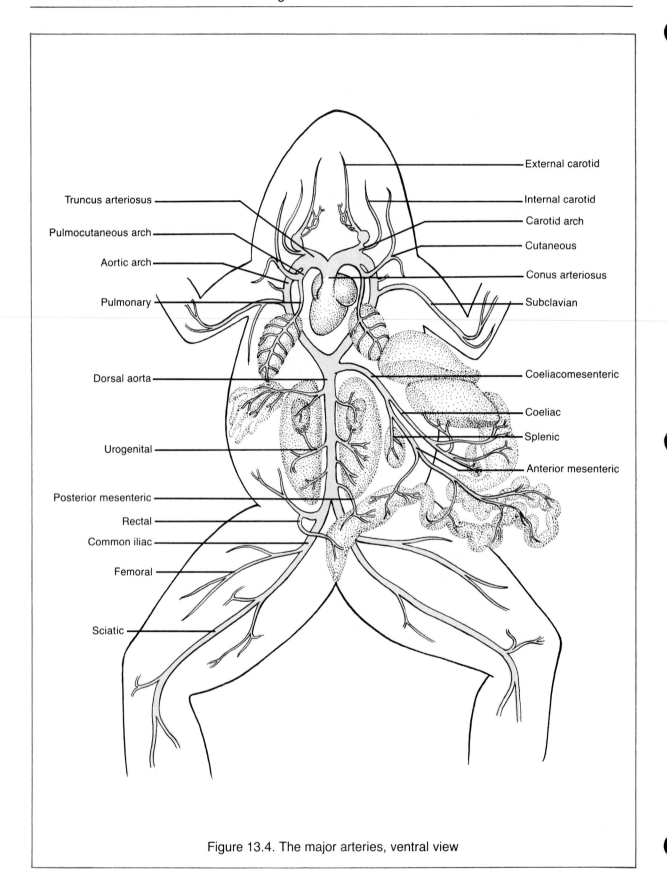

Figure 13.4. The major arteries, ventral view

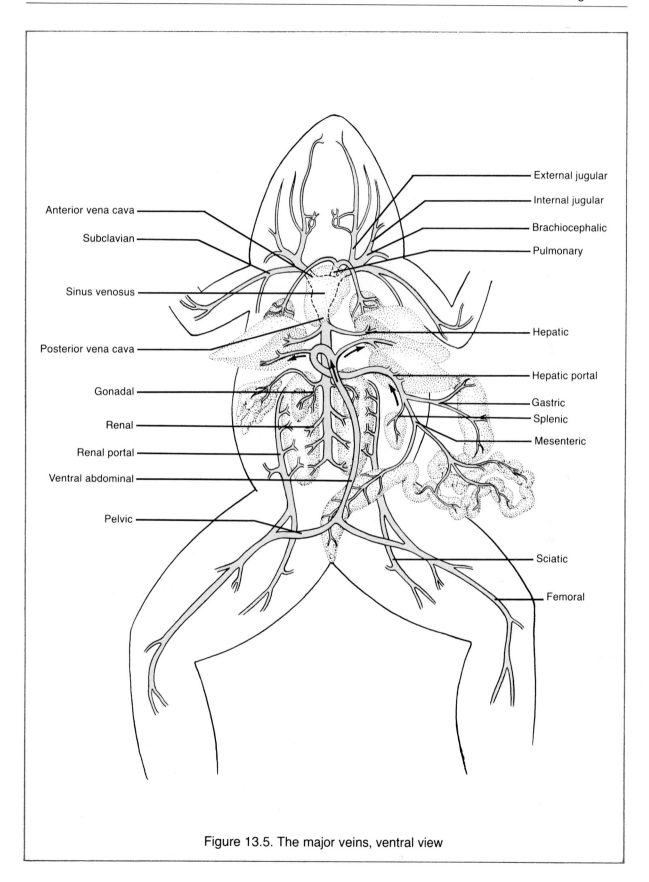

Figure 13.5. The major veins, ventral view

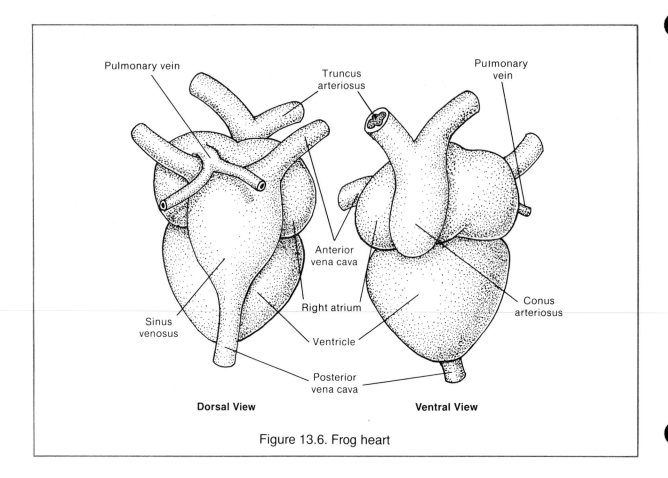

Figure 13.6. Frog heart

branches to form the left and right **common iliacs** that enter the hind legs. Each common iliac forms two major branches. The smaller **femoral artery** serves the thigh, while the larger **sciatic artery** gives off smaller branches and continues to the lower leg.

Major Veins

Return to the heart. Lift the tip of the heart carefully to see the thin-walled, saclike **sinus venosus**, located on the dorsal surface of the heart. See Figure 13.6. It opens into the right atrium and receives blood from three large veins, the venae cavae. The **posterior vena cava** drains the body posterior to the heart and enters the sinus venosus posteriorly. The two **anterior vena cavae** enter the sinus venosus laterally, one from each side. Each drains the head, foreleg, body wall, and skin on its side of the body.

Observe the right anterior vena cava. It is formed by the union of three veins arranged clockwise from posterior to anterior. (1) The **subclavian vein** drains the foreleg, body wall, and skin. (2) The **brachiocephalic vein** drains the shoulder and deeper portions of the head. (3) The **external jugular** drains the more superficial parts of the head.

Locate the smaller **pulmonary veins** that return oxygenated blood from the lungs into the left atrium.

Locate the posterior vena cava. It receives blood from the liver via two **hepatic veins** located just posterior to the heart and at its posterior origin from the **renal veins** draining the kidneys.

Blood from the digestive tract is carried to the liver by the **hepatic portal vein** for processing and then exits via the hepatic veins. The hepatic portal vein will not be injected with blue latex, so it may be difficult to locate.

The **sciatic** and **femoral veins** drain the hind legs and unite to form the **renal portal veins** that carry blood to the kidneys. (Renal portal veins are not found in higher vertebrates.) Blood flows through the kidneys and enters the posterior vena cava via the **renal veins**.

The unpaired **ventral abdominal vein** is formed by the union of the **pelvic veins** posteriorly, and it divides anteriorly to enter the liver lobes.

Now, return to the heart and refresh your understanding of the major arteries and veins associated with it. Then, cut these vessels near the heart, leaving short stubs, and remove the heart. Compare it with Figure 13.6. Make a coronal section through the atria and ventricle to observe the interior structures.

UROGENITAL SYSTEM

The urinary and reproductive systems are closely related and are commonly studied as the urogenital system. See Figures 13.7 and 13.8.

Urinary System

Locate the paired **kidneys**. They are the elongate, reddish-brown (blue in injected frogs) organs located on each side of the posterior vena cava against the dorsal body wall. The **fat bodies**, finger-shaped fatty masses attached to the anterior portions of the kidneys, nourish the frog during hibernation. A slender **adrenal body** (difficult to find) lies along the ventral surface of each kidney.

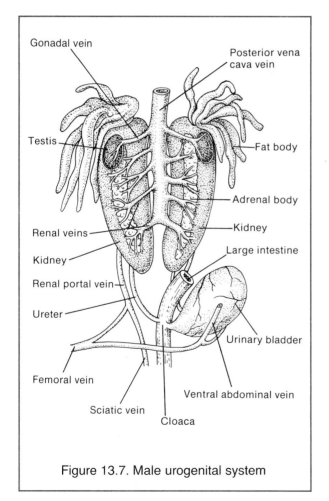

Figure 13.7. Male urogenital system

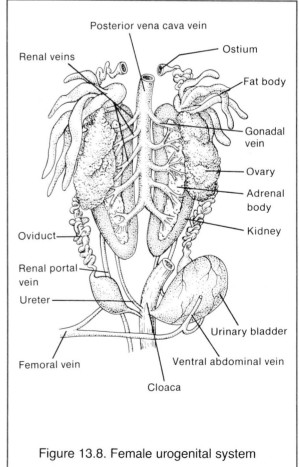

Figure 13.8. Female urogenital system

Ureters carry urine formed by the kidneys to the cloaca for removal from the body. Try to find a slender white ureter near the lateral posterior margin of a kidney.

To observe the cloaca and urinary bladder, it is necessary to cut through the muscles and bones of the pelvic region at the ventral midline until the cloaca is exposed. Perform this step with care.

Locate the cloaca and find the **urinary bladder**, a ventral pouch attached to the cloaca. Urine in the cloaca backs up into the urinary bladder. The cloaca receives urinary wastes, digestive wastes, and gametes (sperm and eggs). Find the junction of the large intestine, ureters, and oviducts with the dorsal surface of the cloaca.

Male Reproductive System

The yellowish, oval **testes** are located on the anterior ventral surface of the kidneys. Sperm pass into the kidneys via several fine tubules, the **vasa efferentia**. The ureters carry both urine and sperm to the cloaca.

Female Reproductive System

The **ovaries** appear as irregular-shaped sacs located on the anterior ventral surface of the kidneys. Ovaries will vary in size and shape in accordance with the breeding season. Locate the coiled whitish **oviducts** that carry eggs from the ovaries to the cloaca. Trace an oviduct to the cloaca.

CONCLUSION

This completes the dissection. Review as necessary so that you can recognize all of the organs. When finished, dispose of your specimen as directed by your instructor. Wash and dry your dissecting instruments and pan. There is no laboratory report for this exercise.

14

DISSECTION OF THE FETAL PIG

OBJECTIVES

On completion of the laboratory session, you should be able to:
1. Identify the major internal organs of the fetal pig and indicate to which organ system each belongs.
2. Describe the relative positions of the major organs.
3. Name the body cavities and the organs each one contains.
4. Define all terms in bold print.

General Dissection Guidelines

The purpose of the dissection is to expose the various organs for study, and it should be done in a manner that causes minimal damage to the specimen. *Follow the directions carefully and in sequence.* Read the entire description of each incision and be sure that you understand it *before* you attempt it. Work in pairs, alternating the roles of reading the directions and performing the dissection. Use scissors for most of the incisions. *Use a scalpel only when absolutely necessary.* Be careful not to damage organs that must be observed later.

Your instructor may have you complete the dissection over several sessions. If so, wrap the

In this exercise, you will study the basic body organization of mammals through the dissection of a fetal (unborn) pig. The dissection will focus on the internal organs located in the ventral body cavity. Your dissection will proceed easier if you understand the common directional terms used to describe the location of organs. These terms are noted in Table 14.1. Figure 14.1 shows the relationship of most of the directional terms, as well as the three common planes associated with a bilaterally symmetrical animal. The integumentary, skeletal, muscular, nervous, and endocrine systems will not be studied. Study the organ systems outlined in Table 14.2 so that you know their functions and the major organs of each.

TABLE 14.1
Anatomical Terms of Direction

Anterior	Toward the head
Posterior	Toward the hind end
Cranial	Toward the head
Caudal	Toward the hind end
Dorsal	Toward the back
Ventral	Toward the belly
Superior	Toward the back
Inferior	Toward the belly
Lateral	Toward the sides
Medial	Toward the midline

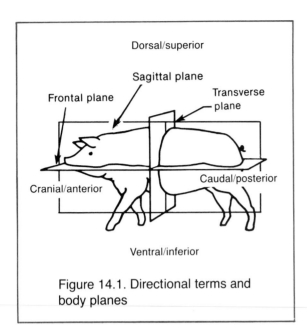

Figure 14.1. Directional terms and body planes

specimen in wet paper towels at the end of each session, and seal it in a plastic bag to prevent it from drying out.

Protect your hands from the preservative by applying a lanolin-based hand cream before starting the dissection or by wearing disposable plastic gloves. Wash your hands thoroughly at the end of each dissection session and reapply the hand cream.

Materials

Dissecting instruments and pins
Dissecting pan, wax bottomed
Dissecting microscope
Hand cream
String
Fetal pig (*Sus scrofa*), preserved and double injected

TABLE 14.2
Major Organs and Functions of the Organ Systems

System	Major Organs	Major Functions
Integumentary	Skin, hair, nails	Protection, cooling
Skeletal	Bones	Support, protection
Muscle	Muscles	Movement
Nervous	Brain, spinal cord, sense organs, nerves	Rapid coordination via impulses
Endocrine	Pituitary, thyroid, parathyroid, testes, ovaries, pancreas, adrenal, pineal	Slower coordination via hormones
Digestive	Mouth, pharynx, esophagus, stomach, intestines, liver, pancreas	Digestion of food, absorption of nutrients
Respiratory	Nasal cavity, pharynx, larynx, trachea, bronchi, lungs	Exchange of oxygen and carbon dioxide
Circulatory	Heart, arteries, capillaries, veins, blood	Circulation of blood, transport of materials
Lymphatic	Lymphatic vessels, lymph, spleen, thymus, lymph nodes	Cleansing and return of extracellular fluid to bloodstream
Urinary	Kidneys, ureters, urinary bladder, urethra	Formation and removal of urine
Reproductive		
Male	Testes, epididymis, vas deferens, seminal vesicles, bulbourethral gland, prostate gland, urethra, penis	Formation and transport of sperm and semen
Female	Ovaries, oviducts, uterus, vagina, vulva	Formation of eggs, sperm reception, intrauterine development of offspring

EXTERNAL ANATOMY

Examine the fetal pig and locate the external features shown in Figure 14.2. Determine the sex of your specimen. The **urogenital opening** in the female is immediately ventral to the anus and has a small **genital papilla** marking its location. A male is identified by the **scrotal sac** ventral to the anus and a **urogenital opening** just posterior to the **umbilical cord**. Two rows of nipples of mammary glands are present on the ventral abdominal surface of both males and females, but the mammary glands later develop only in maturing females. Mammary glands and hair are two distinctive characteristics of mammals.

Make a transverse cut through the umbilical cord and examine the cut end. Locate the two **umbilical arteries** that carry blood from the fetal pig to the placenta, and the single **umbilical vein** that returns blood from the placenta to the fetal pig.

POSITIONING THE PIG FOR DISSECTION

Position your specimen in the dissection pan as follows:

1. Tie a piece of heavy string about 20 in. long around each of the left feet.

2. Place the pig on its back in the pan. Run the strings under the pan, and tie them to the corresponding right feet to spread the legs and expose the ventral surface of the body. This also holds the pig firmly in position.

HEAD AND NECK

This portion of the dissection will focus on structures of the mouth and pharynx. If your instructor wishes for you to observe the salivary glands, a demonstration dissection will be prepared for you to observe.

Dissection Procedure

To expose the organs of the mouth and pharynx, start by inserting a pair of scissors in the angle of the lips on one side of the head and cut posteriorly through the cheek. Open the mouth as you make your cut and follow the curvature of the tongue to avoid cutting into the roof of the mouth. Continue through the angle of the mandible (lower jaw). Just posterior to the base of the tongue, you will see a small whitish projection, the **epiglottis**, extending toward the roof of the mouth. Hold down the epiglottis and surrounding tissue and continue your incision

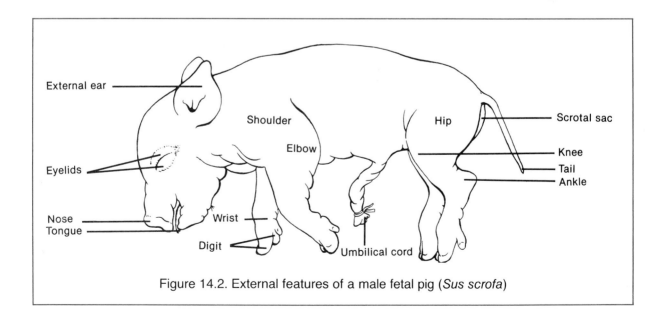

Figure 14.2. External features of a male fetal pig (*Sus scrofa*)

dorsal to it and on into the opening of the **esophagus**. Now, repeat the procedure on the other side so that the lower jaw can be pulled down to expose the structures as shown in Figure 14.3.

Observations

Only a few deciduous teeth will have erupted, usually the third pair of incisors and the canines. Other teeth are still being formed and may cause bulging of the gums. Make an incision in one of these bulges to observe the developing tooth.

Observe the tongue. Note that it is attached posteriorly and free anteriorly as in all mammals. Locate the numerous **papillae** on its surface, especially near the base of the tongue and along its anterior margins. Papillae contain numerous microscopic taste buds.

Observe that the roof of the mouth is formed by the anteriorly located **hard palate**, supported by bone and cartilage, and the posteriorly located **soft palate**. Paired nasal cavities lie dorsal to the roof of the mouth. The space posterior to the nasal cavities is called the **nasal pharynx**, and it is contiguous with the **oral pharynx**, the throat, located posterior to the mouth. Locate the opening of the nasal pharynx posterior to the soft palate. The oral pharynx may be difficult to visualize because your incision has cut through it on each side, but it is the region where the glottis, esophageal opening, and opening of the nasal pharynx are located.

Make a midline incision through the soft palate to expose the nasal pharynx. Try to locate the openings of the **eustachian tubes** in the dorsolateral walls of the nasal pharynx. Eustachian tubes allow air to move into or out of the

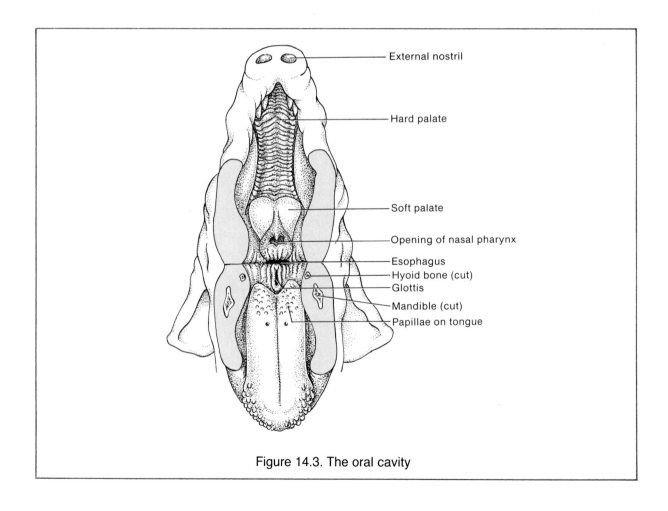

External nostril

Hard palate

Soft palate

Opening of nasal pharynx

Esophagus

Hyoid bone (cut)

Glottis

Mandible (cut)

Papillae on tongue

Figure 14.3. The oral cavity

middle ear to equalize the pressure on the eardrums.

When you have completed your observations, close the lower jaw by tying a string around the snout.

GENERAL INTERNAL ANATOMY

In this section, you will open the **ventral cavity** to expose the internal organs. The ventral cavity consists of the **thoracic cavity** that is located anterior to the diaphragm, and the **abdominopelvic cavity** that is located posterior to the diaphragm.

Dissection Procedure

Using a scalpel, carefully make an incision, through the skin only, from the base of the throat to the umbilical cord along the ventral midline. This is incision 1 in Figure 14.4. Separate the skin from the body wall along the inci-

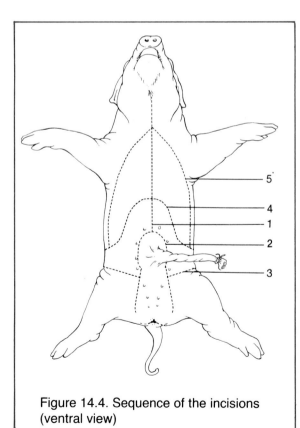

Figure 14.4. Sequence of the incisions (ventral view)

sion just enough to expose the body wall. Do this by lifting the cut edge of the skin with forceps while separating the loose fibrous connective tissue that attaches the skin to the body wall with the handle of a scalpel or a blunt probe.

While lifting the body wall at the ventral midline with forceps, use scissors to make a small incision through the body wall about halfway between the **sternum** and the **umbilical cord**. (Always cut away from your body when using scissors. This gives you better control. Rotate the dissecting pan as necessary.) Once the incision is large enough, insert the forefinger and middle finger of your left hand and lift the body wall from below with your fingertips as you make the incision. This position gives you good control of the incision and prevents unnecessary damage.

Extend the incision along the ventral midline anteriorly to the tip of the sternum and posteriorly to the umbilical cord. Insert the blunt blade of the scissors under the body wall as you make your cuts. Be careful not to cut any underlying organs.

In this manner, continue cutting through both skin and body wall around the anterior margin of the umbilical cord and posteriorly along each side as shown for incision 2. Then, make the lateral incisions just in front of the hind legs (incision 3). Make incision 4 from the midline laterally following the lower margin of the ribs as shown. This will allow you to fold out the resulting lateral flaps.

The **umbilical vein**, which runs from the umbilical cord to the liver, must be cut to lay out the umbilical cord and attached structures posteriorly. Tie a string on each end of the umbilical vein so you will recognize the cut ends later. You will have an unobstructed view of the abdominal organs when this is done.

Extend incision 1 anteriorly through the sternum using heavy scissors. Locate the **diaphragm** and cut it free from the ventral body wall. Lift the left half of the ventral thoracic wall as you cut the connective tissue forming the mediastinal septum that extends from the ventral wall to and around the heart. While lifting the left ventral wall, cut through the thoracic wall (ribs and all) at its lateral margin (incision 5), and cut through the muscle tissue at the anterior end of the sternum so that the left thoracic ventral wall can be removed. Repeat the process for the right thoracic ventral wall. The heart and lungs are now exposed.

Continue incision 1 anteriorly to the chin, and separate the neck muscles to expose the thymus gland, thyroid gland, larynx, and trachea. Remove muscle tissue as necessary to expose these structures.

Wash out the cavities of the pig in a sink while being careful to keep the organs in place.

THE THORACIC ORGANS

The thoracic cavity is divided into left and right pleural cavities containing the **lungs**. See Figure 14.5. The left lung has three lobes, and the right lung has four lobes. The inner thoracic wall is lined by a membrane, the **parietal** **pleura**, and each lung is covered by a **visceral** **pleura**. In life, fluid secreted into the potential space between these membranes, the **pleural cavity**, reduces friction as the lungs expand and contract during breathing.

At the midline, the parietal pleurae join with fibrous membranes to form a connective tissue partition between the pleural cavities called the **mediastinum**. The **heart** is located within the mediastinum. The heart is enclosed in a pericardial sac formed of the **parietal pericardium** and an outer fibrous membrane that is attached to the diaphragm. Remove this sac to expose the heart, if you have not already done so. The **visceral pericardium** tightly adheres to the surface of the heart.

In the neck region, locate the **larynx** (voice box) that is composed of cartilage and contains

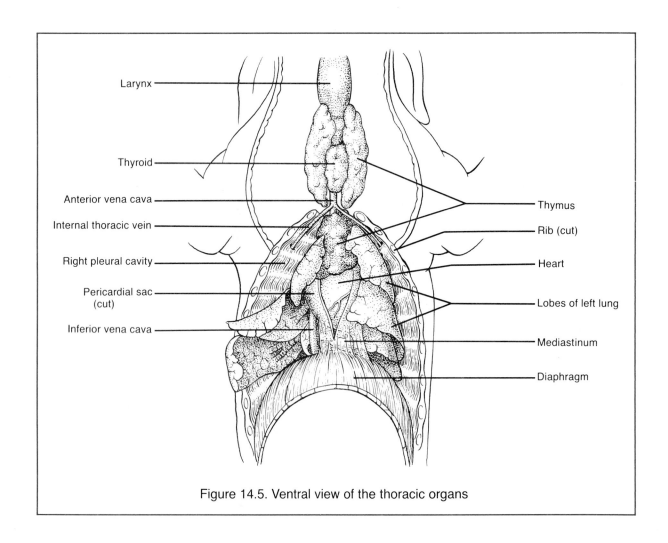

Figure 14.5. Ventral view of the thoracic organs

the vocal folds (cords). The **trachea** (windpipe) extends posteriorly from the larynx and divides dorsal to the heart to form the **bronchi** that enter the lungs. You will see these structures more clearly later after the heart has been removed.

Locate the whitish **thymus gland** that lies near the anterior margin of the heart and extends into the neck on each side of the trachea. The thymus is an endocrine gland, and it is the maturation site for T-lymphocytes that are important components of the immune system.

Locate the small, dark-colored **thyroid gland** located on the anterior surface of the trachea at the base of the neck. The thyroid is an endocrine gland that controls the rate of metabolism in the body.

Remove the thymus and thyroid glands as necessary to get a better view of the trachea and larynx, but don't cut major blood vessels. Carefully remove the connective tissue supporting the trachea so that you can move it to one side to expose the **esophagus** located dorsal to it. The esophagus is a tube that carries food from the pharynx to the stomach. In the thorax, it descends through the mediastinum dorsal to the trachea and heart and then penetrates the diaphragm to open into the stomach. You will get a better view of it later on.

DIGESTIVE ORGANS IN THE ABDOMEN

The inner wall of the abdominal cavity is lined by the **parietal peritoneum**, while the internal organs are covered by the **visceral peritoneum**. The internal organs are supported by thin membranes, the **mesenteries**, that consist of two layers of peritoneum between which are located blood vessels and nerves that serve the internal organs.

The digestive organs are the most obvious organs within the abdominal cavity, especially the large liver that fits under the dome-shaped diaphragm. See Figure 14.6.

STOMACH

Lift the liver anteriorly on the left side to expose the **stomach**. The stomach receives food from the esophagus and releases partially di-

gested food into the small intestine. Food is temporarily stored in the stomach, and the digestion of proteins begins there. The saclike stomach is roughly J shaped. The longer curved margin on the left side is known as the **greater curvature**. The shorter margin between the openings of the esophagus and small intestine on the right side is the **lesser curvature**.

A saclike fold of mesentery, the **greater omentum**, extends from the greater curvature to the dorsal body wall. The elongate, dark organ supported by the greater omentum is the **spleen**, a lymphatic organ that stores and filters the blood in an adult. Locate the **lesser omentum**, a smaller fold of mesentery extending from the lesser curvature of the stomach and small intestine to the liver. The peritoneum also attaches the liver to the diaphragm.

Cut open the stomach along the greater curvature from the esophageal opening to the junction with the small intestine. The greenish material within it and throughout the digestive tract is the **meconium**, which is composed mostly of epithelial cells sloughed off from the lining of the digestive tract, mucus, and bile from the gall bladder. The meconium is passed in the first bowel movement of a newborn pig.

Wash out the stomach and observe the interior of the stomach. Note the numerous folds in the stomach lining that allow the stomach to expand when filled with food. Observe that the stomach wall is thicker just anterior to the junction with the small intestine. This thickening is the **pyloric sphincter** muscle that usually is closed, but it opens to release material from the stomach into the small intestine. Locate a similar, but less obvious, thickening at the esophagus—stomach junction. This is the **cardiac sphincter** muscle that opens to allow food to enter the stomach, but it usually remains closed to prevent regurgitation.

SMALL INTESTINE

The small intestine consists of two parts: duodenum and jejunum-ileum. The first part of the small intestine is the short **duodenum**. It extends from the stomach, curves posteriorly, then turns anteriorly toward the stomach. The duodenum ends and the **jejunum-ileum** begins where the small intestine again turns poste-

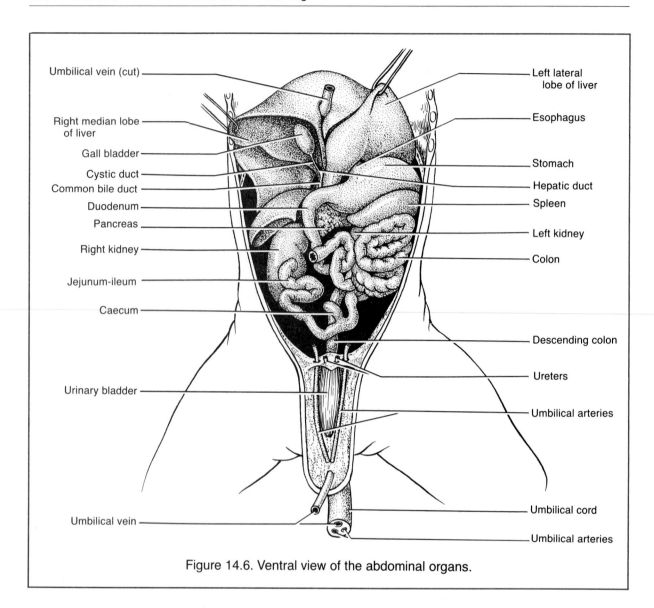

Figure 14.6. Ventral view of the abdominal organs.

Labels (left side, top to bottom): Umbilical vein (cut), Right median lobe of liver, Gall bladder, Cystic duct, Common bile duct, Duodenum, Pancreas, Right kidney, Jejunum-ileum, Caecum, Urinary bladder, Umbilical vein

Labels (right side, top to bottom): Left lateral lobe of liver, Esophagus, Stomach, Hepatic duct, Spleen, Left kidney, Colon, Descending colon, Ureters, Umbilical arteries, Umbilical cord, Umbilical arteries

riorly. In life, most of the digestion of food and absorption of nutrients occurs within the small intestine.

Lift out the small intestine and observe how it is supported by mesenteries that contain blood vessels and nerves. Without damaging blood vessels, remove a 1-in. segment of the jejunum-ileum, cut through its wall lengthwise, open it up, and wash it out. The inner lining has enormous numbers of microscopic projections, the **villi**, that give it a velvety appearance. Place a small flattened section under water in a small petri dish and examine the inner lining under a dissecting microscope for a better look. The absorption of nutrients occurs through the villi.

PANCREAS AND LIVER

Locate the pennant-shaped **pancreas** located in the mesentery within the curve of the duodenum. It secretes the hormone insulin as well as digestive enzymes that are emptied into the duodenum via a tiny **pancreatic duct**. Try to find this small duct by carefully dissecting away the peritoneum from a little finger of pancreatic tissue that follows along the descending duodenum.

The **liver** is divided into five lobes. It carries out numerous vital metabolic functions and stores nutrients. Nutrients absorbed from the digestive tract are processed by the liver before being released into the general circulation. In addition, the liver produces **bile**. Bile consists of bile pigments, by-products of hemoglobin breakdown, and bile salts. Bile salts help emulsify fats which facilitates their digestion in the small intestine.

Lift the right median lobe of the liver to locate the **gall bladder** on its undersurface just to the right of where the umbilical vein enters the liver. Careful dissection of the peritoneum will reveal the **hepatic duct** that carries bile from the liver and the **cystic duct** that carries bile to and from the gall bladder. These two ducts merge to form the **common bile duct** that carries bile into the anterior part of the duodenum. A sphincter muscle at the end of the common bile duct prevents bile from entering the duodenum except when food enters the duodenum from the stomach. When this sphincter is closed, bile backs up and enters the gall bladder via the cystic duct where it is stored until needed. Food entering the duodenum triggers a hormonal control mechanism that causes contraction of the gall bladder forcing bile into the duodenum.

LARGE INTESTINE

Follow the small intestine to where it joins with the **large intestine** or **colon**. At this juncture, locate the small side pouch, the **caecum**, that is small and nonfunctional in pigs and humans. The **ileocaecal valve** is a sphincter muscle that prevents food material in the colon from reentering the small intestine. Make a longitudinal incision through this region to observe the ileocaecal opening and valve.

In life, the large intestine contains large quantities of intestinal bacteria that decompose the nondigested food material that enters it. Water is reabsorbed back into the blood forming the feces that are expelled in defecation.

The pig's large intestine forms a unique, tightly coiled spiral mass. From the spiral mass, it extends anteriorly to loop over the duodenum before descending posteriorly against the dorsal wall into the pelvic region where its terminal portion, the **rectum**, is located. The external opening of the rectum is the **anus**. The rectum will be observed in a later portion of the dissection.

CIRCULATORY SYSTEM

The circulatory system consists of the heart, arteries, capillaries, veins, and blood. **Blood** is the carrier of materials that are transported by the circulatory system. It is pumped by the **heart** through **arteries** that carry it away from the heart and into the **capillaries** of body tissues. Blood is collected from the capillaries by **veins** that return the blood to the heart.

In this section, you are to (1) locate the major blood vessels, noting their location and function, and (2) identify the external features of the heart. In a double-injected fetal pig, the arteries are injected with red latex and the veins are injected with blue latex. A few exceptions exist, however. The pulmonary trunk, ductus arteriosus, and pulmonary arteries are injected with blue latex to indicate that they carry deoxygenated blood after birth, and the pulmonary veins are injected with red latex to indicate that they carry oxygenated blood after birth.

As you study the heart, major veins, and major arteries, carefully remove tissue as necessary to expose the vessels. This is best done by separating tissues with a blunt probe and by picking away connective tissue from the blood vessels with forceps. Make your observations in accordance with the descriptions and sequence that follows.

Before starting your dissection of the circulatory system, it is important to understand the basic pattern of circulation in an adult mammal, as shown in Figure 14.7, and contrast that with the circulation in a fetal mammal, as shown in Figure 14.8.

Adult Circulation

In the adult mammal, deoxygenated blood is returned from the body to the **right atrium** of the heart by two large veins. The **anterior vena cava** returns blood from regions anterior to the heart. The **posterior vena cava** returns blood from areas posterior to the heart. Blood from the digestive tract is carried by the **hepatic portal vein** to the liver whose metabolic processes modifies the nutrient content of the blood before it is carried by the **hepatic vein** into the posterior vena cava.

At the same time that deoxygenated blood enters the right atrium, oxygenated blood from the

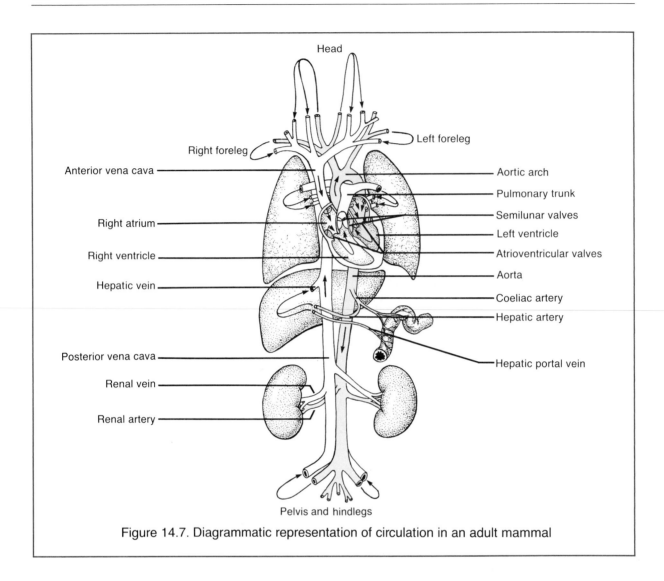

Figure 14.7. Diagrammatic representation of circulation in an adult mammal

lungs is carried by the **pulmonary veins** into the **left atrium** of the heart.

When the atria contract, blood from each atrium is forced into its corresponding ventricle. Immediately thereafter, the ventricles contract. During ventricular contraction, blood pressure in the ventricles closes the **atrioventricular valves**, preventing a backflow of blood into the atria, and opens the **semilunar valves** at the bases of the two large arteries that exit the heart. Deoxygenated blood in the **right ventricle** is pumped through the **pulmonary trunk** that branches into the **pulmonary arteries** carrying blood to the lungs. Oxygenated blood in the **left ventricle** is pumped into the **aorta**

that divides into numerous branches to carry blood to all parts of the body except the lungs.

At the end of ventricular contraction, the semilunar valves close, preventing a backflow of blood into the ventricles, and the atrioventricular valves open, allowing blood to flow from the atria into the ventricles in preparation for another heart contraction.

Fetal Circulation

In a fetus, the placenta is the source of oxygen and nutrients, and it removes metabolic wastes from the blood. The lungs, digestive tract, and kidneys are nonfunctional. Circulatory adapta-

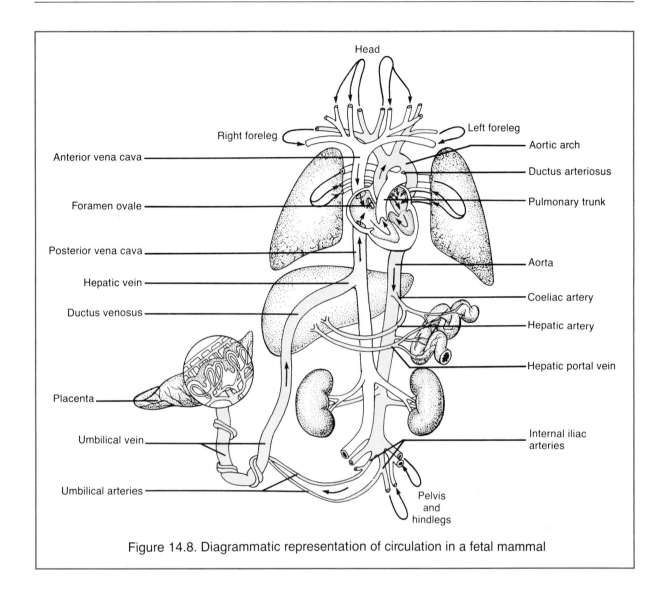

Figure 14.8. Diagrammatic representation of circulation in a fetal mammal

Labels (left, top to bottom): Anterior vena cava, Foramen ovale, Posterior vena cava, Hepatic vein, Ductus venosus, Placenta, Umbilical vein, Umbilical arteries

Labels (top): Head, Right foreleg, Left foreleg

Labels (right, top to bottom): Aortic arch, Ductus arteriosus, Pulmonary trunk, Aorta, Coeliac artery, Hepatic artery, Hepatic portal vein, Internal iliac arteries

Labels (bottom): Pelvis and hindlegs

tions to this condition make the circulation of blood in the fetus quite different from that in an adult.

Deoxygenated blood is returned from the body via the anterior and posterior vena cavae as in the adult. However, blood rich in oxygen and nutrients is carried from the placenta by the **umbilical vein** directly through the liver by a segment called the **ductus venosus** and on into the posterior vena cava. Therefore, the posterior vena cava returns blood to the right atrium that has a high oxygen concentration, but less than that in the umbilical vein.

The fetal heart contains an opening between the right and left atria called the **foramen ovale** that allows much of the blood entering the right atrium from the posterior vena cava to pass directly into the left atrium. This enables blood high in oxygen to enter the left ventricle which pumps it through the aorta to the body.

Much of the blood pumped from the right ventricle through the pulmonary trunk passes through a fetal vessel called the **ductus arteriosus** into the aorta to bypass the lungs and augment the blood supply to the body. Very little blood is carried by the pulmonary arteries to the nonfunctional lungs and returned to the heart by the pulmonary veins.

In the pelvic region, the **umbilical arteries** arise from the internal iliac arteries and carry blood to the placenta where wastes are removed and oxygen and nutrients are picked up from the maternal blood.

At birth, the lungs are inflated by breathing, and the umbilical blood vessels, ductus venosus, and ductus arteriosus constrict, causing blood pressure changes that close the foramen ovale. These changes produce the adult pattern of circulation. Growth of fibrous connective tissue subsequently seals the foramen ovale and converts the constricted vessels into ligamentous cords.

The Heart and Its Great Vessels

Study the structure of the heart and the great vessels shown in Figure 14.9, 14.10, and 14.11. Be sure that you know the relative positions of these vessels before trying to locate them in your specimen.

If you haven't done so, carefully cut away the pericardial sac from the heart and its attachment to the great vessels. Locate the **left** and **right atria**, blood receiving chambers, and the **left** and **right ventricles**, blood pumping chambers. Note the **coronary arteries** and **cardiac veins** that run diagonally across the heart at the location of the ventricular septum, a muscular partition separating the two ventricles. Coronary arteries supply blood to the heart muscle, and blockage of these arteries results in a heart attack.

While viewing the ventral surface of the heart, locate the large **anterior vena cava** that returns blood from the head, neck, and forelegs to the right atrium. By lifting up the apex of the heart you will see the large **posterior vena cava** that returns blood from regions posterior to the heart into the right atrium. Later, when you remove the heart, you will see where these veins enter the right atrium.

Now, locate the **pulmonary trunk** that exits the right ventricle and, just dorsal to it, the **aorta** that exits the left ventricle. These large arteries appear whitish due to their thick walls. Move the apex of the heart to your left and locate the whitish **ductus arteriosus** that carries blood from the pulmonary trunk into the aorta,

bypassing the nonfunctional fetal lungs. You will see the **pulmonary arteries** and **veins** later when you remove the heart.

Major Vessels of the Head, Neck, and Thorax

As you read the description of each major vessel, locate it first in Figures 14.10 and 14.11; then locate it in your dissection specimen.

As noted earlier, the large **anterior vena cava** returns blood from the head, neck, and forelegs into the right atrium. It is formed a short distance anterior to the heart by the union of the left and right brachiocephalic veins. The **brachiocephalic veins** are large but very short. Each one is formed by the union of two jugular veins and a subclavian vein.

There are two jugular veins on each side of the neck. The **external jugular** is more laterally located and drains superficial tissues. The **internal jugular** is more medially located and drains deep tissues including the brain. External and internal jugular veins may join just before their union with the subclavian vein. A **cephalic vein** drains part of the shoulder and joins with the external jugular near the union of the jugular veins.

The paired **subclavian veins** drain most of the shoulders and forelegs. Follow a subclavian vein laterally to locate the entrance of a **subscapular vein** that drains part of the shoulder. Distal to the subscapular vein the subclavian becomes the **axillary vein** in the axilla or armpit area and then becomes the **brachial vein** as it enters the foreleg.

Once you have located these veins, cut the superior vena cava leaving a stub at the heart, and lift it and the attached veins anteriorly to expose the underlying arteries. Locate the **aorta** and note that it forms the aortic arch as it curves posteriorly dorsal to the heart. Two major arteries branch from the aortic arch. The first and larger branch is the **brachiocephalic artery**. The second and smaller branch is the **left subclavian artery**.

Follow the brachiocephalic anteriorly where it branches to form the **right subclavian artery** and a **carotid trunk** that divides to form a pair of **common carotid arteries** that parallel the internal jugular veins to carry blood to the head and neck. At the base of the head, each common

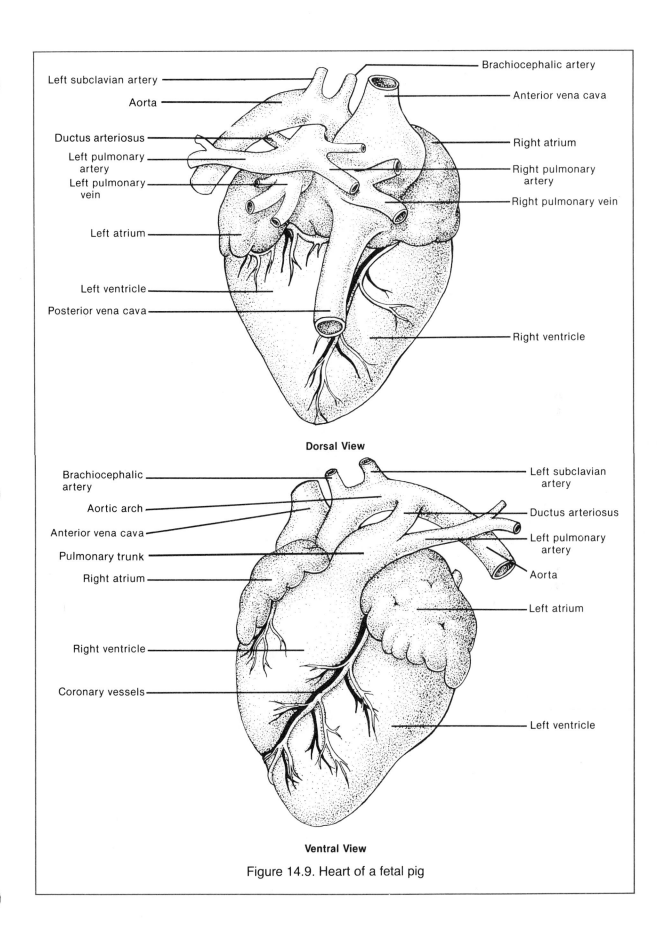

Left subclavian artery

Aorta

Ductus arteriosus

Left pulmonary artery

Left pulmonary vein

Left atrium

Left ventricle

Posterior vena cava

Brachiocephalic artery

Anterior vena cava

Right atrium

Right pulmonary artery

Right pulmonary vein

Right ventricle

Dorsal View

Brachiocephalic artery

Aortic arch

Anterior vena cava

Pulmonary trunk

Right atrium

Right ventricle

Coronary vessels

Left subclavian artery

Ductus arteriosus

Left pulmonary artery

Aorta

Left atrium

Left ventricle

Ventral View

Figure 14.9. Heart of a fetal pig

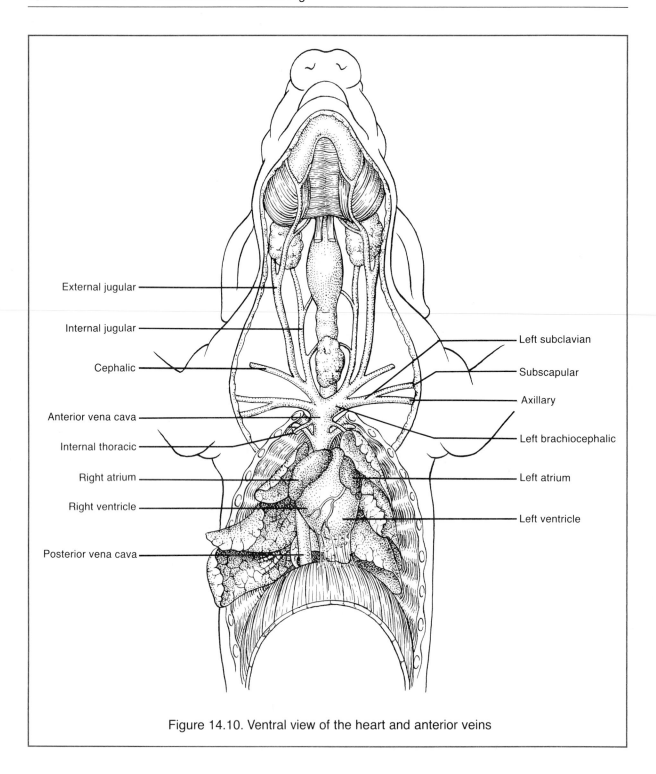

External jugular

Internal jugular

Cephalic

Anterior vena cava

Internal thoracic

Right atrium

Right ventricle

Posterior vena cava

Left subclavian

Subscapular

Axillary

Left brachiocephalic

Left atrium

Left ventricle

Figure 14.10. Ventral view of the heart and anterior veins

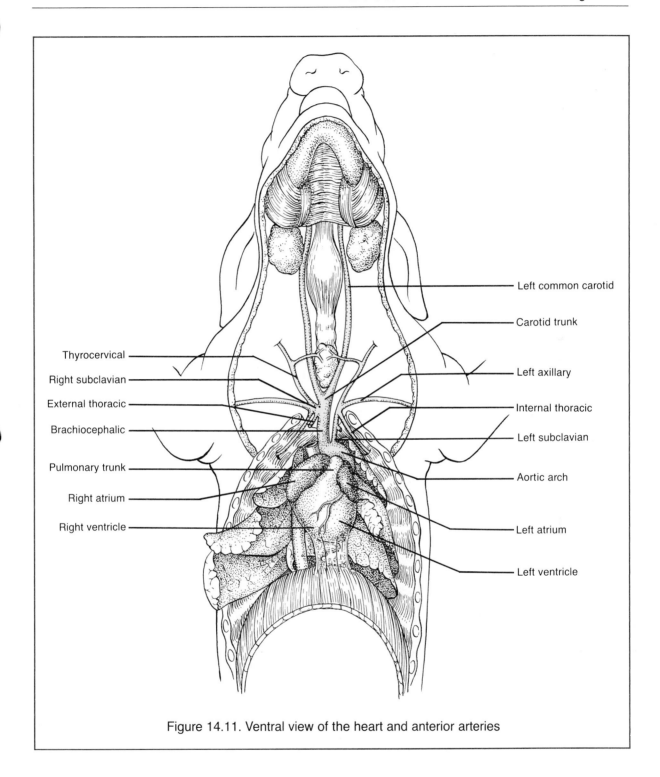

Figure 14.11. Ventral view of the heart and anterior arteries

carotid branches into internal and external carotid arteries.

The right subclavian artery gives off (1) an anterior branch, the **thyrocervical artery** that supplies organs of the neck, and (2) posterior branches, the **internal** and **external thoracic arteries** to the thorax wall. The continuation of the subclavian becomes the **right axillary artery** that enters the foreleg to become the **brachial artery**.

The **left subclavian artery** forms four major branches that match the branches of the right subclavian artery. From anterior to posterior they are (1) a thyrocervical artery, (2) a left axillary artery, (3) an external thoracic artery, and (4) an internal thoracic artery.

After you have located the major anterior arteries and veins, carefully cut through the major vessels to remove the heart. Leave stubs of the vessels on the heart and identify them by comparing your preparation with Figure 14.9. Especially locate the **pulmonary arteries** and **veins**.

Major Vessels Posterior to the Diaphragm

Follow the aorta and posterior vena cava posteriorly through the diaphragm. Compare your observations with Figures 14.12 and 14.13.

Just posterior to the diaphragm, locate the **hepatic vein** from the liver where enters the posterior vena cava. Immediately posterior to this union, the **ductus venosus** enters the posterior vena cava. Recall that the ductus venosus is an extension of the **umbilical vein** in the fetal pig. Dissect away liver tissue as necessary to expose the unions of the hepatic vein and ductus venosus with the posterior vena cava. Save the liver tissue adjacent to the hepatic vein and ductus venosus, but remove and discard the rest. Cut through the left side of the diaphragm, and push the abdominal organs to your left (specimen's right) to expose the posterior vena cava and aorta.

Lift the stomach anteriorly to expose the first branch from the aorta posterior to the dia-

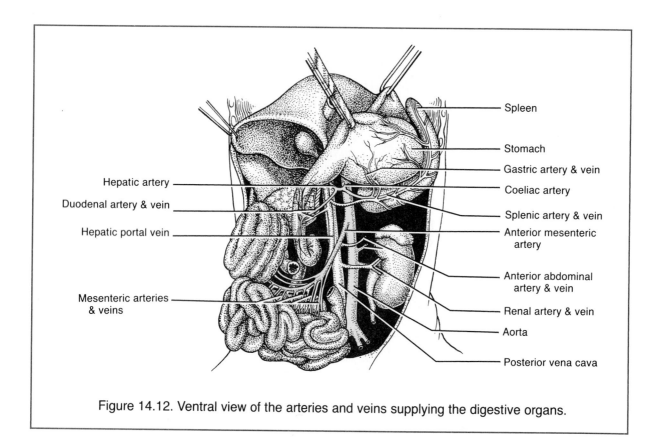

Figure 14.12. Ventral view of the arteries and veins supplying the digestive organs.

Labels in figure:
- Hepatic artery
- Duodenal artery & vein
- Hepatic portal vein
- Mesenteric arteries & veins
- Spleen
- Stomach
- Gastric artery & vein
- Coeliac artery
- Splenic artery & vein
- Anterior mesenteric artery
- Anterior abdominal artery & vein
- Renal artery & vein
- Aorta
- Posterior vena cava

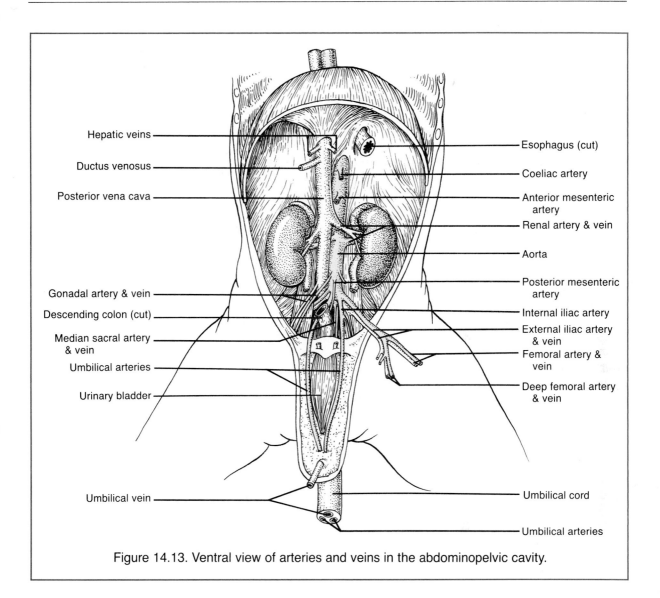

Figure 14.13. Ventral view of arteries and veins in the abdominopelvic cavity.

phragm. This is the **coeliac artery**. It gives off branches to the stomach, spleen, and liver (**hepatic artery**). Posterior to the coeliac is a large branch, the **anterior mesenteric artery**, that gives off numerous branches to the intestine. Note that many small veins from the intestine join to form the **hepatic portal vein** that runs anteriorly to enter the liver. This vein carries nutrient-rich blood from the intestine to the liver for processing prior to being released into the general circulation.

Now remove most of the intestines but save a small section associated with the hepatic portal vein. This will expose the lower portion of the aorta and posterior vena cava.

Locate the **renal arteries** and **renal veins** serving the kidneys. Since the kidneys are located dorsal to the parietal peritoneum, you will have to strip away some of this tissue to observe the vessels and kidneys clearly. The **adrenal glands** (endocrine glands), narrow strips of tissue, should be visible on the anterior margin of the kidneys.

Carefully dissect away the connective tissue to expose the posterior portions of the aorta, posterior vena cava, and the attached vessels. Locate the ureters that carry urine from the kidneys to the urinary bladder, which is located in the reflected tissue containing the umbilical arteries and umbilical cord. Free them from surrounding tissue so they are movable, but don't cut through them.

The small arteries and veins posterior to the renal vessels serve the gonads, ovaries or testes as the case may be. Just below the gonadal arteries, the **posterior mesenteric artery** arises from the ventral surface of the aorta and supplies most of the large intestine.

Locate the pair of **external iliac arteries** that branch laterally from near the posterior end of the aorta. After giving off small branch arteries to the body wall, each external iliac artery extends into the thigh, where it divides into the medially located **deep femoral artery** and the laterally located **femoral artery**. Each of these arteries is associated with a corresponding vein, the **deep femoral vein** and the **femoral vein** that drain the leg and merge to form the **external iliac vein**.

Now locate the medial terminal branches of the aorta, the **internal iliac arteries**. The major extensions of these arteries form the **umbilical arteries** that pass over the urinary bladder to enter the umbilical cord. The other branches of the internal iliac arteries supply the pelvic area and lie alongside the **internal iliac veins** that drain the pelvic area.

The external and internal iliac veins join to form a **common iliac vein** on each side of the body. The union of the common iliac veins form the posterior beginning of the posterior vena cava.

Located between the common iliac veins you will see the **median sacral artery** and **vein** serving the dorsal wall of the pelvic area.

THE RESPIRATORY SYSTEM

The respiratory system functions to bring air into the lungs where blood releases its load of carbon dioxide and receives oxygen for transport to body cells. Air enters the lungs via a series of air passageways. Air enters the **nostrils** and flows through the **nasal cavity**, where it is warmed and moistened. It passes into the **nasal pharynx** and on into the **oral pharynx**, where it enters the **larynx** via a slitlike opening called the **glottis**. You observed these components when you dissected the oral cavity and pharynx. From the larynx, air passes down the **trachea** that branches to form the **primary bronchi**, which carry air into the **lungs**.

Return now to the neck and thoracic cavity to observe the respiratory organs. Locate the larynx, trachea, and primary bronchi. Since the heart has been removed, the trachea and primary bronchi may be readily seen by dissecting away some of the surrounding connective tissue and blood vessels.

Make a midventral incision in the larynx, open it, and locate the **vocal folds** located on each side. Note how the trachea branches to form the primary bronchi. Each primary bronchus divides within the lung to form smaller and smaller air passages until tiny microscopic **bronchioles** terminate in vast numbers of saclike **alveoli**, the sites of gas exchange. Dissect along a primary bronchus to locate the **secondary bronchi** that enter the lobes of a lung. Remove a section of trachea and observe the shape of the **cartilagenous rings** that hold it open. Make a section through a lung to observe its spongy nature and the air passages within it.

UROGENITAL SYSTEM

The urinary and reproductive organs compose the urogenital system. They are considered together because the organs are closely interrelated. Refer to Figures 14.14 or 14.15 depending on the sex of your specimen. You are responsible for knowing the urogenital system of each sex, so prepare your dissection carefully and exchange it when finished for a specimen of the opposite sex.

Urinary System

If you haven't removed the intestines, do so now to expose the urinary organs. Leave a stub of the large intestine because you will want to locate the rectum later. Dissect away the parietal peritoneum to expose the **kidneys** located dorsal to it and against the dorsal body wall. The kidneys are held in place by connective tissue. An **adrenal gland** is located on the anterior surface of each kidney.

Locate the origin of a **ureter** on the medial surface of a kidney near where the renal vessels enter it. As urine is formed by the kidney, it

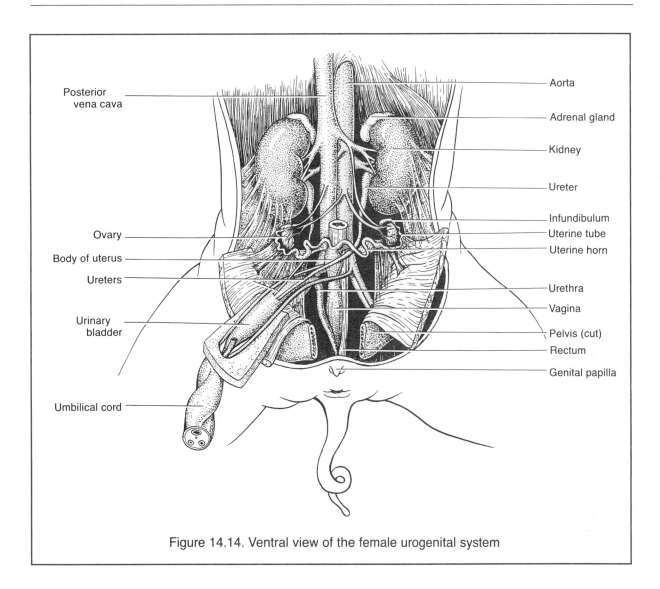

Figure 14.14. Ventral view of the female urogenital system

passes into the ureter for transport by peristalsis (wavelike contractions) to the urinary bladder. Trace the ureter posteriorly to its dorsally located entrance into the **urinary bladder**. The urinary bladder lies between the umbilical arteries on the reflected portion of the body wall containing the umbilical cord. Note that the posterior portion of the urinary bladder narrows to form the **urethra** that enters the pelvic cavity. It will be observed momentarily.

If your instructor wants you to dissect a kidney, make a coronal section to expose its interior. Compare the internal structure with Figure 21.2 in Exercise 21 as you read the accompanying description.

Female Reproductive System

The **uterus** is located dorsal to the urinary bladder. It consists of two **uterine horns** that join posteriorly at the midline to form the **body of the uterus**. Locate the uterine horns and the body of the uterus.

Follow the uterine horns anteriolaterally to the small, almost nodulelike, **ovaries** just posterior to the kidneys where they are supported by mesenteries. The uterine horns end at the posterior margin of the ovaries where they are continuous with the tiny, convoluted **uterine tubes** that continue around the ovaries to their anterior margins. The anterior end of a uterine

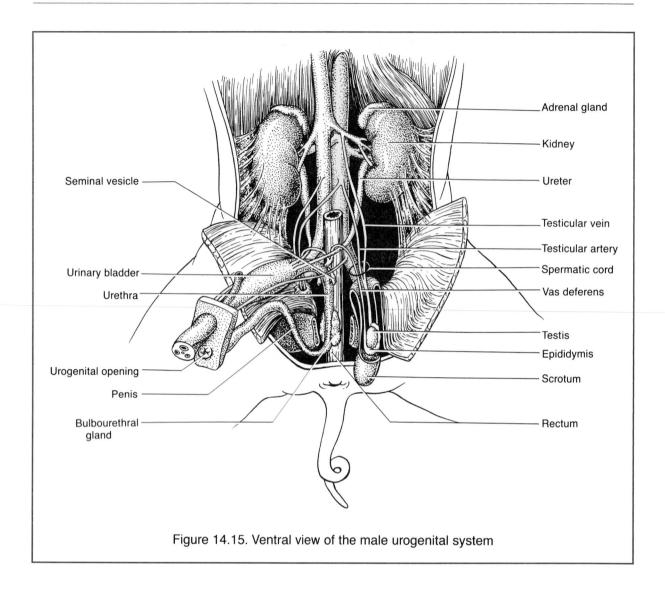

Seminal vesicle

Urinary bladder

Urethra

Urogenital opening

Penis

Bulbourethral
gland

Adrenal gland

Kidney

Ureter

Testicular vein

Testicular artery

Spermatic cord

Vas deferens

Testis

Epididymis

Scrotum

Rectum

Figure 14.15. Ventral view of the male urogenital system

tube forms an expanded funnel-shaped opening, the **infundibulum**, that receives the immature ovum when it is released from an ovary. Fertilization occurs in the uterine tubes, and the fetal pigs develop in the uterine horns.

Now, remove or reflect the skin from the ventral surface of the pelvis. Use your scalpel to cut carefully at the midline through the muscles and bones of the pelvic girdle. Spread the legs and open the pelvic cavity to expose the **urethra** and, just dorsal to it, the **vagina** extending posteriorly from the body of the uterus. You can now see the **rectum**, the terminal portion of the large intestine, located dorsally to the vagina. Note that the vagina and urethra unite to form

the **vaginal vestibule** a short distance from the external urogenital opening that is identified externally by the genital papilla.

Male Reproductive Organs

The **testes** develop within the body cavity just posterior to the kidneys. Later in fetal development, the testes descend through the **inguinal canals** into the **scrotum**, an external pouch. The scrotum provides a temperature for the testes that is slightly less than body temperature. The lower temperature is necessary for the production of viable sperm. Follow a testicular artery and vein to locate the inguinal canal.

On one side, cut open the scrotum to expose a testis. Locate the **epididymis**, a tortuous mass of tiny tubule that begins on the anterior margin of the testis and extends along the lateral margin to join posteriorly with the **vas deferens** or sperm duct. Trace the vas deferens anteriorly from the scrotum, through the inguinal canal to where it loops over a ureter to enter the urethra.

Locate the **urogenital opening** just posterior to the umbilical cord in the reflected flap of body wall containing the urinary bladder. The **penis** extends posteriorly from this point. Make an incision alongside the penis and free it from the body wall. Push it to one side, and use your scalpel to make a midline incision through the pelvic muscles and bones. Spread the legs and open the pelvic cavity so you can dissect out the pelvic organs. Locate the **urethra** at the base of the urinary bladder and carefully remove connective tissue around it, separating it from the **rectum** as it continues posteriorly into the pelvic cavity.

Where the vas deferentia enter the urethra, locate the small glands located on each side of the urethra. These are the **seminal vesicles**. Between the seminal vesicles on the dorsal surface of the urethra is the small **prostate gland**. Follow the urethra posteriorly to locate the pair of **bulbourethral glands** on each side of the urethra where it enters the penis. At ejaculation, these accessory glands secrete the fluids that transport sperm.

CONCLUSION

This completes the dissection. If you have done it thoughtfully, you have gained a good deal of knowledge about the body organization of mammals, including humans. Dispose of your specimen as directed by your instructor. There is no laboratory report for this exercise.

15

PHOTOSYNTHESIS

OBJECTIVES

On completion of the laboratory session, you should be able to:
1. Write the summary equation for photosynthesis and identify the role of each reactant.
2. Identify materials or conditions necessary for photosynthesis.
3. Describe the pathway that each reactant and product take to reach or leave the photosynthetic cells of a leaf.
4. Separate and identify the chloroplast pigments.
5. Describe the effect of light intensity on the rate of photosynthesis.
6. Define all terms in bold print.

Living organisms require a constant source of energy to operate their metabolic (living) processes. The ultimate source of energy is the sun. Except for the few chemosynthetic bacteria, all organisms depend on the conversion of **light energy** into the **chemical energy** of organic nutrients by **photosynthesis**. The photosynthesizers are plants, plantlike protists, and cyanobacteria. All of these organisms contain **chlorophyll**, the green pigment that captures light energy and converts it into the chemical energy of organic molecules. Photosynthetic organisms are called **autotrophs** (self-feeders) since they are able to synthesize organic nutrients from inorganic materials. **Heterotrophs** (feed on others) must obtain their organic nutrients by feeding on other organisms.

The summary equation for photosynthesis is shown in Figure 15.1. The process is not as simple as the summary equation suggests, however. It actually is a complex series of chemical reactions that may be separated into two parts: (1) a light reaction and (2) a dark reaction.

During the **light reaction**, chlorophyll absorbs light energy and converts it into chemical energy, and in the process, water is split releasing oxygen molecules. The **dark reaction** normally occurs in the presence of light, but it does not *require* light energy, hence its name. In this reaction, chemical energy, which was formed in the light reaction, is used to synthesize glucose by combining (1) carbon dioxide from the atmosphere and (2) hydrogen split from water in the light reaction. Consult your text for a complete discussion of photosynthesis.

When **glucose**, a six-carbon sugar, is formed in a photosynthesizing cell, much of it is converted into **starch** for temporary storage. Starch is a polysaccharide composed of many glucose units. Using glucose as the base material, plants can synthesize all other organic compounds required for their metabolic and structural needs.

In this exercise, you will use the presence of starch in chlorophyll-containing cells as indirect evidence that photosynthesis has occurred. The presence of reducing sugars, notably, glucose and fructose, is not evidence of photosynthesis since they may result from carbohydrates transported from other parts of the plant.

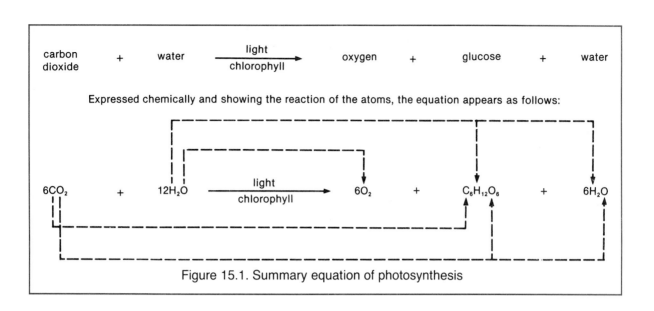

Figure 15.1. Summary equation of photosynthesis

Materials

Fluorescent lamps
Hot plates
Beakers, 100 ml, 250 ml
Bell jars
Boiling chips
Cotton, nonabsorbent
Dissecting instruments
Leaf shields
Petri dishes
Alcohol, isopropyl 70%
Dropping bottles of:
   Benedict's solution
   iodine (I_2 + KI) solution
Soda lime, 8 mesh
Coleus plants, green and white, in 4-in. pots
Corn (*Zea*) seedlings, green and albino
Fairy primose (*Primula malacoides*) or tomato
   (*Lycopersicum*) in 4-in. pots
Prepared slides of lilac (*Syringa*) leaf

THE LEAF AND PHOTOSYNTHESIS

Although photosynthesis occurs in all cells containing chlorophyll, the leaf is the primary photosynthetic organ in vascular plants. Most leaves consist of two major parts: (1) a flattened **blade** where most of the photosynthesis occurs and (2) a **petiole**, a slender stalk that attaches the leaf to the stem.

The internal structure of a blade is shown in Figure 9.12 on page 75. The **epidermal layers** provide protection for internal tissues. Excessive water loss by evaporation is prevented by the waxy **cuticle** secreted on the outer surface of the epidermis. Specialized cells, **guard cells**, control the size of tiny pores called **stomata** and further protect against excessive water loss since water vapor, as well as oxygen and carbon dioxide, passes through the stomata.

Guard cells have walls of unequal thickness. Since they are the only cells of the epidermis that contain chlorophyll, they are active in photosynthesis. When they are photosynthesizing, water tends to diffuse into them, which increases the intracellular pressure and causes the cells to bend and open the stomata. When the guard cells are not photosynthesizing and the products of photosynthesis are removed, water diffuses out of the cells. This causes the cells to flatten out and close the stomata. In this way the stomata of a leaf are opened and closed.

The **mesophyll** is the main photosynthesizing tissue, especially the upper palisade layer. The lower spongy layer allows free movement of gases among the cells. The **veins** of the leaf are embedded in the mesophyll. They provide support for the blade as well as the transport of water and minerals by the **xylem** and organic nutrients by the **phloem**.

Assignment 1

1. ***Complete items 1a and 1b on Laboratory Report 15 that begins on page 403***.
2. Examine a prepared slide of *Syringa* leaf, x.s., and compare it with Figure 9.12. Note the adaptations of the leaf for photosynthesis and the tissues involved directly or indirectly in photosynthesis.
3. ***Complete items 1c and 1d on the laboratory report***.

CARBON DIOXIDE AND PHOTOSYNTHESIS

In this experiment, you will test the null hypothesis "carbon dioxide is not necessary for photosynthesis" by determining the presence or absence of starch in leaves of plants exposed to air with and without carbon dioxide (CO_2). Two healthy plants were placed in darkness for 48 hr and then placed under separate bell jars. Plant A was exposed to air containing CO_2, and plant B was exposed to air from which the CO_2 had been removed by soda lime. Both plants were exposed to light for 24 hr. The setup is shown in Figure 15.2. The presence or absence of CO_2 is the only independent variable.

Assignment 2

1. Remove a leaf from each plant. Cut off the petiole of the leaf from plant A so that you can distinguish the leaves.
2. Carefully follow the procedures in Figure 15.3 to test each leaf for the presence of starch. A gray to blue-black color indicates the presence of starch. *Caution: Alcohol is flammable; do not get it near an open flame.*
3. ***Complete item 2 on the laboratory report***.

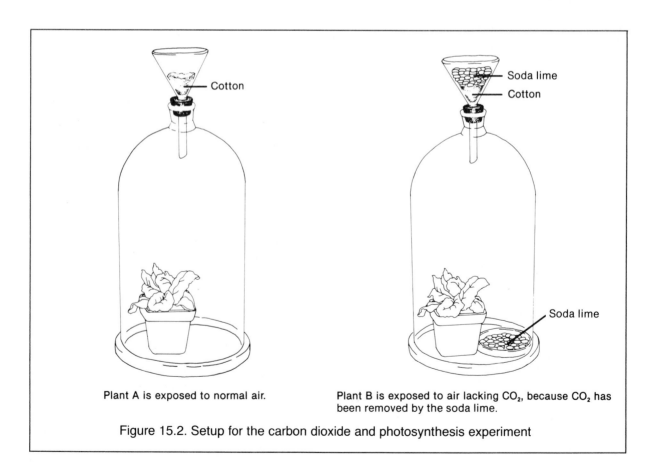

Plant A is exposed to normal air.

Plant B is exposed to air lacking CO_2, because CO_2 has been removed by the soda lime.

Figure 15.2. Setup for the carbon dioxide and photosynthesis experiment

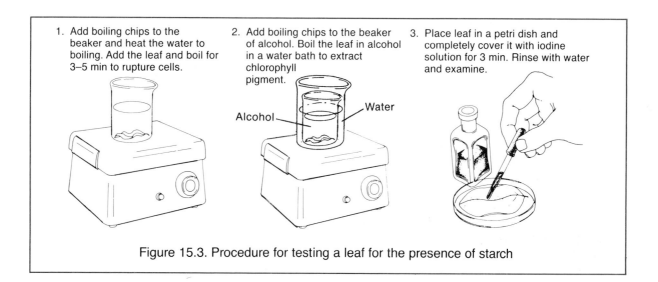

1. Add boiling chips to the beaker and heat the water to boiling. Add the leaf and boil for 3–5 min to rupture cells.

2. Add boiling chips to the beaker of alcohol. Boil the leaf in alcohol in a water bath to extract chlorophyll pigment.

Alcohol   Water

3. Place leaf in a petri dish and completely cover it with iodine solution for 3 min. Rinse with water and examine.

Figure 15.3. Procedure for testing a leaf for the presence of starch

LIGHT AND PHOTOSYNTHESIS

In this section, you will test the null hypothesis "light is not necessary for photosynthesis." The presence or absence of light is the only independent variable. Healthy green plants were placed in darkness for 48 hr. Leaf shields were placed on the leaves, and the plants were exposed to light and normal air for 24 hr prior to the laboratory session.

Assignment 3

1. Remove a leaf from one of the plants. Trace the leaf on a piece of paper, showing the position of the leaf shield.
2. Remove the leaf shield and test the leaf for the presence of starch following the procedures in Figure 15.3. Compare the position of the leaf shield and the distribution of starch.
3. ***Complete item 3 on the laboratory report***.

CHLOROPHYLL AND PHOTOSYNTHESIS

Variegated leaves have unequal distributions of chlorophyll, and occasionally, certain leaves of a plant may lack chlorophyll due to a mutation. Such leaves may be used to determine whether chlorophyll is necessary for photosynthesis. On the stock table are *Coleus* plants and normal and albino corn seedlings that were exposed to light for 24 hr prior to the laboratory session.

Assignment 4

1. Remove a variegated leaf from a *Coleus* plant on the stock table. Cut the leaf in half along the midrib. Sketch half of the leaf showing the distribution of chlorophyll.
2. Test one half of the leaf for the presence of starch as shown in Figure 15.3, and compare the starch and chlorophyll distributions.
3. Use scissors to separate the green and nongreen portions of the other half of the leaf, and test *each portion* separately for the presence of sugar as shown in Figure 15.4. A light green, yellow, or orange coloration indicates the presence of glucose. Compare your results with the distribution of chlorophyll.
4. Observe the corn seedlings, noting that about 25% of them are albinos. Are the albino leaves carrying on photosynthesis?
5. Test green and albino leaves of the corn seedlings for the presence of starch and sugar as shown in Figures 15.3 and 15.4.
6. ***Complete item 4 on the laboratory report***.

1. Remove a leaf from a healthy plant exposed to light for 24 hr.

2. With scissors cut the leaf into small pieces.

3. Place leaf fragments in a test tube with two droppers of water. Place test tube in a beaker half filled with boiling water and boil 5 min.

4. Pour off the fluid into another test tube.

5. Add 5 drops of Benedict's solution to the fluid.

6. Place test tube in boiling water for 5 min. Yellow or orange color is positive test for glucose.

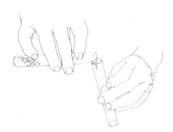

Figure 15.4. Procedure for testing a leaf for the presence of glucose

CHLOROPLAST PIGMENTS

When white light is passed through a prism, the component wavelengths are separated to form a rainbowlike spectrum that we perceive as colors ranging from violet (380 nanometers, nm) through blue, green, yellow, and orange to red (760 nm). Light energy decreases as the wavelengths increase.

When white light strikes a leaf, some wavelengths are absorbed and others are reflected or transmitted. The greatest absorption of light by chlorophyll is in the red, blue, and violet wavelengths. See Figure 15.5. Most leaves appear green because the green wavelengths are reflected and not absorbed by chlorophyll, which is usually the dominant pigment in the leaf. Most leaves also have other light-absorbing pigments in the chloroplasts, notably the yellow **carotenes** and the yellow-orange **xanthophylls**. These pigments absorb different wavelengths of light than chlorophyll and pass the captured energy to chlorophyll.

Chloroplast pigments may be separated by **paper chromatography**, a technique that takes advantage of slight differences in the relative solubility of the pigments. Your instructor has prepared a concentrated solution of chloroplast pigments by using a blender to homogenize spinach leaves in acetone, filtering the homogenate, and concentrating the filtrate by evaporation.

Materials

Chromatography jar
Chromatography paper, 1-in. width
Cork stopper for the jar
Paintbrush, finetipped
Paper clip
Pliers, pointed nosed
Scissors
Straight pin
Chloroplast pigment solution
Developing solvent

Assignment 5

1. Obtain the jar, cork, straight pin, paper clip, and chromatography paper for the setup shown in Figure 15.6. Handle the paper by the edges since oils from your fingers can adversely affect the results. Cut small notches about 2 cm from one end of the paper, and cut off the corners of this end.
2. Insert a pin into the cork stopper and bend it to form a hook. Attach the paper clip to the paper and suspend it in the jar *before adding the solvent*. The end of the paper should be *just above the bottom of the jar*. Adjust the paper clip or cut off the paper to achieve this length. Then remove the paper and clip for step 3.
3. Use a small paintbrush to paint a line of chloroplast pigment solution across the pa-

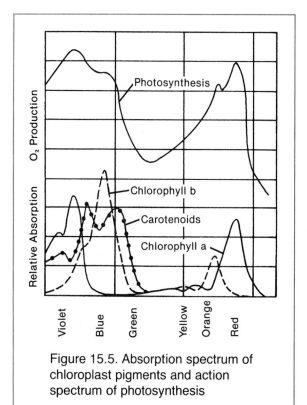

Figure 15.5. Absorption spectrum of chloroplast pigments and action spectrum of photosynthesis

per between the notches. Let it dry. Repeat this step six times to obtain a dark pigment line.
4. Pour the developing solvent into your jar *under a fume hood* to a depth of about 1 cm.

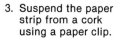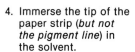

1. Paint a thin line of pigment solution across the paper strip between the notches.
2. Pour petroleum ether into a clean jar to a depth of 0.5 cm.
3. Suspend the paper strip from a cork using a paper clip.
4. Immerse the tip of the paper strip (*but not the pigment line*) in the solvent.

Repeat six times, letting pigment line dry between applications.

Figure 15.6. Procedure for the paper chromatography setup

Caution: The developing solvent and the pigment solution are volatile and highly flammable. Keep them away from open flames.

5. Return to your workstation, and insert the cork and suspended chromatography paper into the jar. The end of the paper, but not the pigment line, should be in the solvent. *Do not move the jar for 30 min.*

6. After 30 min, remove the paper and place it on a paper towel. Pour the developing solvent into the waste jar under the fume hood.

7. Examine your chromatogram. Locate the bright yellow carotene at the top, the blue-green chlorophyll a, and the yellow-green chlorophyll b. The remaining yellow bands are xanthophylls.

8. **Complete item 5 on the laboratory report**.

LIGHT INTENSITY AND THE RATE OF PHOTOSYNTHESIS

As noted in the summary equation of photosynthesis (Figure 15.1), oxygen molecules are one of the products. For each molecule of oxygen produced, a molecule of carbon dioxide is used in the formation of glucose. Some of the oxygen is used by the plant cells, and the excess is released into the atmosphere. This excess oxygen forms the oxygen in the atmosphere.

In the next experiment, the volume of oxygen produced in a 3-min interval is used as an indicator of the rate of photosynthesis at different light intensities. The light intensity is varied by changing the distance between the plant and a constant light source. See Figure 15.7. You will place the light source 30 cm, 60 cm, 90 cm, and 120 cm from the plant in this experiment. The experiment is best performed by groups of two to four students.

Materials

Heat shield (glass container of water)
Meterstick
Pipette, 1 ml
Plastic tubing, 7-cm lengths
Ring stand

Ring stand test-tube clamps
Scalpel or razor blade
Spot lamp, 150 watt, or fluorescent lamp
Syringe and needle, 5 ml
Test tube
Test-tube rack
Tubing clamp, screw-type
Sodium bicarbonate solution, 2%
Elodea shoots

Assignment 6

1. *Complete items 6a to 6e on the laboratory report.*

2. Fill a test tube about three-fourths full of 2% sodium bicarbonate ($NaHCO_3$) solution, which will provide an adequate concentration of CO_2.

3. Make a diagonal cut through the stem of a leafy shoot of *Elodea* about 10 to 13 cm from its tip. Use a sharp scalpel, being careful not to crush the stem. Insert the *Elodea* shoot, tip down, into the test tube so that the cut end is 1 to 2 cm below the surface of the solution. Place the tube in a test-tube holder on a ring stand. If bubbles of oxygen are released too slowly from the cut end of the stem, recut the stem at an angle to obtain a good production of bubbles.

4. Place a short piece of plastic tubing snugly over the tip (pointed end) of a 1-ml pipette as shown in Figure 15.7. Place a screw-type tubing clamp on the tubing and tighten it until it is *almost closed*. Place the pipette in a ring-stand clamp with the tip up, and lower the base of the pipette into the test tube so that the cut end of the *Elodea* shoot is inserted into the pipette. Secure the pipette in this position. See Figure 15.7. Study the 0.01-ml graduations on the pipette to ensure that you know how to read them.

5. If using a spot incandescent light source, fill a rectangular glass container with water to serve as a heat filter. The heat filter must be placed between the *Elodea* shoot and the spot lamp, about 10 cm in front of the *Elodea*. A heat filter is not necessary if using a fluorescent lamp.

6. You will vary the light intensity by plac-

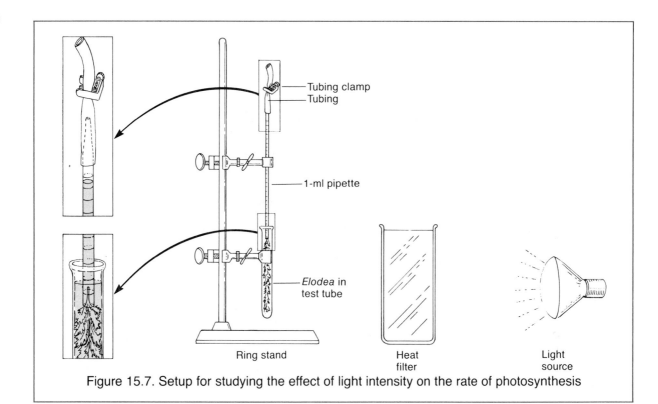

Figure 15.7. Setup for studying the effect of light intensity on the rate of photosynthesis

ing the light source at 30 cm, 60 cm, 90 cm, and 120 cm from the tube containing the *Elodea* shoot. Start at the 30-cm distance and arrange your setup as shown in Figure 15.7. Turn off the room lights.

7. Insert the needle of a 5-ml syringe into the tubing and gently pull out the syringe plunger to raise the level of the fluid in the pipette to the 0.9-ml mark. Tighten the screw clamp to close the tubing on the tip of the pipette to hold the fluid level. Then remove the syringe.

8. Bubbles of oxygen should start forming at the cut end of the *Elodea* stem and rise into the pipette displacing the solution.

9. After allowing 10 minutes for equilibration, read and record the fluid level in the pipette, and record the time of the reading. Be sure to take your readings at the bottom of the meniscus as shown in Figure 15.8. ***Record your data in the chart in item f on the laboratory report***.

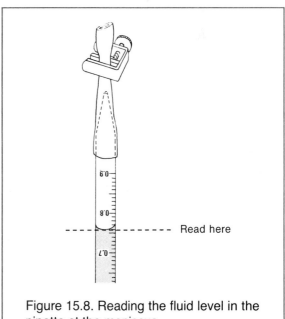

Figure 15.8. Reading the fluid level in the pipette at the meniscus

Exactly 3 min later, record another reading. Then determine the volume of oxygen produced within the 3-min interval. Repeat the process two more times to determine oxygen production during three separate 3-min intervals at this light intensity. Then calculate the average volume of oxygen produced per minute, milliliters of oxygen per minute (ml/O$_2$/min), and record this figure in the chart in item f on the laboratory report.

10. Use the same procedure as in 9 to determine the rate of oxygen production at 60 cm, 90 cm, and 120 cm. Move the light source to obtain the new distance and change the water in the heat filter at each new distance. Allow 10 minutes for equilibration to the new distance before starting your readings.

11. **Complete the laboratory report**.

THE COLOR OF LIGHT AND THE RATE OF PHOTOSYNTHESIS

Procedures similar to those used to determine the effect of light intensity on the rate of photosynthesis may be used to determine the effect of different wavelengths (colors) of light on the rate of photosynthesis. If time permits, design an experiment to test the effect of different wavelengths of light on the rate of photosynthesis. Obtain the colored filters from your instructor and perform the experiment. Write up your experiment, results, and conclusions on a separate sheet of paper using the format preferred by your instructor.

CELLULAR RESPIRATION

All organisms must have a continuous supply of energy from an external source to operate their metabolic functions. The sun is the ultimate energy source, and photosynthesis converts light energy into the **chemical bond energy** of organic nutrients. The three major classes of organic nutrients are **carbohydrates**, **fats**, and **proteins**. For the energy stored in these molecules to be used for cellular work, the chemical bonds must be broken by **cellular respiration**.

Cellular respiration and combustion are both oxidation processes. Combustion is an *uncontrolled oxidation* that releases large amounts of heat energy due to the simultaneous breaking of many chemical bonds. For burning to occur, the substance must be heated (the addition of energy) to its combustion temperature.

In contrast, cellular respiration is an *enzymatically controlled oxidation* that breaks bonds sequentially and releases energy in small amounts so that about 40% of the energy is "captured" in high-energy phosphate bonds ($\sim$P). Although nutrients must be energized with $\sim$P before oxidation occurs, they do not have to be heated to their combustion temperatures. The enzymes are what make the difference.

The captured energy is carried in the form $\sim$P, which combines with **adenosine diphosphate (ADP)** to form **adenosine triphosphate (ATP)**. ATP is the immediate source of energy for cellular work. It transfers $\sim$P to power the chemical reactions within the cell. Usable energy is always transferred as $\sim$P. Study Figure 16.1.

The two types of cellular respiration are aerobic and anaerobic. Compare the summary equation in Figure 16.2.

Anaerobic respiration does not require molecular oxygen. A nutrient, such as glucose, is degraded by **glycolysis** to give a net yield of two ATP molecules. Only certain fungi and bacteria can live by anaerobic respiration alone.

Fermentation is a synonym for anaerobic respiration in bacteria and yeasts and is a commercially important process in the production of alcoholic beverages, industrial chemicals, and bakery products.

Aerobic respiration requires molecular oxygen. A nutrient, such as glucose, is first broken down by glycolysis to form pyruvic acid, which is then completely oxidized to carbon dioxide via the **tricarboxylic acid cycle**. Most of the ATP is produced by **oxidative phosphorylation** in which energy-rich electrons removed during glycolysis and the tricarboxylic acid cycle are passed through a series of carriers to molecular

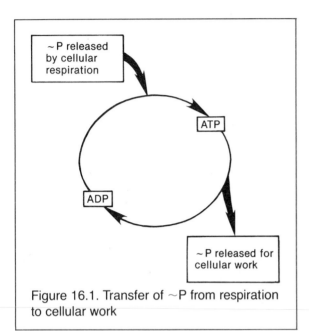

Figure 16.1. Transfer of ~P from respiration to cellular work

---

Aerobic respiration: 36 ATP, net

glucose + oxygen ⟶ carbon dioxide + water

$C_6H_{12}O_6$ + $6O_2$ ⟶ $6CO_2$ + $6H_2O$

Anaerobic respiration: 2 ATP, net

In Yeast

glucose ⟶ carbon dioxide + ethyl alcohol

$C_6H_{12}O_6$ ⟶ $2CO_2$ + $2C_2H_5OH$

In Animals

glucose ⟶ lactic acid

$C_6H_{12}O_6$ ⟶ $2C_3H_6O_3$

Figure 16.2. Summary equations for aerobic and anaerobic respiration

---

oxygen, along with hydrogen ions, to form **metabolic water**. Aerobic respiration of a glucose molecule yields a net of 36 ATP molecules. Thus, organisms using aerobic respiration have a tremendous energy advantage over anaerobic organisms.

In accordance with the summary equation, the occurrence of aerobic respiration may be detected by the consumption of the **reactants** or the accumulation of the **products**.

Assignment 1

Complete item 1 on Laboratory Report 16 that begins on page 407.

Materials

Aluminum foil
Celsius thermometers
Cotton, nonabsorbent
Drinking straws
Glass marking pen
Glass tubing, 2-cm lengths
Medicine droppers
Rubber-bulb air syringe
Rubber stoppers for test tubes
Test tubes
Test-tube rack
Vacuum bottles

Bromthymol blue, 0.004%, in dropping bottles
Crickets or isopods
Pea seeds, germinating
Primula or *Viola* plants

RESPIRATION AND CARBON DIOXIDE

The accumulation of carbon dioxide may be used as an indicator of cellular respiration. A dilute bromthymol blue solution, a pH indicator, can be used to detect an increase in carbon dioxide within a closed container. As carbon dioxide accumulates, it causes a decrease in pH (an increase in acidity) by combining with water to form carbonic acid which, in turn, releases hydrogen ions (H+) as it dissociates. The hydrogen ions react with the bromthymol blue and cause it to turn yellow. See Figure 16.3.

In this section, you will determine if cells of living organisms produce carbon dioxide by cellular respiration.

Assignment 2

1. Perform experiment 1 to determine if air exhaled from your lungs contains an accumulation of CO_2. If body cells produce CO_2, it will be carried by blood to the lungs for removal from the body.

Step 1: CO_2 + H_2O ⟶ H_2CO_3

carbon dioxide + water ⟶ carbonic acid

Step 2: H_2CO_3 ⇌ H^+ + HCO_3^-

carbonic acid ⇌ hydrogen ion + bicarbonate ion

Figure 16.3. Reaction of carbon dioxide and water

a. Place 3 drops of 0.004% bromthymol blue into each of two numbered test tubes.
b. Exhale your breath through a drinking straw into tube 1 for 3 minutes.
c. Use a rubber-bulbed syringe to pump atmospheric air into tube 2 for 3 minutes.
d. ***Record any color change in the bromthymol blue solutions in the table in item 2 on the laboratory report.***

2. Perform experiment 2 to determine if germinating seeds and live invertebrates produce CO_2.
   a. Place 3 drops of 0.004% bromthymol blue into each of 3 numbered test tubes.
   b. Place several short segments of glass tubing into each tube so that seeds and invertebrates will be kept out of the solution.
   c. Place 6 to 10 germinating seeds into tube 1, place 3 crickets or 6 to 10 isopods into tube 2, and place nothing else in tube 3.
   d. Gently (to prevent a sudden increase in air pressure from injuring the invertebrates) insert a rubber stopper into each tube. Place the tubes in a test-tube rack for 20 minutes. Then observe any color change in the bromthymol blue.
   e. ***Record your results on the laboratory report and complete item 2.***

Assignment 3

You have learned that living cells continuously produce carbon dioxide via cellular respi-

ration and that the summary equation for aerobic cellular respiration is essentially the reverse of the summary equation for photosynthesis. What happens in chlorophyll-containing cells when photosynthesis is occurring and when it is not?

1. Place 3 drops of 0.004% bromthymol blue into three numbered test tubes.
2. Obtain two leaves from a healthy plant. Insert one leaf into tubes 1 and 2 so that the upper surface of the leaf is against the glass. Do not put the leaves into the solutions. Place nothing in tube 3.
3. Insert rubber stoppers into each tube. Wrap tube 1 in aluminum foil to exclude light.
4. Place the tubes in a test-tube rack and expose them to light, but not heat, for 30 min. After 30 min, note the color of the bromthymol blue solution in each tube.
5. ***Complete item 3 on the laboratory report.***

RESPIRATION AND HEAT PRODUCTION

Normal body temperature in birds and mammals is produced by heat that is lost as a by-product of cellular respiration. Do simpler organisms produce heat in a similar way?

A few hours before the start of the laboratory session, your instructor set up three vacuum bottles. Bottle 1 contains germinating pea seeds, bottle 2 contains live invertebrates, and bottle 3 contains nothing. A one-hole cork stopper with a thermometer was placed in each bottle to measure the temperature inside the bottle.

Assignment 4

1. Read and record the temperatures in the bottles.
2. ***Complete item 4 on the laboratory report.***

TEMPERATURE AND RESPIRATION RATE

You will now determine the effect of temperature on the rate of cellular respiration. The rate will be determined by measuring oxygen consumption. You will use pea seeds, crickets or isopods, and mice as the test specimens. This will allow you to detect any basic differences in the response of plants, poikilothermic animals, and homeothermic animals to changes in temperature.

These experiments are best done by groups of four students with each group assigned to a particular temperature: 10°C, room temperature, or 40°C. If time is limited, your instructor may assign a different experiment to each group with all groups sharing the data.

When your group has been formed, read through the experiments and establish a division of labor to cover the tasks. Once the experiments have been set up, one person should keep the time, one or two persons should take the readings, and one person should record the data.

Materials

Balance
Beakers, 500 ml
Celsius thermometer
Colored water for manometer
Cotton, nonabsorbent
Glass marking pen
Pipettes, 1 ml
Plastic tubing
Respirometer (Figure 16.4)
Ring stand and clamps
Rubber stoppers for test tubes, one hole
Soda lime
Test tubes
Test-tube racks, wire type
Tubing clamps
Water baths
Wire screen
Sodium chloride solution, 10%, colored, in
    dropping bottles
Crickets or isopods
Frogs
Pea seeds, germinating
White mice

Plants and Invertebrates

Assignment 5

1. Set up three numbered test tubes at the assigned temperature. The interior of each tube must be dry. Place soda lime in each tube to a depth of about 2 cm. Then insert

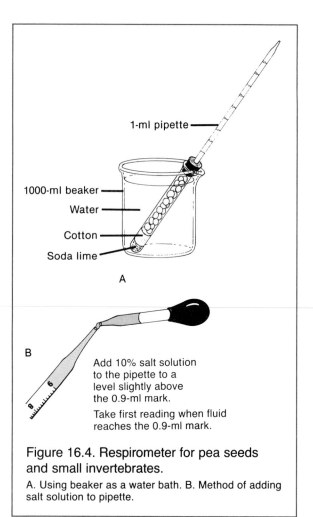

1-ml pipette

1000-ml beaker

Water

Cotton

Soda lime

A

B

Add 10% salt solution to the pipette to a level slightly above the 0.9-ml mark.

Take first reading when fluid reaches the 0.9-ml mark.

Figure 16.4. Respirometer for pea seeds and small invertebrates.

A. Using beaker as a water bath. B. Method of adding salt solution to pipette.

a loose pad of cotton to separate the specimens from the soda lime. Place the tubes in a test-tube rack.

2. Determine the number of pea seeds and invertebrates that are to be placed in the test tubes. Blot any water from them and weigh them to the nearest 0.1 g on the balance.

3. Add specimens to the tubes as follows:
    Tube 1: germinating pea seeds
    Tube 2: invertebrates
    Tube 3: nothing

4. Insert the base of a 1-ml pipette into the one-hole stopper so that it is flush with the inner surface of the stopper. Insert the stopper into a test tube. Repeat for each tube.

5. Place the respirometer at the assigned temperature for 10 min for temperature

equilibration. It is important that the respirometer is placed at an angle (not vertical) as shown in Figure 16.4 to prevent the salt solution from running down into the pipette. After 10 min, place a small drop of 10% sodium chloride solution at the tip of each pipette, and note how it is drawn into the pipettes of the experimental tubes.

6. You are to determine the movement of the fluid for each tube during five 3-min test intervals. Record the reading on the pipette at the beginning and end of each 3-min test period. Be sure to take your readings at the front edge of the fluid. Discard the lowest and highest readings (Why?) and calculate the average of the remaining three readings. Subtract the average movement in the control tube (tube 3) from the average movement in the experimental tubes. Calculate the average respiration rate (ml O_2/ hr/g) for each specimen as shown below. ***Record your data in item 5b on the laboratory report***.

$$\frac{\text{ml } O_2}{3 \text{ min}} \times \frac{60 \text{ min}}{1 \text{ hr}} = \text{ml } O_2/\text{hr}$$

$$\frac{\text{ml } O_2/\text{hr}}{\text{wt in grams}} = \text{ml } O_2/\text{hr/g}$$

7. Clean the apparatus and your workstation.
8. Exchange data with groups doing the experiment at different temperatures.
9. ***Complete item 5 on the laboratory report***.

Frog and Mouse

In this experiment, you will use a respirometer similar to the one in Figure 16.5 to determine the oxygen consumption of a frog (poikilotherm)

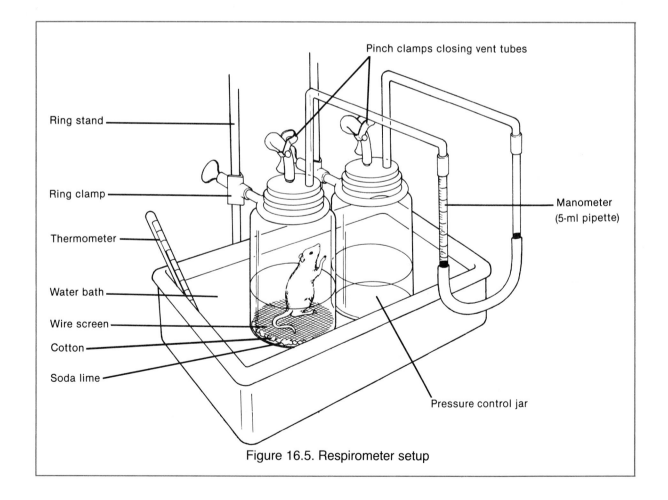

Figure 16.5. Respirometer setup

and a mouse (homeotherm) at three different temperatures. Your group will be assigned to perform a part of the experiment. Exchange your data with student groups performing other parts of the experiment.

The manometer (U-tube) of the respirometer connects the experimental and control chambers. A pressure change in one will cause the movement of the manometer fluid toward the chamber with the lowest pressure. Since carbon dioxide is absorbed by the soda lime, oxygen consumption can be measured by the movement of fluid toward the experimental chamber. Note that any change in pressure in the control chamber is automatically reflected in the level of the fluid.

Perform the experiment as described below. Record your results and those of other student groups in item 6b on the laboratory report.

Assignment 6

1. Set up a respirometer as shown in Figure 16.5. Place soda lime in the bottom of each chamber to a depth of about 2 cm. Cover the soda lime with about 1 cm of cotton and add a wire screen. Place the respirometer in a water bath at the assigned temperature.

2. Weigh the animal assigned to you to the nearest 0.1 g. (Mice should be picked up by their tails.)

3. Place the animal in the experimental chamber. *With the vent tubes open,* loosely replace the stopper. *Caution: Failure to keep the vent tubes open when inserting the stopper may injure the test animal due to a sudden increase in air pressure.*

4. After 5–10 min for temperature equilibration, *with the vent tubes open,* insert the stopper snugly into the jar. Close the vent tubes, record the time, and take the first reading from the manometer. Exactly 3 min later, take the second reading. *Open the vent tubes, remove the stopper, and place it loosely on top of the jar.*

5. After 3–5 min, insert the stopper, close the vent tubes, and take the first reading of the second replica. Proceed as before. Repeat to make at least five replicas. **Record your data in item 6b on the laboratory report.**

6. Discard the lowest and highest values, and calculate the average oxygen consumption (milliliters per 3-min interval) and the ml O_2/hour/gram of body weight.

7. Return the animal to its cage, and clean the respirometer and your workstation.

8. Exchange data with groups doing the experiment at different temperatures.

9. Plot the respiration level (ml O_2/hr/g) for each of the organisms studied in *item 6f on the laboratory report.*

10. **Complete the laboratory report.**

DIGESTION

Large food molecules of carbohydrates, proteins, and fats must be broken down by **digestion** into smaller, absorbable, nutrient molecules that can be used by cells in cellular respiration. The degradation of complex food molecules into absorbable nutrients occurs by **enzymatic hydrolysis**, the process of splitting large molecules by the addition of water and the catalytic action of enzymes.

Intracellular digestion is commonplace in cells, even photosynthetic cells. Heterotrophs use intracellular digestion or **extracellular digestion** to convert food molecules into absorbable nutrient molecules.

Bacteria and fungi secrete digestive enzymes into the surrounding substrate, and nutrients resulting from the digestion are absorbed. Protozoans and sponges engulf food into vacuoles where digestion occurs. Coelenterates and flatworms use both extracellular and intracellular digestion. Higher animals have a digestive tract that is used to (1) ingest and digest food extracellularly, (2) absorb nutrients, and (3) remove the nondigestible wastes.

ENZYME ACTION

When reduced to its simplest form, life is a series of chemical reactions. These reactions are complex and are controlled by **enzymes**, organic catalysts that greatly increase the rate of the reactions. Life could not exist without enzymes. Enzymes are usually proteins, and the sequence of their amino acids is determined by the genetic code of DNA molecules.

Most enzymes are specific in the type of chemical reactions that they control. This specificity is determined by the shape of the enzyme, which allows it to fit onto a particular **substrate molecule**. The shape of an enzyme is determined by the kind and sequence of amino acids that compose it. Weak **hydrogen bonds**, which form between amino acids of the chain, are responsible for the three-dimensional shape of the enzyme. Since these bonds are easily broken, the shape of the enzyme may be altered by changes in temperature and pH. Such changes inactivate the enzyme. Most enzymes function best in narrow temperature and pH ranges. Later in the exercise, you will investigate the action of a digestive enzyme to learn more about the nature of enzymes and specifically about the role of enzymes in digestion.

Digestive enzymes are only one type of the thousands of enzymes present in living organisms. The role of digestive enzymes is to speed up the hydrolysis of food molecules. Many different digestive enzymes are required to complete the digestion of food since a particular enzyme acts only on a single type of food molecule. The basic pattern of digestion may be summarized as follows:

$$\begin{array}{ccc} \text{Large} & & \text{small} \\ \text{food} & \xrightarrow[+\ H_2O]{\text{enzymes}} & \text{nutrient} \\ \text{molecules} & & \text{molecules} \end{array}$$

The interaction of an enzyme and a food molecule is shown in Figure 17.1. The enzyme (E) combines with the **substrate** (food) molecule (S) to form an **enzyme-substrate complex** (ES) where hydrolysis occurs. Then, the **product** molecules (P) separate from the enzyme and the unchanged enzyme is recycled to react with another substrate molecule. Note that the enzyme is not altered in the reaction and may be used repeatedly so that relatively few enzymes can catalyze a great number of reactions. The interaction of enzyme and substrate may be summarized as follows:

$$E + S \longrightarrow ES \longrightarrow E + P$$

Assignment 1

Complete item 1 on Laboratory Report 17 that begins on page 411.

DIGESTION IN *PARAMECIUM*

Paramecium exhibits intracellular digestion as found in protozoans. Food particles are collected by the oral groove and inserted into a food

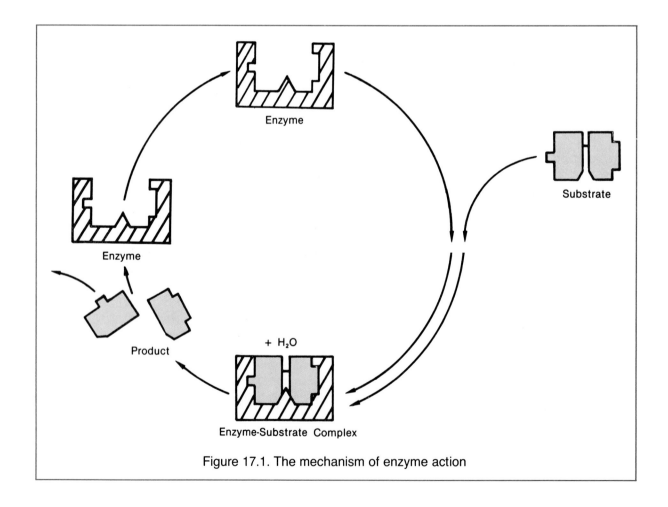

Figure 17.1. The mechanism of enzyme action

vacuole where digestion occurs. The vacuole circulates through the cell in a definite pattern, and the nondigestible parts are expelled through the cell membrane at a site called the anal pore. See Figure 17.2.

In this section, you will observe the digestion of yeast cells by a *Paramecium*. The yeast cells have been stained with congo red, a pH indicator. As digestion proceeds, the yeast cells will change from red to purple and finally to blue. This color change will allow you to follow the progress of digestion.

Materials

Colored pencils
Medicine droppers
Microscope slides and cover glasses
Protoslo
Paramecium culture
Yeast cells stained with congo red

Assignment 2

1. Make a slide of the *Paramecium* culture containing yeast cells stained with congo red. Add Protoslo to slow down the specimens, and add a cover glass so that you can observe them at $100\times$ and $400\times$. Correlate the location of the vacuoles with the color of the yeast cells.
2. Color the vacuoles in Figure 17.2 to show the progress of digestion.
3. ***Complete item 2 on the laboratory report***.

THE HUMAN DIGESTIVE SYSTEM

Refer to Figures 17.3 and 17.4 as you study the human digestive system as an example of a vertebrate digestive system. Table 17.1 lists the major divisions of the digestive tract and their

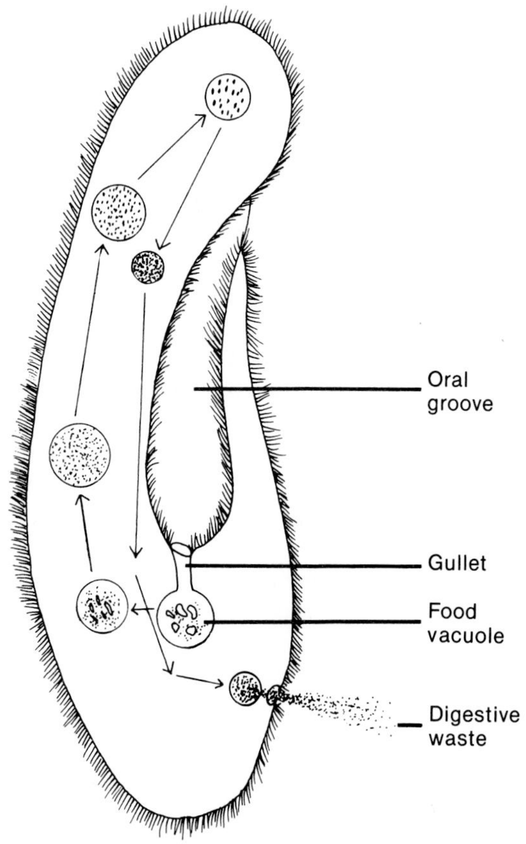

Oral groove

Gullet

Food vacuole

Digestive waste

Figure 17.2. Digestion in *Paramecium*

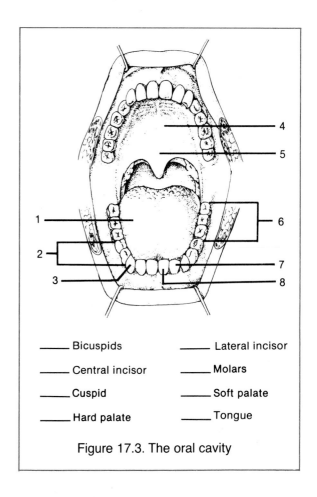

_____ Bicuspids   _____ Lateral incisor

_____ Central incisor   _____ Molars

_____ Cuspid   _____ Soft palate

_____ Hard palate   _____ Tongue

Figure 17.3. The oral cavity

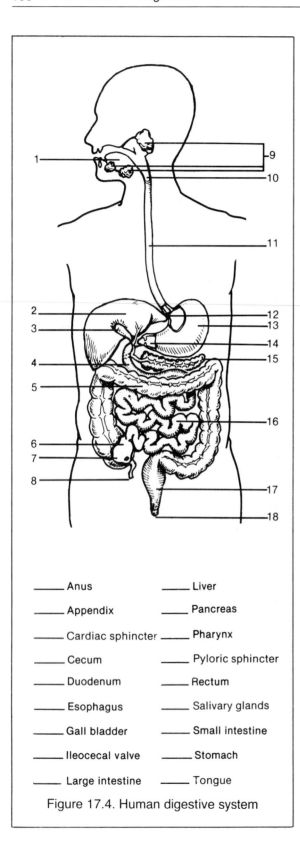

_____ Anus   _____ Liver

_____ Appendix   _____ Pancreas

_____ Cardiac sphincter _____ Pharynx

_____ Cecum   _____ Pyloric sphincter

_____ Duodenum   _____ Rectum

_____ Esophagus   _____ Salivary glands

_____ Gall bladder   _____ Small intestine

_____ Ileocecal valve   _____ Stomach

_____ Large intestine   _____ Tongue

Figure 17.4. Human digestive system

TABLE 17.1
Major Functions of the Digestive Tract Divisions

Structure	Major Function
Mouth	Eating, chewing, and swallowing food; digestion of starch begins here
Pharynx	Carries food to esophagus
Esophagus	Carries food to stomach by peristalsis
Stomach	Mixes gastric juice with food to form chyme; protein digestion begins here
Small intestine	Mixes chyme with bile and intestinal and pancreatic juices; digestion and absorption of nutrients completed here
Large intestine	Decomposition of undigested materials by bacteria; reabsorption of water to form feces
Anus	Defecation

functions. Table 17.2 indicates the end products of digestion for the major food groups.

The Oral Cavity

The mouth contains a number of structures that assist in the digestive process. During chewing, the **teeth** break the food into smaller pieces, while the **tongue** manipulates the food and mixes it with saliva. **Saliva** is produced by three pairs of **salivary glands**. The tongue pushes the food mass back into the **pharynx** to initiate the swallowing reflex, which carries the food into the esophagus.

The roof of the mouth, formed by the anterior **hard palate** and the posterior **soft palate**, divides the oral and nasal cavities. This arrangement allows you to breathe while chewing.

There are 32 teeth in a complete set of teeth in an adult. From front to back in one half of each jaw and starting at the anterior midline, they are **central incisor**, **lateral incisor**, **cuspid**, first and second **bicuspids** (premolars), and the first, second, and third **molars**. The

TABLE 17.2
End Products of Digestion

Food	End Products
Carbohydrates	Monosaccharides
Proteins	Amino acids
Fats	Fatty acids and monoglycerides

third molars (wisdom teeth) often become impacted due to the evolutionary shortening of the jaws.

Esophagus to Anus

Food is moved through the digestive tract by **peristalsis**, wavelike contraction of the muscles in its walls. The **esophagus** carries food from the pharynx to the **stomach**. The **cardiac sphincter**, a circular muscle located at the esophagus-stomach junction, opens to allow the passage of food and closes to prevent regurgitation. In the stomach, food is mixed with **gastric juice** and converted to a semiliquid mass called **chyme**. Enzymes in gastric juice begin the digestion of proteins and certain fats. Chyme is then released in small amounts into the small intestine. The **pyloric sphincter** controls the passage of chyme into the small intestine.

In the small intestine, chyme is mixed with bile, pancreatic juice, and intestinal juice. Bile and pancreatic juice enter the **duodenum**, the first portion of the small intestine. The bile and pancreatic ducts join to form a common opening into the duodenum.

Bile is secreted by the **liver**, the large gland in the upper right portion of the abdominal cavity. It is temporarily stored in the **gall bladder**. Bile emulsifies fats to facilitate fat digestion. **Pancreatic juice** is produced by the **pancreas**, a pennant-shaped gland located between the stomach and the duodenum. **Intestinal juice** is secreted by the inner lining (mucosa) of the small intestine. Enzymes in pancreatic juice and intestinal juice act sequentially to complete the digestion of food. Digestion of food and absorption of nutrients into the blood are completed in the small intestine.

The nondigestible material passes into the **large intestine** via the **ileocecal valve**, an-

other sphincter muscle. The **appendix** is a small vestigial appendage attached to the pouchlike **cecum**, the first portion of the large intestine. The large intestine includes ascending, transverse, descending, and sigmoid regions. It ends with the **rectum**, a muscular portion that expels the feces through the **anus**. Decomposition of nondigestible materials by bacteria and the reabsorption of water are the major functions of the large intestine.

Histology of the Small Intestine

Examine the structure of the small intestine in cross section in Figure 17.5. Note the four layers of the intestinal wall.

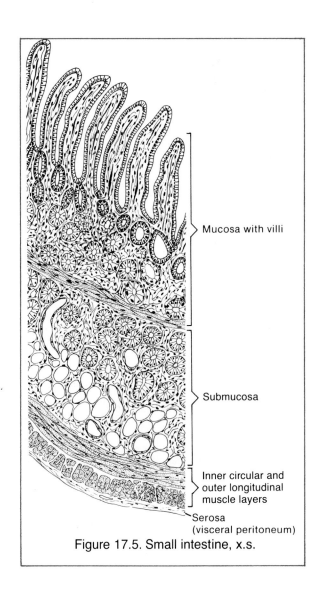

Mucosa with villi

Submucosa

Inner circular and outer longitudinal muscle layers

Serosa (visceral peritoneum)

Figure 17.5. Small intestine, x.s.

Peritoneum. The outer protective membrane that lines the coelom and covers the digestive organs.

Muscle layers. Outer longitudinal and inner circular layers whose contractions mix food with digestive secretions and move the food mass by peristalsis.

Submucosa. Connective tissue containing blood vessels and nerves serving the digestive tract.

Mucosa. The inner epithelial lining that secretes intestinal juice and absorbs nutrients.

Locate the **villi**, fingerlike extensions of the mucosa that project into the lumen of the small intestine. Villi increase the surface area of the mucosa, which facilitates the absorption of nutrients. See Figure 17.6.

Materials

Colored pencils
Human torso model with removable organs
Prepared slides of small intestine, x.s.

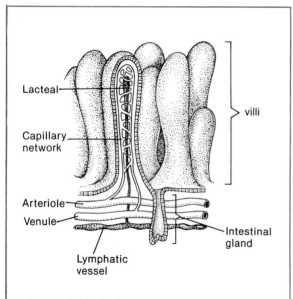

Figure 17.6. A diagrammatic representation of villi in the small intestine

Assignment 3

1. Label Figures 17.3 and 17.4 and color-code the digestive organs.
2. Locate the organs of the digestive system on the human torso model. Note their relationships.
3. Examine a prepared slide of small intestine, x.s., and locate the four layers of the intestinal wall.
4. ***Complete item 3 on the laboratory report***.

DIGESTION OF STARCH

In this section, you will assess the effect of temperature and pH on the action of **pancreatic amylase**. This enzyme accelerates the hydrolysis of **starch**, a polysaccharide, to **maltose**, a disaccharide and reducing sugar. See Figure 17.7. The experiments are best done by groups of four students, with each group assigned a different temperature and with all groups sharing their results.

You will dispense solutions from dropping bottles and use a number of test tubes. To avoid contamination, the droppers must not touch other solutions. All glassware must be clean and rinsed with distilled water.

You will dispense "droppers" and "drops" of solutions. As noted earlier, a "dropper" means *one dropper full* of solution (about 1 ml). When dispensing solutions by the "drop," hold all droppers at the same angle to dispense equal-sized drops. All drops should land squarely in the bottom of the test tube. If they run down the side of the tube, your results may be affected.

Thoroughly mix the solutions placed in a test tube. If a vortex mixer is not available, shake the tube vigorously from side to side to mix the contents.

You will use the iodine test for starch and Benedict's test for reducing sugars to determine if digestion has occurred. Examine the demonstration of positive and negative tests prepared by your instructor and use these as standards for comparison.

Iodine Test

1. Add 2 drops of iodine solution to one to two droppers of solution to be tested.

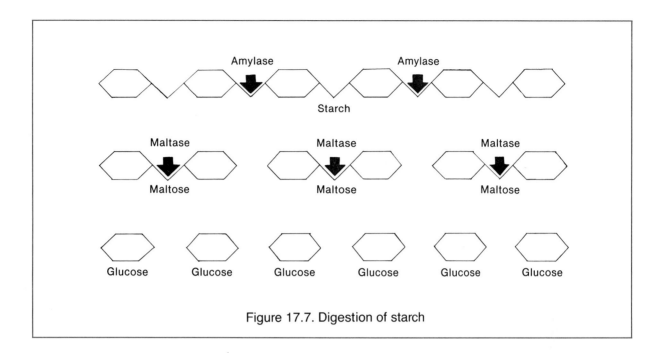

Figure 17.7. Digestion of starch

2. A blue-black color indicates the presence of starch.

Benedict's Test

1. Add 5 drops of Benedict's solution to one dropper of solution to be tested, and heat the mixture to near boiling in a water bath for 3 min.
2. A light green, yellow, orange, or red color indicates the presence of a reducing sugar in that order of increasing concentration.

Materials

Hot plate
Vortex mixer
Water baths at 5°, 37°, and 70°C
Beaker, 400 ml
Celsius thermometer
Glass marking pen
Test tubes
Test-tube holders
Test-tube racks, submersible type
Dropping bottles of:
 Benedict's solution
 buffers, pH 5,8,11
 iodine ($I_2 + KI$) solution
 pancreatic amylase, 0.1%
 soluble starch, 1.0%

Assignment 4

1. Label nine test tubes 1 to 9.
2. Add the pH buffer solutions to the test tubes as follows.
    Tube 1: 2 droppers pH 5
    Tube 2: 2 droppers pH 5
    Tube 3: 3 droppers pH 5

    Tube 4: 2 droppers pH 8
    Tube 5: 2 droppers pH 8
    Tube 6: 3 droppers pH 8

    Tube 7: 2 droppers pH 11
    Tube 8: 2 droppers pH 11
    Tube 9: 3 droppers pH 11
3. Add 5 drops of pancreatic amylase to tubes 1, 3, 4, 6, 7, and 9. Shake the tubes to mix well.
4. Place the test tubes in a test-tube rack and place the rack in the water bath at your assigned temperature for 10 min.
5. While leaving the test tubes in the water bath, add one dropper of starch to tubes 1, 2, 4, 5, 7, and 8. Shake the tubes to mix well. Leave the test tubes in the water bath for 15 min. Table 17.3 shows the contents of each tube for easy reference.
6. After 15 min, remove the test-tube rack and tubes and return to your workstation.
7. Number another set of nine tubes 1B to 9B.

8. Pour half the liquid in tube 1 into tube 1B, pour half the liquid in tube 2 into tube 2B, and so on until you have divided the liquid equally between the paired tubes. You now have nine paired tubes with the members of each pair containing liquid of identical composition.

9. Test all the original nine tubes (1–9) for the presence of starch by adding 2 drops of iodine solution to each tube. ***Record your results in the table in item 4a on the laboratory report***.

10. Test all of the B tubes (1B–9B) for the presence of maltose, a reducing sugar as follows. Add 5 drops of Benedict's solution to each tube, and place the tubes in a water bath (400-ml beaker half full of water) on your hot plate. Place some boiling chips in the beaker and heat the tubes to near boiling for 3 to 4 min. Watch for any color change in the solution. ***Record your results in item 4a on the laboratory report***.

11. Exchange results with other groups using different temperatures. ***Record their results in item 4a on the laboratory report***.

12. ***Complete item 4 on the laboratory report***.

TABLE 17.3
Summary of Tube Contents for the Starch Digestion Experiment

| Tube | Droppers of Solution | | | | |
| | Buffers | | | | |
	pH 5	pH 8	pH 11	Starch	Drops of Amylase
1	2			1	5
2	2			1	
3	3				5
4		2		1	5
5		2		1	
6		3			5
7			2	1	5
8			2	1	
9			3		5

GAS EXCHANGE

Aerobic cellular respiration requires gaseous oxygen (O_2) in the cell and produces carbon dioxide (CO_2). Thus, the continuous exchange of O_2 and CO_2 must occur between the cell and its environment. This exchange always occurs by diffusion across the moist cell membrane. Gases can diffuse across the membrane only if they are in solution.

In unicellular, colonial, and many small multicellular organisms, each cell is either in direct contact with the environment or only a few cells away so that the rate of diffusion is sufficient to provide O_2 and to remove CO_2. In larger animals, however, diffusion through the body surface is too slow to provide adequate gas exchange for interior cells. This is especially true for animals.

GAS EXCHANGE IN ANIMALS

The evolution of special organs for gas exchange in animals has taken two basic directions: evaginations (outpocketings), such as **gills**, and invaginations (inpocketings), such as **lungs**. Many variations of these basic patterns exist, but each exhibits certain essential characteristics. The respiratory surface must be (1) moist, (2) protected from injury, (3) of adequate surface area, and (4) thin for rapid diffusion of gases. Depending on the habitat of the organism, a mechanism exists to pass either water or air over the respiratory surface.

Gases can enter or leave the cells of a gas-exchange organ only through the extracellular fluid that bathes the cells. In aquatic animals, the water of the environment is in contact with the cells and provides this moisture. In terrestrial forms, the ever-present problem of dessication must be overcome while the surface of the organ is kept moist. This problem has been solved in vertebrates by the evolution of internal lungs that maintain a saturated humidity and moist membrane surfaces.

In most cases, the evolution of a circulating internal fluid, either blood or hemolymph, has occurred to transport O_2 and CO_2 between the respiratory organs and the body cells. A close physical relationship exists between the internal fluid and the respiratory surface. The larger the respiratory surface, the faster is the rate of gas exchange between the environment (air or water) and the internal fluid.

Materials

Colored pencils
Breathing mechanics model
Biohazard bag
Cotton
Human torso model
Propper spirometer with sterile, disposable mouthpieces
Dissecting instruments and pans
Syracuse dishes
Alcohol swabs
Amphyl solution, 0.5%
Clams, crayfish, grasshoppers, and fish, living and preserved
Frog, pithed
Sheep pluck
Prepared slides of:
  human louse, cleared to show tracheae
  lung tissue, normal and emphysematous
  trachea, x.s.

Gills

A variety of aquatic organisms have evolved gills as gas-exchange organs. The methods used to increase the surface area and to pass water over the gill surface varies from group to group.

Tracheae

Insects have a unique respiratory system. **Tracheae**, tiny chitinous tubules, extend inward from the **spiracles**. They usually unite with two longitudinal trunks that have smaller branches which carry air *directly* to body tissues. The hemolymph is not involved. Oxygen diffuses from the tracheae into the tissues, and carbon dioxide diffuses from the tissues into the tracheae. Some insects contract and expand the abdomen to "pump" air into and out of the tracheae.

Lungs

The lungs of amphibians are quite simple and little more than thin, air-filled sacs that have an abundant blood supply. In many amphibians,

the skin serves as an additional gas-exchange organ. In higher vertebrates, lungs are more complex.

Mammalian lungs are highly developed with many internal subunits that provide a large surface for rapid gas exchange. The combined respiratory surface is many times greater than the external surface area of the lung.

Assignment 1

1. Examine a prepared slide of a human louse that has been cleared to reveal the tracheal system. Note the arrangement of the tracheae and the location of the spiracles. Compare your slide with Figure 18.1. Color the tracheal system blue in Figure 18.1.

2. Examine the demonstration dissections of a preserved or freshly killed clam, crayfish, and fish. Note the location, structure, and arrangement of the gills. Compare your observations with Figure 18.2. Color the gills red in Figure 18.2.

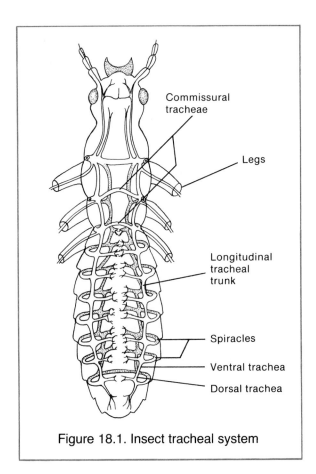

Figure 18.1. Insect tracheal system

3. Observe the living specimens in the aquarium and watch for respiratory movements.
4. Remove a small section of a gill from a clam, crayfish, and fish. Place them in water in a Syracuse dish and examine them with a dissecting microscope. Note the gill structure and how a large surface area is provided in each.
5. ***Complete item 1 on Laboratory Report 18 that begins on page 413***.

THE MAMMALIAN RESPIRATORY SYSTEM

Refer to Figure 18.3 which illustrates the basic components of the human respiratory system, as you study this section.

During **inspiration**, air enters the **nasal cavity** through the **nostrils**. The air is filtered

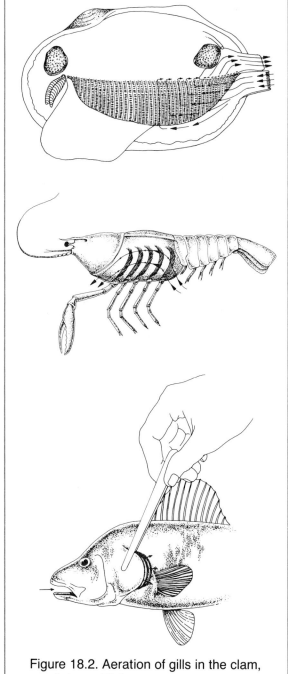

Figure 18.2. Aeration of gills in the clam, crayfish, and fish

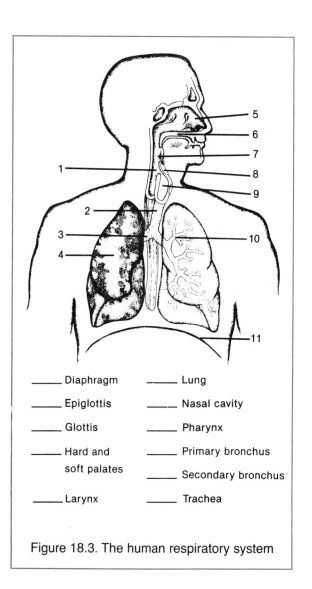

_____ Diaphragm   _____ Lung

_____ Epiglottis   _____ Nasal cavity

_____ Glottis   _____ Pharynx

_____ Hard and soft palates   _____ Primary bronchus

_____   _____ Secondary bronchus

_____ Larynx   _____ Trachea

Figure 18.3. The human respiratory system

and warmed by the moist membranes of the nasal cavity. The surface area of the nasal cavity is increased by the turbinates projecting into the cavity from each side. Air then passes through the **pharynx**, which serves as the passageway for both food and air, and flows through the **glottis** into the larynx. The **larynx** is a cartilaginous box containing the vocal folds (cords). During swallowing, the **epiglottis** folds over and closes the glottis to prevent food from entering the larynx. From the larynx, air passes down the **trachea** (windpipe), branches into the **primary bronchi**, and enters the lungs. The primary bronchi branch into **secondary bronchi** that branch into smaller and smaller tubules (**bronchioles**), and finally terminate in the **alveoli**, clusters of tiny sacs filled with air. Each human lung has about 300 million alveoli that increase the surface area of the lung to about 75 m^2. Capillary networks surround the alveoli, and gas exchange occurs between the air in the alveoli and the capillary blood. See Figure 18.4. Air exits the respiratory system in reverse order during **expiration**.

Most of the respiratory passages are lined with **pseudostratified ciliated columnar epithelium**. As the name implies, the ciliated, columnar cells forming this tissue appear to be arranged in layers, but they actually are not. See Figure 18.5. The **goblet cells** produce a layer of mucus that coats the surface of the cells. Foreign particles, such as bacteria, viruses, pollens, and dust, are trapped in the mucus, which

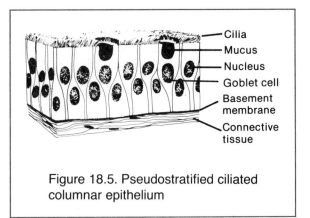

Figure 18.5. Pseudostratified ciliated columnar epithelium

is swept toward the pharynx, upward from the bronchi and trachea and downward from the nasal cavity, by the beating cilia. The mucus and entrapped particles are then swallowed. In this way, foreign materials are removed from the air passages to prevent infection and to keep the membranes free from contaminants.

Assignment 2

1. Observe the demonstration dissection of a pithed frog exposing an inflated lung. Examine it under a dissecting microscope and note the blood cells flowing through capillaries in the thin wall of the lung.
2. Label and color-code the respiratory organs in Figure 18.3. Locate the parts of the respiratory system on a human torso model or chart.
3. Examine the demonstration of the respiratory system of a sheep. Locate the parts that you labeled in Figure 18.3. Observe the **cartilaginous rings** that hold open the trachea. What is their real shape? Cut open the larynx and find the vocal folds. Locate the **epiglottis**, a cartilaginous flap that folds over the glottis during swallowing. Feel the lungs to detect their spongy consistency. If the lungs are not preserved, cut off a small piece and place it in a beaker of water. How do you explain what happens? Examine the cut surface of the lung tissue with a dissecting microscope. Can you see small air passages?
4. Examine prepared slides of normal and emphysematous lung tissue. Note how the

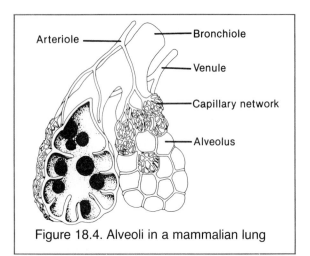

Figure 18.4. Alveoli in a mammalian lung

alveoli have broken down to form large spaces in the emphysematous lung tissue. Few people other than smokers develop **emphysema**.

5. Examine a prepared slide of trachea, x.s., and observe the ciliated epithelium composing the inner lining. Compare your observations with Figure 18.5.

6. ***Complete item 2 on the laboratory report***.

Mechanics of Breathing

Mammals are unique in that the thoracic and abdominal cavities are separated by a thin sheet of muscle, the **diaphragm**, which is the primary muscle involved in breathing. The **intercostal muscles** between the ribs also are involved but play a secondary role.

Assignment 3

1. Examine the breathing mechanics model. See Figure 18.6. The balloons represent the lungs, and the glass tubing represents the trachea and primary bronchi. The glass jar corresponds to the thoracic wall, and the rubber sheet simulates the diaphragm. Note that the balloons are in an enclosed space, representing the thoracic cavity.

2. Observe what happens when the rubber sheet is pulled downward and pushed upward. Determine how this works.

3. ***Complete item 3 on the laboratory report***.

Lung Capacity in Humans

Lung capacities vary among males and females, primarily due to variations in the size of the thoracic cavity and the lungs. Capacities also vary with age. Average capacities are shown in Figure 18.7.

In this section, you will use a spirometer to determine your lung volumes. *Caution: Wipe the spirometer stem with an alcohol swab or cotton soaked in 0.5% Amphyl before and after using the spirometer and use a fresh, sterile, disposable mouthpiece. Place all used mouthpieces in the biohazard bag after use.*

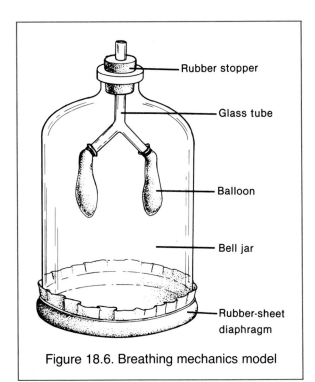

Figure 18.6. Breathing mechanics model

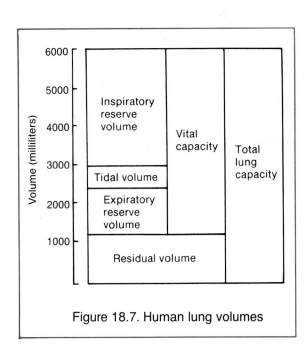

Figure 18.7. Human lung volumes

Assignment 4

Determine your lung volumes and record your results in item 4 on the laboratory report.

1. The volume of air exchange during normal quiet breathing is the **tidal volume** (TV). Determine it as described below.
   a. Rotate the dial of the spirometer so that the needle is at zero. See Figure 18.8.
   b. Disinfect the stem and add a sterile mouthpiece. Place the mouthpiece in your mouth, keeping the spirometer dial upward.
   c. Inhale through your nose and exhale through the spirometer for five normal quiet breathing cycles. Record the dial reading and divide by 5 to determine your tidal volume.
2. The amount of air that can be forcefully exhaled after a maximum inhalation is the **vital capacity** (VC). It is often used as an indicator of respiratory function.
   a. Rotate the dial face of the spirometer to place the needle at zero.
   b. Take two deep breaths and exhale completely after each one. Then take a

breath as deeply as possible and exhale through the spirometer. A slow, even expiration is best. Record the reading.
   c. Repeat step b two more times, resetting the dial face after each measurement.
   d. Record the average of the three measurements as your vital capacity. Compare your vital capacity with those shown in Table 18.1.

TABLE 18.1
Expected Vital Capacities (ml) for Adult Males and Females*

Males

Height (inches)	Age in Years					
	20	30	40	50	60	70
60	3885	3665	3445	3225	3005	2785
62	4154	3925	3705	3485	3265	3045
64	4410	4190	3970	3750	3530	3310
66	4675	4455	4235	4015	3795	3575
68	4940	4720	4500	4280	4060	3840
70	5206	4986	4766	4546	4326	4106
72	5471	5251	5031	4811	4591	4371
74	5736	5516	5296	5076	4856	4636

Females

Height (inches)	Age in Years					
	20	30	40	50	60	70
58	2989	2809	2629	2449	2269	2089
60	3198	3018	2838	2658	2478	2298
62	3403	3223	3043	2863	2683	2503
64	3612	3432	3252	3072	2892	2710
66	3822	3642	3462	3282	3102	2922
68	4031	3851	3671	3491	3311	3131
70	4270	4090	3910	3730	3550	3370
72	4449	4269	4089	3909	3729	3549

Data from Propper Mfg. Co., Inc.

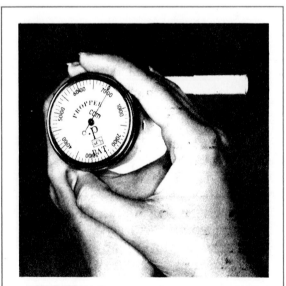

Figure 18.8. Propper spirometer.
Slip on a sterile mouthpiece and rotate the dial face to zero before exhaling through the spirometer.

3. The volume of air that can be exhaled *after* a normal tidal volume expiration is the **expiratory reserve volume** (ERV).

   a. Rotate the dial face of the spirometer so that the needle is at 1000. This is to compensate for the space on the dial between 0 and 1000.

   b. *After* a normal quiet expiration, forcefully exhale as much air as possible through the spirometer. Be sure not to take an extra breath at the start. Subtract 1000 from the reading to determine your expiratory reserve volume.

   c. Based on your determinations above, calculate your inspiratory reserve volume. IRV = VC − (TV + ERV)

   d. ***Complete item 4 on the laboratory report***.

THE CHEMICAL CONTROL OF BREATHING

You are aware that a direct relationship exists between the degree of exercise and the rate of breathing. To determine the chemical factor responsible for this relationship, the following experiments were done. Study the results carefully so that you can form valid conclusions.

Assignment 5

Study each experiment carefully and complete item 5 on the laboratory report.

Experiment 1. The concentration of O_2, CO_2, and hydrogen ions (H+) in arterial blood of a human subject was determined before and after exercise. The results are shown below.

Results

	Before Exercise	After Exercise
pO_2 (mm Hg)	100	80
pCO_2 (mm Hg)	40	45
pH*	7.4	7.3
Breathing rate (cycles/min)	15	45

*Recall that as the pH decreases, the acidity increases.

Complete item 5a on the laboratory report.

Experiment 2. A human subject breathed air from a closed container for 3 min, and the breathing rate was determined as shown below.

Results

	1 Min	2 Min	3 Min
Breathing rate (cycles/min)	16	21	37

Complete item 5b on the laboratory report.

Experiment 3. The same human subject breathed air from a closed container through a tube containing soda lime that removed the CO_2 from the air. The results are shown below.

Results

	1 Min	2 Min	3 Min
Breathing rate (cycles/min)	15	16	15

Complete item 5c on the laboratory report.

Experiment 4. When ammonium chloride (NH_4Cl) was injected into a human subject so that the blood pH was lowered by 0.08 (that is, the acidity was increased), the breathing rate was only slightly increased. Subsequently, when the same subject inhaled sufficient CO_2 to lower the pH only 0.04 (that is, increased acidity only half as much as the NH_4Cl injection), the breathing rate was greatly increased.

Recall that CO_2 unites with water in the blood to form carbonic acid that ionizes to release hydrogen (H^+) and bicarbonate (HCO_3^-) ions. *Therefore, the pH of the blood may be lowered (that is, acidity increased) by excessive amounts of CO_2.*

Complete item 5d on the laboratory report.

GAS EXCHANGE IN VASCULAR PLANTS

Cellular respiration in plants generally occurs at a much slower rate than in animals so that the exchange of O_2 and CO_2 does not require special gas-exchange organs or a transport system. The relatively active living tissues are near the surface of the plant so that diffusion is sufficient to meet the gas-exchange needs.

Gas exchange occurs at a rapid rate between leaf cells and the atmosphere, especially when the cells are photosynthesizing. The fact that green plant cells produce O_2 and absorb CO_2 in photosynthesis does not mean that plant cells do not require O_2 and produce CO_2 in cellular respiration. The O_2 released from plant tissues during photosynthesis is the excess available after respiration needs have been met. Plant tissues that are not photosynthesizing absorb O_2 and release CO_2 like all cells in the process of aerobic cellular respiration.

The size of the stomata controls the rate of gas exchange, and guard cells control the size of a stoma. Guard cells are the only cells of the epidermis that contain chloroplasts and are photosynthetically active. An increase in the intracellular pressure of the guard cells occurs when they are photosynthesizing and causes them to bend and open the stoma due to the unequal thickness of their cell walls. A decrease in intracellular pressure closes the stoma. Stomata are usually open in daylight hours and closed at night. The stomata also are closed when the plant is suffering from excessive water loss.

Materials

Prepared slides of:
  Syringa leaf, x.s.
  leaf epidermis

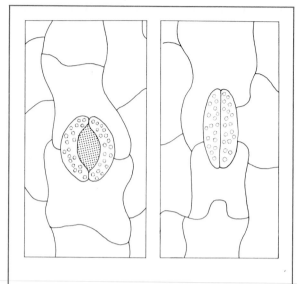

Figure 18.9. Open and closed stomata

Assignment 6

1. Examine a prepared slide of leaf epidermis. Locate the stomata and the guard cells around each stoma. Note the shape of the guard cells and stomata. See Figure 18.9. Color the guard cells green.
2. Examine a prepared slide of *Syringa* leaf, x.s., and locate a stoma in the lower epidermis. Note the space between the stoma and the interior leaf tissues and how the spongy mesophyll cells are arranged to allow movement of gases between the cells. Mesophyll cells provide the primary gas-exchange surfaces in plants. How do they exhibit the characteristics noted earlier for gas-exchange organs in animals?
3. ***Complete item 6 on the laboratory report.***

19

INTERNAL TRANSPORT
IN PLANTS

OBJECTIVES

On completion of the laboratory session, you should be able to:
1. Describe the structure and function of vascular tissues in plants.
2. Describe the processes of root pressure, transpiration, and translocation.
3. Define all terms in bold print.

Each cell carries out its own metabolic processes whether it exists as a single cell or as a part of a multicellular organism. Therefore, each cell must obtain the necessary raw materials from its environment and rid itself of the wastes produced by the metabolic activity.

In unicellular, colonial, and simple multicellular organisms, the cells are either in direct contact with the external environment or only a few cells from it. The rate of diffusion is sufficient for the transport of substances for such cells. In more complex organisms, however, internal cells are located far from the external environment. In such organisms, special tissues or organs or both are necessary to transport substances to and from the cells.

The successful colonization of the land by vascular plants depended on the evolution of conducting tissues. This adaptation enabled the evolution of large plant bodies with a greater specialization of parts than in the less advanced plants. Thus, leaves are specialized for photosynthesis and roots for the uptake of water and dissolved minerals, while the vascular tissue forms a continuous pathway for the transport of substances from one portion of the plant to another.

VASCULAR TISSUES IN PLANTS

Xylem and **phloem** are the vascular (conducting) tissues in plants. Water and dissolved minerals are absorbed by the roots, transported to the xylem, and conducted upward in the root and stem to the leaves, flowers, and fruits. Water and mineral transport in the xylem is always upward.

Organic nutrients produced in the leaves by photosynthesis are transported via the phloem to the various parts of the plant. The transport of organic nutrients is called **translocation**. The process is not clearly understood, although diffusion is involved in part. The translocation of organic nutrients may be either upward or downward within the phloem.

Materials

Net- and parallel-veined leaves
Prepared slides of:
  pine wood (*Pinus*), l.s. and x.s.
  basswood (*Tilia*), mascerated wood
  pumpkin (*Curcubita*) stem, l.s.

corn (*Zea*) stem, x.s.
alfalfa (*Medicago*) stem, x.s.
oak (*Quercus*) stem, 1 yr, x.s.
privet (*Ligustrum*) and corn (*Zea*) leaves, x.s.

Xylem

Xylem tissue is composed of **vessel cells** in flowering plants or **tracheids** in conifers that are attached end to end to form continuous, minute channels. These channels transport water and dissolved minerals from the root tip to the shoot tip. See Figure 19.1. Mature xylem cells are nonliving, lack protoplasm, and have perforated end walls (or no end walls at all), which facilitate the vertical movement of the fluid. Lateral fluid movement occurs through minute holes in the side walls. Xylem also contains elongated, thick-walled fiber cells that provide additional support. In woody plants, xylem tissue is commonly called wood.

Phloem

Phloem tissue is composed of two types of cells: **sieve tube cells** and **companion cells**. See Figure 19.2. Sieve tubes contain strands of cytoplasm but no nucleus. The cytoplasm of one cell is continuous with that of the adjacent cells in the series. This is possible because of the perforations in the sieve plate forming the end cell wall. A number of sieve tube cells joined end to end form the sieve tube through which organic nutrients are transported.

The companion cells are smaller cells that are adjacent to the sieve tube cells. They also are joined end to end but do not function in transport. Some biologists believe that the nucleus of the companion cells also controls the cytoplasm of the sieve tube cells.

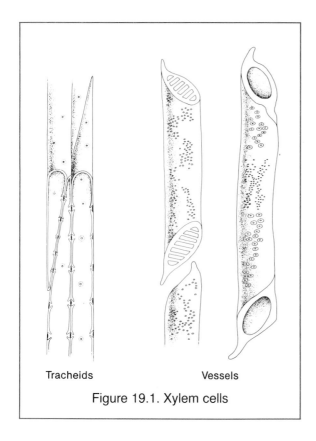

Tracheids   Vessels

Figure 19.1. Xylem cells

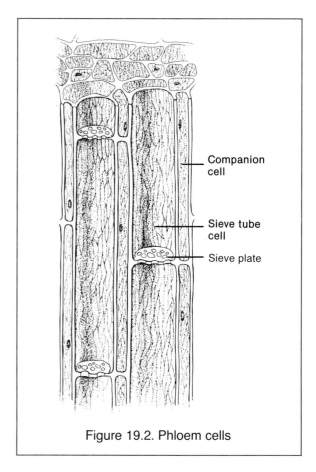

Companion cell

Sieve tube cell

Sieve plate

Figure 19.2. Phloem cells

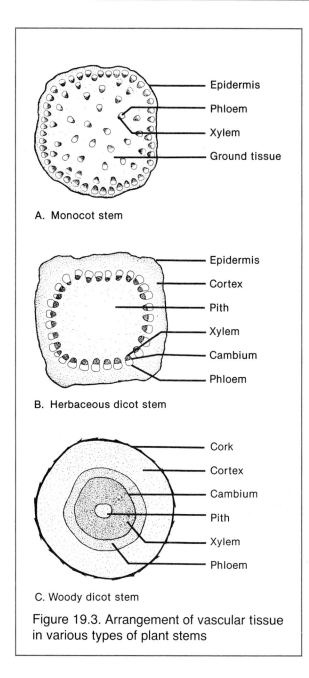

A. Monocot stem

- Epidermis
- Phloem
- Xylem
- Ground tissue

B. Herbaceous dicot stem

- Epidermis
- Cortex
- Pith
- Xylem
- Cambium
- Phloem

C. Woody dicot stem

- Cork
- Cortex
- Cambium
- Pith
- Xylem
- Phloem

Figure 19.3. Arrangement of vascular tissue in various types of plant stems

Organization of Vascular Tissue

Both xylem and phloem form minute channels that are continuous from root tip to shoot tip, and they are always arranged in close association with each other. When observed in the cross section of a stem, the vascular tissue may be arranged in (1) scattered **vascular bundles** in

monocots (e.g., corn), (2) vascular bundles in a broken ring near the stem circumference in herbaceous dicots (e.g., alfalfa), or (3) a continuous vascular ring in woody dicots (e.g., oak).

Figure 19.3 illustrates patterns of vascular tissue distribution. Note that the xylem is always located toward the center of the stem and that the phloem is always oriented toward the periphery. The cambium is located between the xylem and the phloem in stems capable of secondary growth. Consult Exercise 9 for the detailed cellular structure of stems.

Assignment 1

1. ***Complete item 1a on Laboratory Report 19 that begins on page 415.***
2. Examine a prepared slide of pine wood, x.s. and l.s., and locate tracheids similar to those in Figure 19.1.
3. Examine a prepared slide of basswood, macerated wood, and locate vessels like those in Figure 19.1.
4. Examine a prepared slide of pumpkin stem, l.s., and locate sieve tubes and companion cells like those in Figure 19.2.
5. Observe the location of vascular tissues and relative positions of xylem and phloem in prepared slides of:
   a. corn (*Zea*) stem, x.s.
   b. alfalfa (*Medicago*) stem, x.s.
   c. oak (*Quercus*) stem, x.s.
   d. privet (*Ligustrum*) leaf, x.s.
   e. corn (*Zea*) leaf, x.s.
6. ***Complete item 1 on the laboratory report.***

Root Pressure

Water and dissolved minerals are absorbed by the root tips, especially the **root hairs**, which greatly increase the surface area of the root in contact with water in the soil. From the epidermal cells, water moves from cell to cell until it enters a xylem vessel. Water also may move along the cell walls until it reaches the endodermis. It must then pass through a living cell to enter the xylem because of the impermeable cell walls of the **endodermal cells**. These impermeable cell walls constitute the **Casparian strip** and allow some control of the amount of

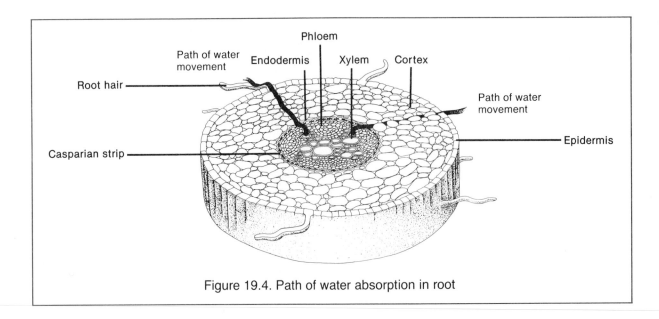

Figure 19.4. Path of water absorption in root

water absorbed. See Figure 19.4. Dissolved minerals may be absorbed by simple diffusion, but experiments have shown that active absorption is usually involved.

The absorptive force of the root cells produces a **root pressure** that forces water upward in the xylem vessels.

Materials

Dissecting instruments
Microscope slides and cover glasses
Root pressure demonstration (see Figure 19.5)
Methylene blue, 0.01%, in dropping bottles
Germinated grass seeds

Assignment 2

1. Examine the root pressure demonstration that has been prepared in accordance with Figure 19.5. Record the level of the fluid in the pipette at 30-min intervals. ***Record your data in item 2a on the laboratory report***.
2. Obtain a germinated grass seed and place it on a microscope slide in a drop of water. Examine it with a dissecting microscope and at 40× and 100× with a compound microscope using reduced light. Note the root hairs. The oldest root hairs are longest.

3. Use a scalpel to cut the seed from the root. Discard the seed. Add a drop of methylene blue, 0.01%, and a cover glass and observe a short root hair at 400×. Observe the junction of the root hair and its epidermal cell. Use reduced light to locate a nucleus.
4. ***Complete item 2 on the laboratory report***.

Transpiration

The evaporation of water from plant tissues is called **transpiration**. Most of the transpiration occurs from leaf tissues, especially the **mesophyll**. The water vapor escapes from the leaf via the open **stomata**. The evaporation of water exerts a "pull" on the columns of water in the xylem vessels. The cohesion of water molecules by hydrogen bonds enables the columns of water to be "pulled" upward by the evaporation of water molecules from the leaves.

The rate of transpiration varies with (1) intrinsic factors, such as the number and size of leaves and the density of stomata, and (2) extrinsic factors, such as available water and climatic and weather conditions.

Materials

Electric fan
Forceps
Graph paper, mm and cm divisions

1. With a sharp blade cut off stem of a well-watered plant.

2. Snugly fit rubber tubing over stem and add a few drops of water.

3. Insert a 1-ml pipette into tubing, forcing water up into pipette. Support with ring stand.

Figure 19.5. Root pressure setup

Microscope slides and cover glasses
Knife, sharp
Pipette, 1 ml
Potometer jar
Two-hole rubber stopper
Nail polish, clear
Branches with leaves

Assignment 3

1. Prepare a potometer setup as shown in Figure 19.6. Insert the pipette into the rubber stopper before cutting the branch from the plant. Do *not* get the leaves wet.
2. Press the stopper firmly into the jar to cause the water to rise near the top of the pipette.
3. Place the potometer on your laboratory table, and determine the rate of transpiration with and without wind created by an electric fan. ***Record your data in item 3b on the laboratory report***.
4. ***Complete items 3c to 3g on the laboratory report***.

Density of Stomata

Since water vapor passes from the leaf via the stomata, the density of stomata and the size of leaves affect the transpiration rate. In what way is the rate affected?

Assignment 4

1. Remove a leaf from a branch. Place it on a sheet of graph paper lined in millimeter and centimeter divisions and trace around it. From the tracing, estimate the surface area of the leaf in square centimeters (1 cm^2 = 100 mm^2). Consider the surface area to be the same for both upper and lower leaf surfaces. ***Record your data in item 4a on the laboratory report***.
2. Coat part of both surfaces of the leaf with clear nail polish. After the nail polish has dried, peel it off, and mount it dry on a slide under a cover glass. Use the $4 \times$ or $10 \times$ objective and *reduced light* to observe the stomata. Record the number of stomata in three fields and calculate the average density per field for each leaf surface. Are stomata on both surfaces of the leaf?
3. Calculate the area of the field in millimeters using the diameter of field determined

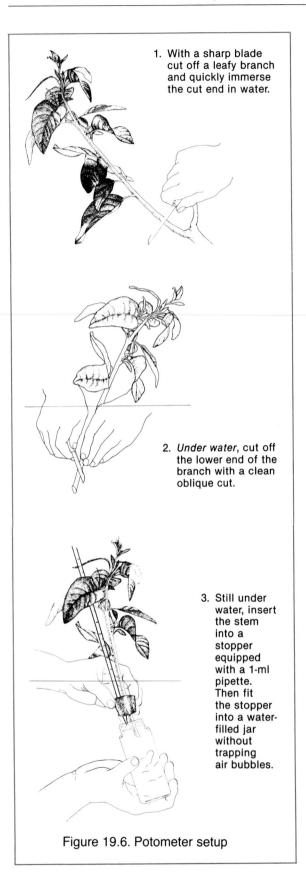

1. With a sharp blade cut off a leafy branch and quickly immerse the cut end in water.

2. *Under water*, cut off the lower end of the branch with a clean oblique cut.

3. Still under water, insert the stem into a stopper equipped with a 1-ml pipette. Then fit the stopper into a water-filled jar without trapping air bubbles.

Figure 19.6. Potometer setup

in Exercise 1 ($A = \pi r^2$). Convert the number of stomata per field to stomata per square centimeter.

$$\frac{\text{Number of stomata}}{\text{Area of field in mm}^2} \times \frac{100 \text{ mm}^2}{1 \text{ cm}^2} = \text{stomata/cm}^2$$

Multiply the area of leaf surface *containing stomata* by the stomata per square centimeter *on that surface* to determine the number of stomata on your leaf.

5. If time permits, determine the distribution and density of stomata on leaves of other plants, and correlate your findings with the normal environmental conditions of the plants.

6. ***Complete the laboratory report***.

20

INTERNAL TRANSPORT IN ANIMALS

OBJECTIVES

On completion of the laboratory session, you should be able to:

1. Describe the components and functions of blood and hemolymph.
2. Describe and identify human blood cells.
3. Explain the basis of blood groups, blood typing, the concept of universal donor and recipient, and the cause of erythroblastosis fetalis.
4. Distinguish between open and closed circulatory systems, and identify and describe the function of their components.
5. Identify the parts of the mammalian heart on charts and a dissected sheep heart, and describe their functions.
6. Trace the path of blood in humans from the heart to the lungs, intestine, head, kidneys, and back to the heart.
7. Define all terms in bold print.

In unicellular animal-like protists and in simple multicellular animals, diffusion alone provides an adequate rate of distribution of materials to meet the metabolic needs of the organism. Larger, more complex animals depend on a **circulatory system** to provide a more rapid transport of materials. Diffusion still plays an important role in the movement of materials between the circulatory fluid and the body cells, however.

Circulatory systems include (1) a circulating fluid (blood or hemolymph), (2) a heart or pulsating vessel, containing one-way valves, that pumps the fluid through the system, and (3) vessels through which the fluid passes. Two types of circulatory systems occur among animal groups: open systems and closed systems.

In **closed circulatory systems**, the circulating fluid is called **blood**, and it is confined within the blood vessels: arteries, capillaries, and veins. Circulation of blood is rapid. Annelids and vertebrates have closed circulatory systems.

In **open circulatory systems**, the circulating fluid is called **hemolymph**, and it is not confined to blood vessels. It is carried from the pulsating heart through one or more open-ended arteries that empty it into the coelom or **hemocoel**. The hemolymph slowly moves around the internal organs on its way back to the heart. Most mollusks and all arthropods have an open circulatory system.

BLOOD AND HEMOLYMPH

Blood and hemolymph are quite variable among animals. In either case, separated cells are carried along in a fluid plasma. At least some of the cells are capable of amoeboid movement. The function of blood and hemolymph is to transport materials from place to place in the animal's body. These materials include organic nutrients, metabolic wastes, hormones, antibodies, mineral ions, oxygen, and carbon dioxide. Most insects are an exception in that respiratory gases are primarily transported by the tracheal system, not the hemolymph.

Two types of **respiratory pigments** occur that facilitate the transport of oxygen. **Hemacyanin**, a blue-green pigment, is dissolved in the plasma of certain arthropods. **Hemoglobin**, a red pigment, is dissolved in the plasma of annelids and some arthropods, and it is confined within red blood cells in vertebrates.

Materials

Anti-sera
  anti-A
  anti-B
  anti-D (Rh)
Biohazard bag
Blood typing box and typing plate
Colored pencils
Cotton, sterile and absorbent
Lancets, sterile and disposable
Microscope slides
Schilling blood charts
Toothpicks, flat
Alcohol swabs
Amphyl solution, 0.5%
Prepared slides of frog, reptilian, and human
  blood

Assignment 1

1. Examine prepared slides of amphibian and reptilian blood. Observe the red blood cells, which are the most numerous cells on the slide. What is their general shape? Do they have a nucleus? Stain has been used to distinguish the blood cells, so the colors you see are artificial.
2. ***Complete item 1 on Laboratory Report 20 that begins on page 419.***

Human Blood

Human blood consists of 45% blood cells and 55% plasma. The formed elements, commonly called blood cells, may be divided into three groups: (1) **erythrocytes**, red blood cells that transport oxygen and carbon dioxide; (2) **leukocytes**, white blood cells that fight infections; and (3) **thrombocytes**, platelets that initiate

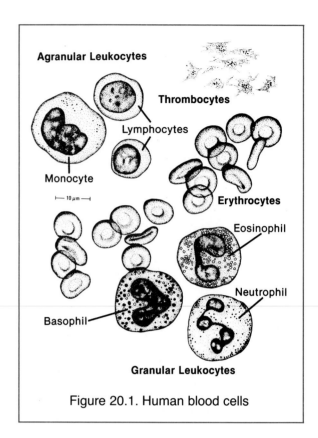

Figure 20.1. Human blood cells

the clotting process. See Figure 20.1. In 1 mm³ of blood, there are about 5,000,000 erythrocytes, 8000 leukocytes, and 350,000 thrombocytes, and these constitute only 45% of the volume! Obviously, blood cells are very small and numerous. They are destroyed and replaced in large numbers daily. The types and distinguishing characteristics of white blood cells when stained with Wright's blood stain are presented in Table 20.1.

Assignment 2

1. Obtain a Schilling blood chart from the stock table. Study the mature blood cells in the bottom row to learn their characteristics and frequency in normal circulating blood. The cells appear as stained with Wright's blood stain. Note that leukocytes are subdivided into **granulocytes** and **agranulocytes** depending on the presence or absence of cytoplasmic granules. Do the mature red blood cells have a nucleus? This condition is true for all mammals. Note the small size of the platelets, which are nonnucleated fragments of larger parent cells.

TABLE 20.1
Leukocytes

Name	Characteristics of Stained Cells	Normal Percentage
Granulocytes		
Neutrophils	Nucleus with three to five lobes, very small lavender cytoplasmic granules	60–70%
Eosinophils	Nucleus with two lobes, red or orange cytoplasmic granules	2–4%
Basophils	Nucleus with two to three lobes, blue or purple cytoplasmic granules	0.5–1%
Agranulocytes		
Lymphocytes	Small cell with a large spherical nucleus and small amount of cytoplasm	20–25%
Monocytes	Large cell with a large irregular or kidney-shaped nucleus	3–8%

2. Color the blood cells in Figure 20.1 as shown on the Schilling blood chart.
3. Examine a prepared slide of human blood, and locate erythrocytes, neutrophils, lymphocytes, and monocytes. **Draw a few red blood cells in item 1c.** Try to find an eosinophil and basophil but don't spend more than 10 min searching. Compare your observations with the blood chart.
4. *Complete items 2 on the laboratory report*.

Human Blood Groups

Human blood may be classified according to the presence or absence of various **antigens** (proteins) on the red blood cells. The presence of these antigens is genetically controlled, so an individual's **blood type** is the same from birth to death. See Table 20.2. The antigens most commonly tested for in blood typing tests are A, B, and D (Rh).

Testing for the presence of A and B antigens determines the ABO blood group. The presence or absence of the D (Rh) antigen determines the Rh blood type. The two factors are combined in designating an individual's blood type (e.g., A,Rh$^+$; A,Rh$^-$; AB,Rh$^+$). Table 20.3 indicates the association of these **antigens** and their corresponding **antibodies** in blood.

All blood types are not compatible with each other. Therefore, knowing the blood types of both donor and recipient in blood transfusions is imperative. The *antigens of the donor* and the

TABLE 20.2
Percentage of ABO Blood Types

Blood Type	% U.S. Blacks	% U.S. Whites
A	25	41
B	20	7
AB	4	2
O	51	50

TABLE 20.3
Antigen-Antibody Associations

Blood Type	Antigen	Antibody
O	None	a,b
A	A	b
B	B	a
AB	AB	None
Rh$^+$	D	None
Rh$^-$	None	d*

*Antibodies are produced by an Rh$^-$ person only after Rh$^+$ red blood cells enter his or her blood.

antibodies of the recipient must be considered in blood transfusions. Transfusion of incompatible blood results in clumping (agglutination) of erythrocytes in the transfused blood, which may plug capillaries and result in death. Clumping of erythrocytes occurs whenever the *antigens of the donor* and the *antibodies of the recipient,* that are designated by the same letter, are brought together. See Table 20.3. The antibodies of the donor are so diluted in the recipient's blood that their effect is inconsequential.

The Rh Factor

About 85% of Caucasians and 99 to 100% of Chinese, Japanese, African blacks, and American Indians possess the D (Rh) antigen and are therefore Rh$^+$. If an Rh$^-$ individual receives a single transfusion of Rh$^+$ blood, antibodies are produced against the D (Rh) antigen. However, no clumping of cells occurs due to the gradual increase in antibodies and the loss of erythrocytes containing the D (Rh) antigen. If a second transfusion of Rh$^+$ blood is received, clumping will occur, and death may result.

Antibodies against the D (Rh) antigen may be produced by an Rh$^-$ woman carrying an Rh$^+$ fetus due to "leakage" of fetal erythrocytes into the maternal blood. The accumulation of antibodies in maternal blood is gradual enough so that no complications result during the first pregnancy with an Rh$^+$ fetus. In subsequent pregnancies with Rh$^+$ fetuses, however, the maternal antibodies may diffuse into the fetal blood and destroy the fetal erythrocytes. This pathological conditions, *erythroblastosis fetalis,* may be fatal to the fetus. The mother suffers no consequences unless she subsequently receives a transfusion of Rh$^+$ blood.

Blood Typing

Testing for the presence of A, B, and D antigens is accomplished by adding **antisera** containing specific antibodies that will combine with these antigens. Because testing for the D antigen requires a temperature of 50°C, you will use a slide warming box and a special blood typing plate to allow simultaneous testing for all three antigens. See Figure 20.2. Each blood typing setup consists of a slide warming box, blood typing plate, small bottles of 70% alcohol and 0.50% Amphyl, a collecting bag, cotton, toothpicks, lancets, glass slides, and one bottle each of anti-A, anti-B, and anti-D sera.

Note: The probability of infection with blood-borne viruses (HIV or hepatitis) is very remote when typing your own blood. If done correctly,

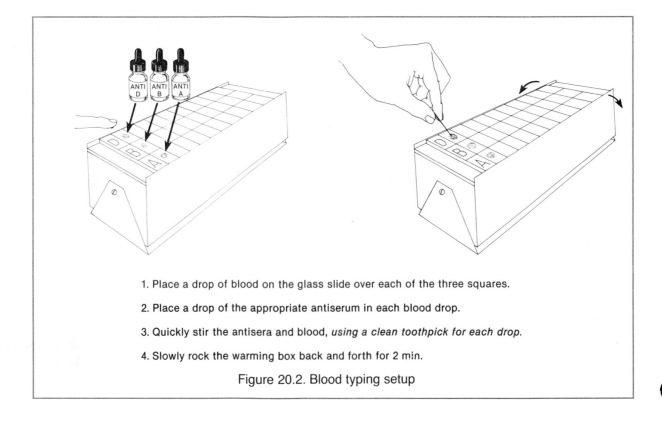

1. Place a drop of blood on the glass slide over each of the three squares.

2. Place a drop of the appropriate antiserum in each blood drop.

3. Quickly stir the antisera and blood, *using a clean toothpick for each drop.*

4. Slowly rock the warming box back and forth for 2 min.

Figure 20.2. Blood typing setup

you will be in contact with your own blood, only. *Be sure to follow the safety precautions and directions given by your instructor as well as the procedures that follow.*

Assignment 3

1. Place a glass slide across the typing plate on the slide warming box and turn on the warming box at least 5 min before use.
2. Wipe a fingertip with the alcohol-saturated cotton. When your fingertip has dried, pierce it with the lancet. Discard the lancet into the collecting bag.
3. Place a drop of blood on the glass slide over each of three squares on the special blood typing plate.
4. Add a drop of anti-D serum to the drop of blood in the D square, a drop of anti-B serum to the drop of blood in the B square,

and a drop of anti-A serum to the drop of blood in the the A square. See Figure 20.2. *Do not touch the blood with the antisera droppers.*

5. Mix the antiserum and blood in each square with a *different* toothpick. (Using the same toothpick for more than one square may invalidate the results.)
6. Rock the warming box back and forth for *exactly 2 min,* and then determine the results. If tiny red granules appear in the blood and anti-D serum mixture within 2 min, the blood is Rh$^+$. If not, it is Rh$^-$. The mixture may have to be examined under a dissecting microscope to determine the type. If clumping occurs in either of the other squares used for ABO blood typing, the red granules will be large and obvious. Determine your blood type by referring to Figure 20.3.

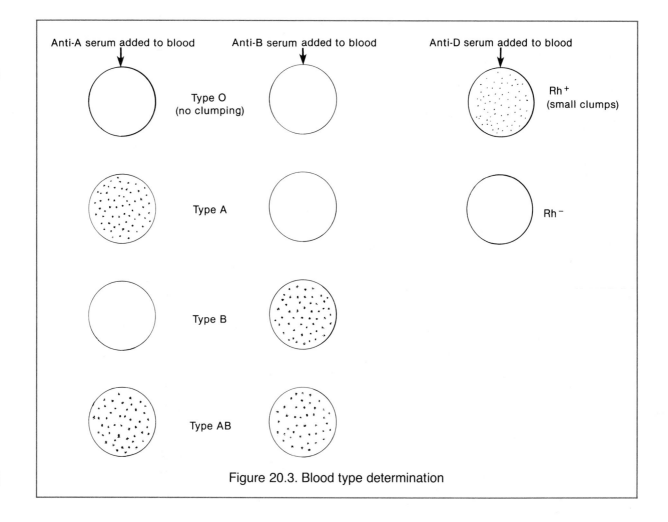

Figure 20.3. Blood type determination

7. *Wipe the pierced fingertip with cotton soaked in 0.5% Amphyl.*
8. *Place anything in contact with blood, such as used toothpicks, cotton, and glass slides in the biohazard bag immediately after use.*
9. **Complete item 3 on the laboratory report.**

Circulation of the Blood

The circulatory system of vertebrates is composed of the (1) **heart**, whose contractions provide the force to circulate the blood; (2) **arteries**, which carry blood away from the heart to all parts of the body; (3) **veins**, which return blood to the heart from the body; and (4) **capillaries**, tiny vessels that connect arteries and veins.

Materials

Heart models of:
  fish
  amphibian
  reptile
  mammal
Colored pencils
Dissecting instruments and pins
Dissecting pan
Frog board and cloth
Frog, live
Sheep heart, fresh or preserved
Prepared slides of:
  cardiac muscle
  artery and vein, x.s.
  atherosclerotic artery, x.s.

The Heart

The vertebrate heart shows a progressive complexity from fish to birds and mammals. Fish have a two-chambered heart composed of an **atrium**, a receiving chamber, and a **ventricle**, a pumping chamber. A valve between the chambers prevents the backflow of blood. Deoxygenated blood is pumped by the ventricle to the gills, where it is oxygenated, before it continues on to the rest of the body.

Amphibians have a three-chambered heart composed of one ventricle and two atria. Blood is pumped to the lungs and body simultaneously by the single ventricle. Deoxygenated blood from the body returns to the right atrium, and oxygenated blood from the lungs enters the left atrium. Blood from each atrium flows into the ventricle, which results in mixing oxygenated and deoxygenated blood in the single ventricle. Thus, the blood nourishing body cells is not highly oxygenated.

Reptiles have a septum that partially divides the ventricle and reduces the mixing of oxygenated and deoxygenated blood. This structure improves the oxygenation of blood supplying the body cells. Birds and mammals have four-chambered hearts with two atria and two ventricles that ensure the circulation of highly oxygenated blood to the body cells. Refer to Figures 20.4 through 20.6 as you study the description of the mammalian heart that follows.

The mammalian heart is a double pump composed of four chambers. The two **atria** receive blood returning to the heart. The two **ventricles** pump blood from the heart. The right atrium and right ventricle compose one pumping unit, and the left atrium and left ventricle compose the other. There is no connection between the two atria or the two ventricles. The **ventricular septum** is a wall of muscle separating the ventricles.

The opening between the atrium and ventricle on each side is guarded by an **atrioventricular valve** that prevents a backflow of blood from the ventricle into the atrium. The **bicuspid valve** separates the left atrium and ventricle. The **tricuspid valve** separates the right atrium and ventricle. The flaps (cusps) of the valves are anchored by tough cords of tissue called **chordae tendineae** to small mounds of muscle, **papillary muscles**, on the inner walls of the ventricles. These cords prevent the valve flaps from being forced into the atria during ventricular contraction and enable the valves to close off the opening. They resemble the lines of a parachute as they restrain the valve flaps.

Semilunar valves are located at the base of both pulmonary and aorta arteries. These valves permit blood to be forced into the arteries but prevent a backflow of blood from the arteries into the ventricles.

The Flow of Blood Through the Mammalian Heart

During the relaxation phase (diastole), blood flows into the right atrium from the **anterior vena cava**, which returns blood from body regions anterior to the heart, and the **posterior**

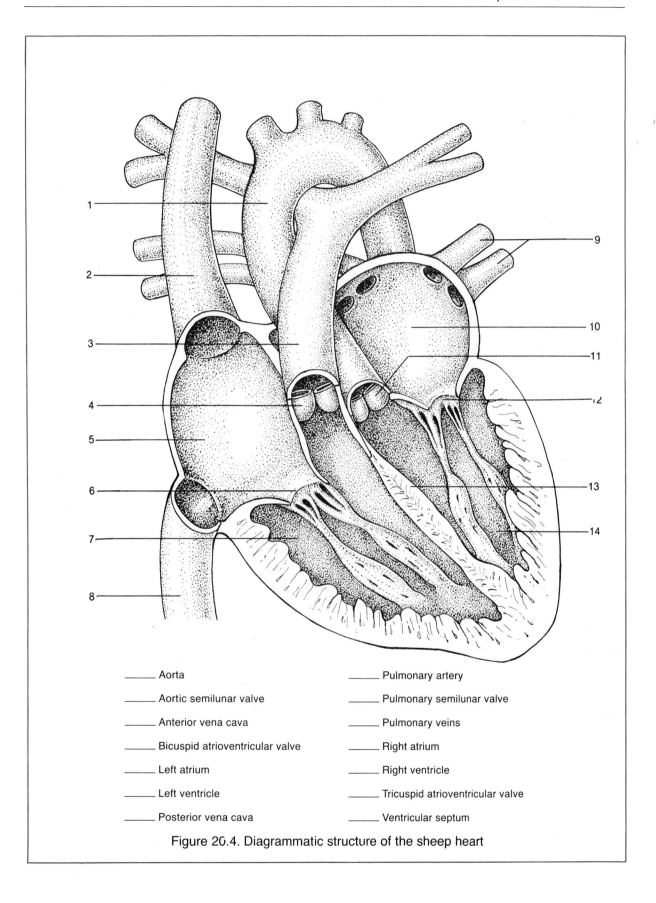

_____ Aorta	_____ Pulmonary artery
_____ Aortic semilunar valve	_____ Pulmonary semilunar valve
_____ Anterior vena cava	_____ Pulmonary veins
_____ Bicuspid atrioventricular valve	_____ Right atrium
_____ Left atrium	_____ Right ventricle
_____ Left ventricle	_____ Tricuspid atrioventricular valve
_____ Posterior vena cava	_____ Ventricular septum

Figure 20.4. Diagrammatic structure of the sheep heart

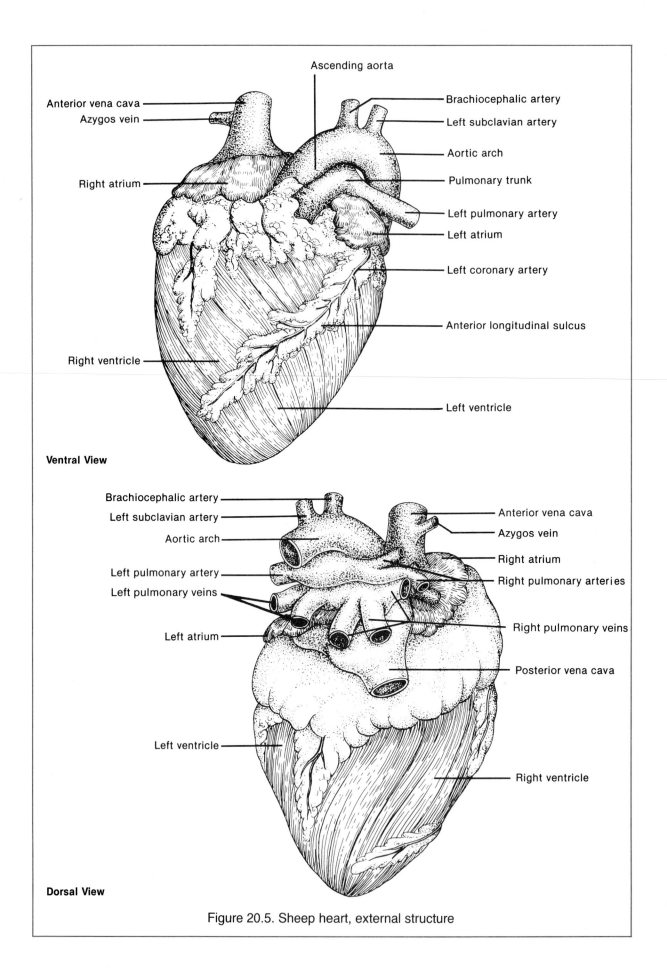

Figure 20.5. Sheep heart, external structure

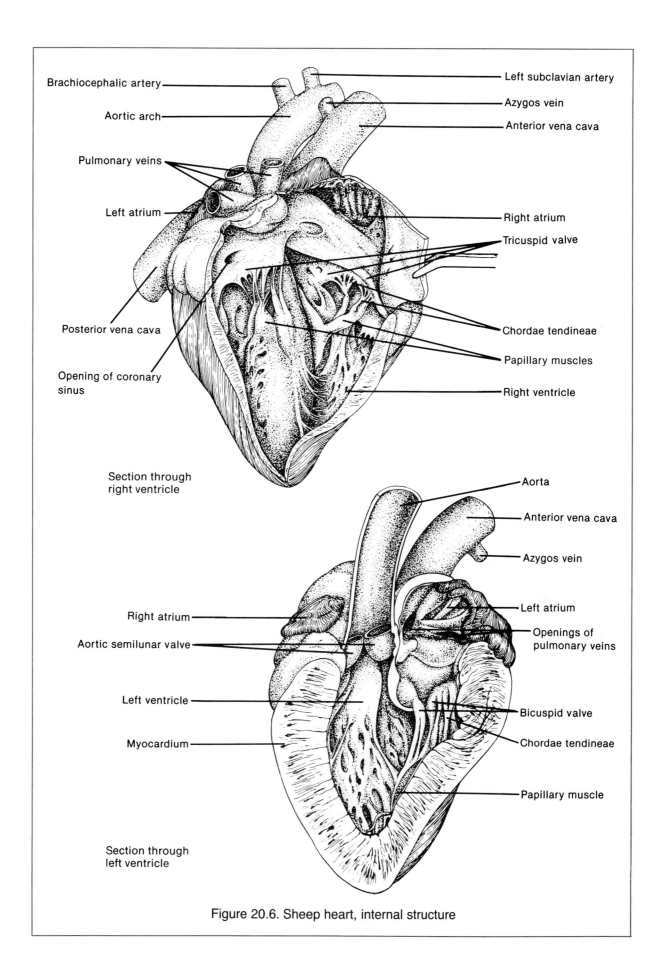

Brachiocephalic artery

Aortic arch

Pulmonary veins

Left atrium

Posterior vena cava

Opening of coronary sinus

Section through right ventricle

Left subclavian artery

Azygos vein

Anterior vena cava

Right atrium

Tricuspid valve

Chordae tendineae

Papillary muscles

Right ventricle

Aorta

Anterior vena cava

Azygos vein

Left atrium

Openings of pulmonary veins

Right atrium

Aortic semilunar valve

Left ventricle

Myocardium

Bicuspid valve

Chordae tendineae

Papillary muscle

Section through left ventricle

Figure 20.6. Sheep heart, internal structure

vena cava, which returns blood from regions posterior to the heart. Simultaneously, the **pulmonary veins** return blood into the left atrium from the lungs. The blood flows from each atrium into the corresponding ventricle.

During contraction (systole), the atria contract, forcing more blood into the ventricles so that they are completely filled. Immediately thereafter, the ventricles contract. The buildup of pressure in the ventricles (1) closes the atrioventricular valves, (2) opens the semilunar valves, and (3) forces blood into the **pulmonary** and **aorta arteries**, which carry blood to the lungs and all other parts of the body, respectively. At the cessation of contraction, the semilunar valves snap shut, and the atrioventricular valves open as the relaxation phase begins. The closing of the heart valves produces the characteristic heart sounds.

Assignment 4

1. Label Figure 20.4.
2. Locate the parts of the mammalian heart on a model or chart.
3. Compare the models of vertebrate hearts and note the differences.
4. ***Complete item 4 on the laboratory report***.

The Sheep Heart

A study of the external and internal structure of a sheep heart will extend your understanding of heart structure and function. Refer to Figures 20.5 and 20.6 as you proceed.

External Features

1. If the **pericardium**, a membranous sac enclosing the heart, is present, note how it envelops the heart and is attached to the heart. Use scissors to remove it.
2. Observe the ventral surface of the heart. See Figure 20.5. Locate a **coronary artery**, which runs along a groove over the ventricular septum dividing the left and right ventricles. Locate the other coronary artery on the dorsal surface.
3. Locate the ventricles and note the difference in size.
4. Observe the atria, thin-walled pouches above the ventricles. Locate the anterior

and posterior vena cavae that carry blood into the right atrium and the four small pulmonary veins that return blood to the left atrium. The pulmonary veins are often embedded in fat on the dorsal surface of the heart. Insert a probe into these veins to determine that they enter the left atrium.
5. Locate the pulmonary artery, which exits the right ventricle and lies next to the left atrium. Insert a probe into it to determine that it comes from the right ventricle.
6. Locate the aorta that exits the left ventricle. A large artery, the brachiocephalic artery, branches from the base of the aorta. Compare the thickness of the walls of the aorta and vena cava.

Dissection Procedures

1. Hold the heart in your left hand with the apex away from you and the dorsal surface up. Insert a blade of the scissors into the anterior vena cava and cut through the right atrium, atrioventricular valve, and to the tip of the right ventricle. If the heart is fresh, you may have to wash out blood clots. Compare the thickness of the atrial and ventricular walls. Observe the atrioventricular valve and the chordae tendineae.
2. Following the procedures in step 1, cut through the left atrium and on to the tip of the left ventricle. Compare the size and wall thickness of the two ventricles.
3. Insert a probe through the aorta and into the left ventricle. Use scissors to cut along the probe from the left ventricle into the aorta. This cut will expose the aortic semilunar valve. Locate the three membranous pockets that form it. Just above the semilunar valve are the two openings of the coronary arteries where they exit the aorta.
4. Following the procedures in step 3, cut through the pulmonary artery into the right ventricle. Examine the pulmonary semilunar valve.

Assignment 5

Complete item 5 on the laboratory report.

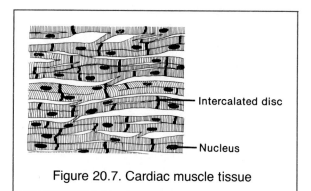

Figure 20.7. Cardiac muscle tissue

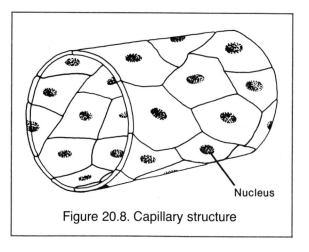

Figure 20.8. Capillary structure

Assignment 6

1. Examine a prepared slide of **cardiac muscle**, the type of muscle tissue composing the heart. Compare your slide with Figure 20.7 and locate the labeled parts.
2. Note that each cell has a centrally located nucleus and striations and is separated from adjoining cells by **intercalated discs**. Observe how the fibers join together to form a meshlike arrangement rather than remaining as individual fibers.
3. ***Complete item 6 on the laboratory report.***

The Pattern of Circulation

In vertebrates, blood flows through a system of closed vessels to transport materials to and from body cells. **Arteries** carry blood from the heart, and they divide into smaller and smaller arteries and ultimately lead to **arterioles**, which are nearly microscopic in size. Arterioles lead to the **capillaries**, the smallest blood vessels. The walls of the capillaries are composed of a single layer of squamous epithelial cells. See Figure 20.8. Exchange of materials occurs between blood in capillaries and the body cells. From capillaries, blood flows into tiny veins called **venules** that combine to form larger **veins** that carry blood back to the heart.

Figure 20.9 shows the relationship between an arteriole, capillaries, and a venule. The diameter of the arterioles affects the flow of blood and blood pressure and is controlled by circular smooth muscles that respond to impulses from the autonomic nervous system. The flow of blood into capillaries is controlled by the **precapillary sphincter muscle** that is governed by im-

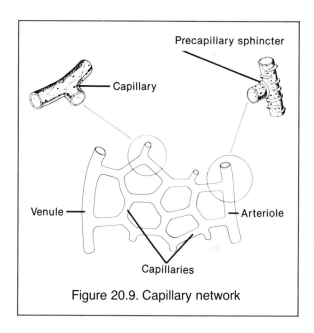

Figure 20.9. Capillary network

pulses from the autonomic nervous system and local chemical stimuli. An increase in the local CO_2 concentration opens the sphincter and increases capillary blood flow. Similarly, a decrease in CO_2 concentration reduces capillary blood flow.

The exchange of materials between the body cells and capillary blood involves **tissue fluid** as an intermediary. Tissue fluid is the thin layer of extracellular fluid that covers all cells and tissues. This is the pattern of the exchange:

$$\text{Blood in} \rightleftharpoons \text{tissue} \rightleftharpoons \text{body}$$
$$\text{capillaries} \qquad \text{fluid} \qquad \text{cells}$$

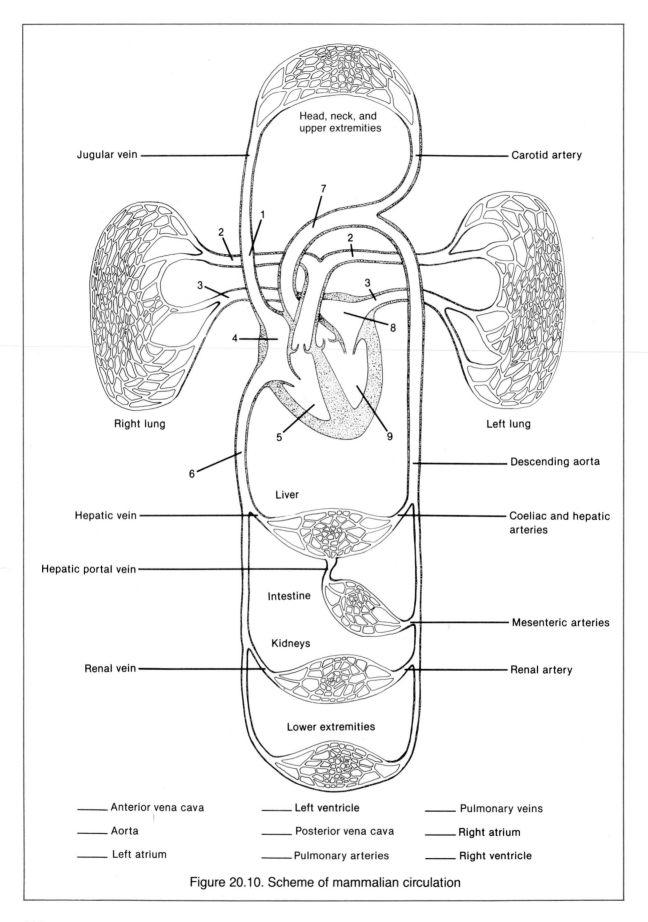

Head, neck, and
upper extremities

Jugular vein ———— Carotid artery

1

2

7

2

3

3

8

Right lung

4

5

9

Left lung

6

Descending aorta

Liver

Hepatic vein ———— Coeliac and hepatic
arteries

Hepatic portal vein ————

Intestine

Mesenteric arteries

Kidneys

Renal vein ———— Renal artery

Lower extremities

_____ Anterior vena cava   _____ Left ventricle   _____ Pulmonary veins

_____ Aorta   _____ Posterior vena cava   _____ Right atrium

_____ Left atrium   _____ Pulmonary arteries   _____ Right ventricle

Figure 20.10. Scheme of mammalian circulation

Since the mammalian heart is a double pump, there are two separate pathways of circulation. The **pulmonary circuit** pumps **deoxygenated blood** to the lungs via pulmonary arteries. After the exchange of O_2 and CO_2, **oxygenated blood** returns to the heart via pulmonary veins. The **systemic circuit** pumps oxygenated blood to the body (except the lungs) where O_2 and CO_2 are exchanged between the body cells and capillary blood. Then the deoxygenated blood is returned to the heart.

Assignment 7

1. *Complete items 7a and 7b on the laboratory report*.
2. Examine the capillary blood in the webbing of a frog's foot that has been set up under a demonstration microscope. Note the pulsating flow in the feeding arteriole and the smooth flow in the collecting venule. Observe the blood cells moving in single file through a capillary. How would you describe the flow of blood in the capillary? Keep the foot wet with water. Turn off the microscope light when you have finished observing. *Complete item 7c on the laboratory report*.
3. Examine prepared slides of artery and vein, x.s. Compare the difference in thickness of the walls due to differences in the amount of smooth muscle and connective tissue. Locate the single layer of squamous endothelial cells forming the interior lining. *Complete item 7d on the laboratory report*.
4. Examine a prepared slide of an atherosclerotic artery, x.s., and note the fatty deposit that partially plugs the vessel. Cholesterol is primarily responsible for this deposit. This type of obstruction in coronary arteries often causes heart attacks and may require coronary bypass surgery.
5. Label Figure 20.10. Add arrows to indicate the direction of blood flow. Color vessels carrying deoxygenated blood blue and those carrying oxygenated blood red.
6. *Complete item 7 on the laboratory report*.

21

EXCRETION

The metabolic processes of animals produce wastes that are harmful and must be removed. In simple animals, this is accomplished by diffusion alone, but more complex animals use special organs or organ systems. The principal metabolic wastes are **carbon dioxide** (CO_2) formed by cellular respiration and **nitrogenous wastes** produced by the breakdown of amino acids composing proteins. Most animals use respiratory organs to expedite the removal of CO_2 and have special excretory organs to get rid of nitrogenous wastes.

In your study of invertebrate animals, you observed three types of excretory organs. Annelids have paired, segmental **nephridia** that remove nitrogenous wastes from the coelomic fluid and blood and excrete them through nephridiopores. Crustaceans, such as the crayfish, possess paired **green glands** located at the base of the first antennae. These glands remove nitrogenous wastes from the hemolymph and excrete them through external openings. Similarly, insects have **Malpighian tubules** that remove nitrogenous wastes from the hemolymph and excrete them into the intestine for passage from the body.

NITROGENOUS WASTES

When amino acids are deaminated, the amine groups tend to form toxic **ammonia**. This substance simply diffuses from small, simple animals, but diffusion is inadequate for its removal in larger, more complex forms. In aquatic vertebrates, nitrogenous wastes, including ammonia, are flushed from the body with copious amounts of water. Terrestrial forms, which must conserve water, convert ammonia into **urea** and **uric acid**. Urea, the primary nitrogenous waste in amphibians and mammals, is excreted in an aqueous solution, but uric acid, the primary nitrogenous waste in insects, reptiles and birds, is excreted with relatively little water as a semisolid mass.

The process of urinary excretion not only removes the nitrogenous waste products of metabolism, but it also regulates the concentration of ions, water, and other substances in body fluids. The urinary systems of marine, freshwater, and terrestrial vertebrates are each uniquely adapted to perform these functions in radically different environments.

THE MAMMALIAN URINARY SYSTEM

The mammalian urinary system consists of (1) a pair of **kidneys** that remove nitrogenous wastes, excess ions, and water from the blood to form **urine**; (2) a pair of **ureters** that carry urine by peristalsis from the kidney to the urinary bladder; (3) a **urinary bladder** that serves as a temporary storage container; and (4) a **urethra** that carries urine from the bladder during urination. See Figure 21.1.

The Kidney

The basic structure of a kidney is shown in coronal section in Figure 21.2. The outer portion is the cortex, which contains vast numbers of capillaries and **nephrons**. The inner **medulla**

contains the **renal pyramids** that are composed mainly of **collecting tubules**. The tip (papilla) of each pyramid is inserted into a funnellike **calyx** that unites with the **renal pelvis**.

The Nephron

The nephron is the functional unit of the kidney. Each human kidney contains about 1 million nephrons. A nephron consists of (1) a **Bowman's** (nephron) **capsule** that surrounds an arteriole capillary tuft, a **glomerulus**, and (2) a tortuous, thin-walled **tubule** that leads to a collecting tubule. The nephron tubule consists of three parts. The **proximal convoluted tubule** leads from the nephron capsule to a U-shaped **loop of Henle** that is contiguous with a **distal convoluted tubule**. Many renal tubules are joined to a single collecting tubule.

Materials

Colored pencils
Model of human kidney
Model of human urinary system
Dissecting instruments and pan
Sheep kidney, fresh or preserved
Sheep kidney, triple injected
Prepared slide of kidney cortex

Assignment 1

Complete item 1 on Laboratory Report 21 that begins on page 423.

Assignment 2

1. Label and color-code parts of the urinary system in Figure 21.1 and parts of the kidney and nephron in Figure 21.2.
2. Study the models of the urinary system and kidney. Locate the parts shown in Figures 21.1 and 21.2.
3. Examine the demonstration whole kidney. Note its distinctive shape and locate the ureter and renal blood vessels.
4. Obtain a sheep kidney that has been coronally sectioned and locate the parts shown in Figure 21.2A. Correlate the structure of the parts with their functions that were previously described.
5. Examine the sectioned triple-injected kidney set up under a dissection microscope. Locate the glomeruli and nephron capsules.

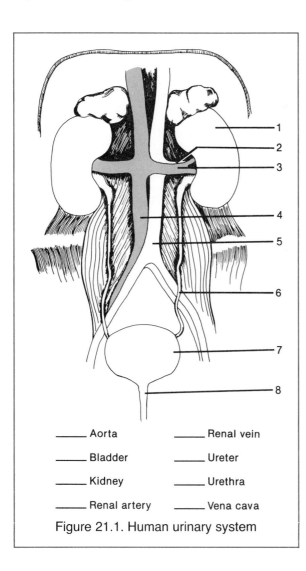

_____ Aorta   _____ Renal vein
_____ Bladder   _____ Ureter
_____ Kidney   _____ Urethra
_____ Renal artery   _____ Vena cava

Figure 21.1. Human urinary system

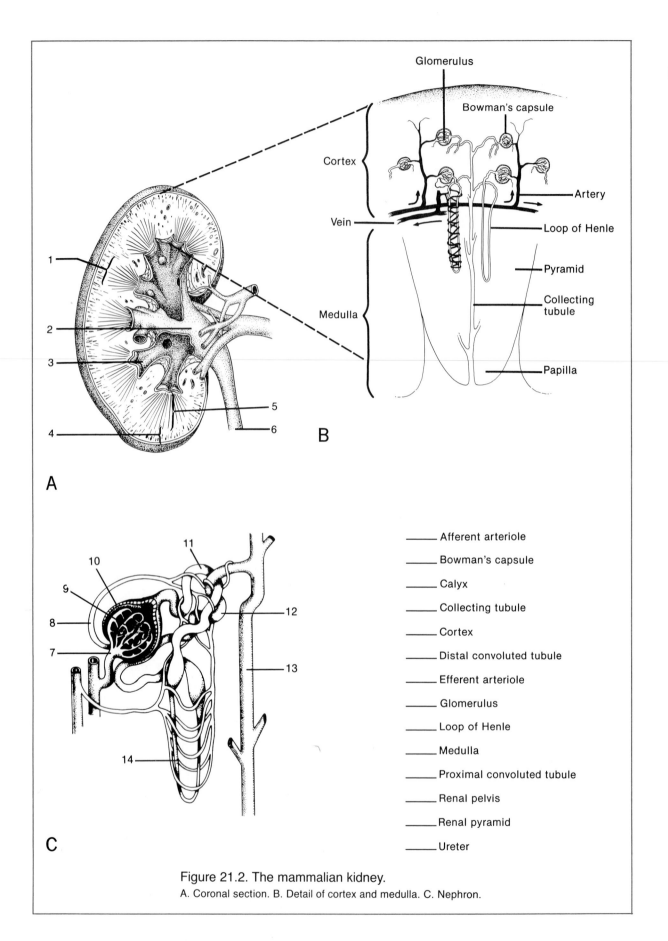

Glomerulus

Bowman's capsule

Cortex

Artery

Vein

Loop of Henle

Medulla

Pyramid

Collecting tubule

Papilla

1

2

3

5

6

4

A

B

11

10

9

8

7

12

13

14

C

_____ Afferent arteriole

_____ Bowman's capsule

_____ Calyx

_____ Collecting tubule

_____ Cortex

_____ Distal convoluted tubule

_____ Efferent arteriole

_____ Glomerulus

_____ Loop of Henle

_____ Medulla

_____ Proximal convoluted tubule

_____ Renal pelvis

_____ Renal pyramid

_____ Ureter

Figure 21.2. The mammalian kidney.
A. Coronal section. B. Detail of cortex and medulla. C. Nephron.

6. Examine a prepared slide of kidney cortex. Locate a glomerulus and nephron capsule. Note the thin wall of the tubules. How many cells thick is the tubule wall?

7. ***Complete item 2 on the laboratory report***.

Urine Formation

Blood is carried from the aorta to the kidneys by renal arteries and returned from the kidneys by renal veins into the vena cava. While passing through the kidneys, the concentration of nitrogenous wastes and other substances in the blood is controlled by (1) the partial removal of diffusable materials in the glomerular blood by **filtration** into Bowman's capsules, (2) **selective reabsorption** of useful substances from the tubules back into the blood, and (3) **secretion** of certain ions from the capillary blood into the tubules.

The renal artery branches into progressively smaller arteries leading ultimately to the afferent arterioles supplying blood to the glomeruli. The **afferent arteriole** leading to a glomerulus is larger in diameter than the **efferent arteriole** carrying blood from the glomerulus. This causes an increase in blood pressure in the glo-

merulus and accelerates the movement (filtration) of diffusable materials from the glomerular blood into the capsule. Thus, some of all diffusable materials pass from the glomerulus into Bowman's capsule.

The fluid in the capsule, the **filtrate**, flows through the nephron tubule, which is enveloped by a capillary network. Selective tubule reabsorption of needed materials—especially water, nutrients, and mineral ions—from the filtrate into the capillary blood occurs as the filtrate moves through the tubule. At the same time, certain excess ions (especially H^+ and K^+) may be secreted from the capillary blood into the filtrate. Both **tubule reabsorption** and **tubule secretion** involve active and passive transport mechanisms. The remaining fluid flows from the nephron tubule into a collecting tubule of a renal pyramid, continues into a calyx, and on into the renal pelvis. This fluid, now called **urine**, leaves the kidney via the ureter and is carried to the urinary bladder by peristalsis.

Assignment 3

1. Study Table 21.1. Note the relationships among the daily volumes of blood circulated through human kidneys, filtrate removed, and urine formed. ***Complete items 3a to 3d on the laboratory report***.

2. Study Table 21.2. Note how the concentration of selected substances varies in different parts of the nephron and blood vessels. Use Figure 21.2 to orient yourself to the anatomical association of the structures involved.

TABLE 21.1
Volume of Blood, Filtrate, and Urine Formed in Humans in a 24-Hr Period (liters)

Blood flow through the kidneys	1700
Filtrate removed from blood	180
Urine formed	1.5

TABLE 21.2
Concentration (mg/100 ml) of Dissolved Substances in Various Regions of the Nephron

Materials in Water Solution	1 Afferent Arteriole	2 Efferent Arteriole	3 Bowman's Capsule (Filtrate)	4 Collecting Tubule (Urine)*	5 Renal Vein
Urea	30	30	30	2000	25
Uric acid	4	4	4	50	3.3
Inorganic salts	720	720	720	1500	719
Protein	7000	8000	0	0	7050
Amino acids	50	50	50	0	48
Glucose	100	100	100	0	98

*Trace amounts are not included.

Note that the concentration of urea is constant in columns 1, 2, and 3 of Table 21.2. This indicates that urea is a small molecule that easily diffuses from the glomerulus into Bowman's capsule. How do you explain the change in urea concentration in columns 4 and 5?

3. **Complete item 3 on the laboratory report**.

URINALYSIS

In humans, the composition of urine is commonly used to assess the general functioning of the body. Diet, exercise, and stress may cause variations in composition and concentration, but significant deviations usually result from malfunctions of the body. See Table 21.3.

TABLE 21.3
Urine Components Evaluated by Urinalysis

Component	Normal	Abnormal*
Color	Straw to amber	Pink, red-brown, or smoky urine may indicate blood in the urine. The higher the specific gravity, the darker is the color. Nearly colorless urine may result from excessive fluid intake, alcohol ingestion, diabetes insipidis, or chronic nephritis.
Turbidity	Clear to slightly turbid	Excess and persistent turbidity may indicate pus or blood in the urine.
pH	4.8 to 8.0 Ave. = 6.0	Acid urine may result from a diet high in protein (meats and cereals) or a high fever; alkaline urine results from a vegetarian diet or bacterial infections of the urinary tract.
Specific gravity	1.003 to 1.035	Low values result from a deficiency of antidiuretic hormone or kidney damage that impairs water reabsorption. High values result from diabetes mellitis or kidney disease, allowing proteins to enter filtrate.
Blood or hemoglobin	Absent	Presence of intact RBCs may result from lower urinary tract infections, kidney disease allowing RBCs to enter filtrate, lupus, or severe hypertension. Presence of hemoglobin occurs in extensive burns and trauma, hemolytic anemia, malaria, and incompatible transfusions.
Protein	Absent or trace	Proteins are present in kidney diseases allowing proteins to enter the filtrate, and they may be present in fever, trauma, anemia, leukemia, hypertension, and other nonrenal disorders. They may occur due to excessive exercise and high-protein diets.
Glucose	Absent or trace	Presence usually indicates diabetes mellitis.
Ketones	Absent or trace	Presence results from excessive fat metabolism as in diabetes mellitis and starvation.
Bilirubin	Small amounts	Excess may indicate liver disease (hepatitis or cirrhosis) or blockage of bile ducts.
Nitrite (bacteria)	Absent	Presence indicates a bacterial urinary tract infection.
Leukocytes (pus)	Absent	Presence indicates a urinary tract infection.

*Only a few causes of abnormal values are noted.

Materials

Biohazard bag
Collecting cup, plastic
Multistix 9 SG reagent strips
Simulated urine samples

Assignment 4

In this section, you will use "dip sticks" to analyze several simulated urine samples to determine if they are normal or if their characteristics suggest possible disease. Table 21.3 indicates normal values and some abnormal characteristics. Study it before proceeding.

1. Obtain six Multistix 9 SG reagent strips (dip sticks) and a color chart that indicates how to read the results. Study the color chart carefully to be sure that you understand the time requirements for reading the dip sticks and the location and interpretation of each reagent band. Note: The sequence of the reagent bands from the tip to the handle of a dip stick matches the sequence of specific tests listed from top to bottom on the color chart.

2. Obtain six numbered test tubes containing simulated urine samples. Place them in a test-tube rack at your workstation.

3. Analyze the urine samples one at a time by dipping a dip stick completely into the sample so that all reagent bands are immersed. Remove the dip stick and place it on a paper towel with the reagent bands facing upward. Read your results after 1 min.

4. If you wish to test your own urine, obtain a plastic collecting cup and a Multistix 9 SG reagent strip from your instructor and perform the test in the restroom. Dispose of the urine in a toilet, place the collecting cup in the biohazard bag and read your results. Then place the dip stick in the biohazard bag.

5. ***Record your results in item 4a on the laboratory report. Compare your results with Table 21.3 and complete item 4 on the laboratory report.***

22

CHEMICAL CONTROL

OBJECTIVES

On completion of the laboratory session, you should be able to:
1. Describe the physical properties and action of IAA on plant growth as determined in the experiments.
2. Describe the effect of thyroxine on body weight and metabolic rate as determined in mice.
3. Describe the effect of acetylcholine and epinephrine on the heart rate of a frog.
4. Define all terms in bold print.

Chemical control is widespread among organisms and is responsible for much of the orderly fashion in which living processes are conducted. Much of the chemical control is carried out by chemical messengers called **hormones**. Five types of hormones have been identified in vascular plants, and the existence of others is suspected. Plant hormones are secreted by certain cells and are transmitted primarily by diffusion to stimulate or inhibit specific activity in *adjacent* or *nonadjacent* cells, depending on the hormone.

In animals, chemical control occurs in three distinct ways. **Neurotransmitters** are chemicals released by neuron axons into synaptic junctions that either stimulate or inhibit impulse formation and transmission in *adjacent* neurons. Thus, neural activity depends on these transmitter substances that are secreted in exceedingly small amounts and exert their effect only at the synapse. Some neurons secrete **neu-

rosecretory hormones** that, like transmitter substances, affect changes in the membrane potential of target cells, but travel greater distances via either diffusion or bloodstream to affect *nonadjacent* cells. This mode of action occurs in simple animals such as coelenterates as well as in higher forms. The so-called **true hormones** occur in mollusks and higher animals and are secreted by **endocrine cells** that may function independently or be organized into **endocrine glands**. Hormones secreted by endocrine cells diffuse into body fluids from which they are picked up by specific *nonadjacent* target cells. These hormones seem to exert their effect by stimulating or inhibiting enzyme activity.

IAA AND PLANT GROWTH

Whenever a planted seed germinates, the shoot grows upward out of the soil, while the root grows downward. Have you wondered what mechanism produces this growth pattern? This behavioral response is called **geotropism**. Roots are said to exhibit positive geotropism because they grow toward the pull of gravity. Shoots exhibit negative geotropism. The response to light, **phototropism**, is another common tropism observed in plants.

These tropisms result primarily from a plant's response to varying concentrations of **indoleacetic acid (IAA)**, the primary auxin hormone that promotes cell elongation. Since experiments with IAA require considerable time and care, you will not perform the experiments but will examine results of a series of experiments using oat coleoptiles. These experiments are similar to those that established the role of IAA in plants.

A **coleoptile** is a hollow sheathlike structure found in corn, oats, and other grasses. It protects the young leaves curled within it while it forces its way through the soil. Once the coleoptile breaks through the ground and is exposed to sunlight, it stops growing. The leaves break through, expand, and start photosynthesizing. The **region of cell division** in a coleoptile is at its base where it is attached to the shoot tip. Above this region is the **region of elongation**, which is responsible for most of the linear growth of the coleoptile. See Figure 22.1.

Materials

Demonstration of phototropism
Demonstrations of stem bending by IAA
Lanolin
Lanolin with 1% IAA
Oat seedlings with coleoptiles

Assignment 1

1. ***Complete item 1 on Laboratory Report 22 that begins on page 427.***

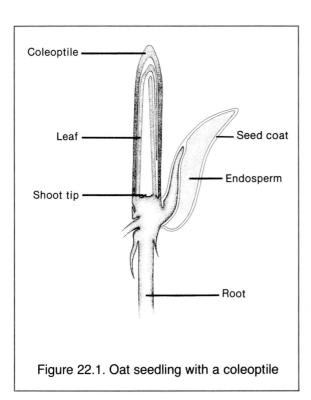

Figure 22.1. Oat seedling with a coleoptile

Assignment 2

1. Study the experiments in Figure 22.2. Note the way each experiment is set up and the results obtained.
2. ***Complete items 2a–2e on the laboratory report***.
3. Examine the geotropism demonstration in vascular plants. The terminal bud has been removed from plant A but not from plant B. How do you explain the response of each plant?
4. Examine the phototropism demonstration in vascular plants. The terminal bud has been removed from plant A but not from plant B. How do you explain the response of each plant?
5. Examine the growth pattern of plants to which either plain lanolin or lanolin + 1% IAA was applied along one side of the stem.
6. ***Complete item 2 on the laboratory report***.

THYROXINE AND METABOLIC RATE

In this section, you will investigate the effect of thyroxine on weight gain and the metabolic rate of immature white mice. **Thyroxine** is a hormone secreted by the **thyroid gland**. It is carried to all parts of the body by the bloodstream.

Metabolism is the overall exchange of materials and transformation of energy in living organisms. Metabolism per unit time is the **metabolic rate**. Metabolic rate may be measured as oxygen (O_2) consumption, carbon dioxide (CO_2) production, or heat (calories) production.

In this experiment, you will determine the consumption of O_2 as an indicator of metabolic rate by using a respirometer (see Figure 16.5 on p. 153). The respirometer consists of two stoppered jars connected by a U-shaped tube (manometer) that is partially filled with fluid. One jar contains the mouse; the other serves as a pressure control. A pressure change in one jar will cause movement of the fluid toward the jar with less pressure. If CO_2 is removed by an absorbent, O_2 consumption can be determined by measuring the movement of fluid in the 5-ml volumetric pipette composing one arm of the U-

Experiment 1

Two groups of oat seedlings were prepared as follows:
  Group A—The coleoptile tips were removed.
  Group B—Seedlings were left in normal state.
After 12 hr in darkness, the results were:
  Group A—No growth
  Group B—Average growth of 4.2 mm

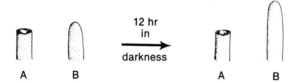

Experiment 2

Three groups of oat seedlings were prepared as follows:
  Group A—Coleoptile tips were removed, and plain agar blocks were placed on the cut surface of the coleoptiles.
  Group B—Coleoptile tips were removed, and agar blocks containing an extract of coleoptile tips were placed on the cut surface of the coleoptiles.
  Group C—Seedlings were left in normal state.
The results after 12 hr in darkness were:
  Group A—No growth
  Group B—Average growth of 3.9 mm
  Group C—Average growth of 4.0 mm

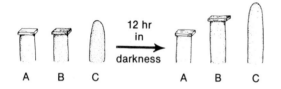

Experiment 3

Three groups of seedlings were prepared as follows:
  Group A—Coleoptile tips were removed, and agar blocks containing an extract of coleoptile tips were placed on the right side of the cut surface of the coleoptiles.
  Group B—Same as group A except that the agar blocks were placed on the left side of the cut surface of the coleoptiles.
  Group C—Seedlings were left in normal state.
The results after 12 hr in darkness were:
  Group A—Growth causing bending to the left
  Group B—Growth causing bending to the right
  Group C—Normal growth

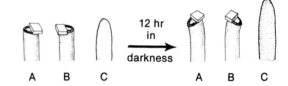

Experiment 4

Two groups of seedlings were placed in a horizontal position and prepared as follows:
  Group A—Coleoptile tips were removed.
  Group B—Seedlings were left in normal state.
After 12 hr in darkness, the results were:
  Group A—No growth
  Group B—Growth causing tips to bend upwards (negative geotropism)

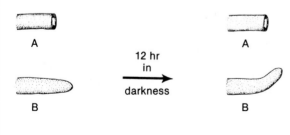

Experiment 5

Three groups of seedlings were prepared as follows:
  Group A—Coleoptile tips were removed.
  Group B—Seedlings were left in normal state.
  Group C—Tinfoil "cap" was placed on coleoptile tip.
After 12 hr exposure to unilateral light, the results were:
  Group A—No growth
  Group B—Growth causing tip to bend toward the light (positive phototropism)
  Group C—Upward growth only, that is, no bending

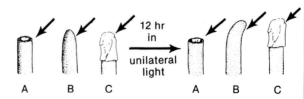

Figure 22.2. Control of plant tropisms by IAA

tube or, if O_2 is available, by injecting, at selected intervals, measured amounts of O_2 into the system with a syringe until the fluid level in the U-tube is returned to the starting point.

Three young male white mice were obtained 10 days ago. Each was weighed and placed in a separate cage. Food was present at all times, and the amount of food consumed was carefully measured. Mouse A was given drinking water containing 0.02% thyroxine, mouse B was given drinking water containing 0.25% $KClO_4$, and mouse C was given plain water. Thus, mouse A is experimentally **hyperthyroid**, mouse B is experimentally **hypothyroid**, and mouse C is normal.

Your instructor will provide the following data for each mouse:

1. The weight (in grams) at the start of the 10-day diet
2. The quantity (in grams) of food consumed

This experiment is best performed by groups of two to four students. If time is limited, your instructor may assign a different mouse to each student group with all groups sharing data.

Materials

Balance for weighing mice
Respirometer (Figure 16.5)
Water baths
Celsius thermometer
Colored water for manometer
Cotton, absorbent
Wire screen
Potassium perchlorate, 0.25%
Soda lime
Thyroxine, 0.02%

Assignment 3

1. Obtain a respirometer or construct one as shown in Figure 16.5. Place soda lime in the bottom of each jar to a depth of about 2 cm, cover it with a layer of absorbent cotton, and add the wire screen.
2. Add colored water to the manometer, if necessary, so that it extends about one third of the way up each arm.
3. Place the jars in a water bath at room temperature. During the experiment, keep the temperature of the water constant by adding warm or cold water. Why is the maintenance of a constant temperature important?
4. From your instructor, obtain the starting weights and food consumption for each mouse. Weigh each mouse prior to placement in the respirometer. *Record these data in item 3a on the laboratory report.*
5. Pick up a mouse by its tail and place it in the experimental chamber of the respirometer, and *with the vent tube open,* loosely replace the stoppers. Allow 5 min for temperature equilibration.
6. *With the vent tubes open,* insert the stoppers. Close the vent tubes with pinch clamps and record your first reading. Take the second reading after 3 min. Open the vent tubes. **Record your readings in item 3b on the laboratory report.**
7. Following the procedure in step 6, measure the O_2 consumption in five replicates, each with a duration of 3 min. Open the vent tubes for 3 min between replicates.
8. Discard the lowest and highest readings and calculate the average O_2 consumption.
9. **Complete item 3 on the laboratory report.**

CHEMICAL CONTROL AND HEART RATE

The heartbeat in vertebrates is controlled by factors both internal and external to the heart. Internal control is by the rhythmic depolarization of the sinuatrial node (S-A node), which causes the rhythmic contractions of the heart. The rate of the contractions is mostly controlled externally by the vagus and accelerator nerves leading to the S-A node. Impulses passing down the **vagus nerve** cause the secretion of **acetylcholine** at the junction of the nerve and heart, while impulses in the **accelerator nerve** cause a secretion of **norepinephrine** by the axon tips. These neurotransmitters affect the rate and force of the heart contractions.

The **adrenal glands** secrete the hormone **epinephrine** (adrenalin), a chemical similar to norepinephrine. Epinephrine is carried throughout the body by the blood. It also affects the rate and force of the heartbeat and diverts blood from the digestive tract into the skeletal muscles and nervous system. Epinephrine is the hormone that prepares animals to respond to an emergency or stressful situation. The "butterflies" that you may have experienced before a game or speech are side effects of epinephrine stimulation.

Observe the effect of epinephrine and acetylcholine on the heart rate of a double-pithed frog. A double-pithed frog has had its brain and spinal cord destroyed mechanically so that no impulses are sent or received by the central nervous system. The heart continues to beat for a while, however. All organs and tissues do not die at the instant the brain is destroyed or at the same rate. *Record your data in item 4a on the laboratory report.*

Materials

Dissecting instruments and pins
Dissecting pan
Stopwatch
Hypodermic needles, 22 gauge
Hypodermic syringes, 0.5 or 1.0 ml
Acetylcholine, 1:10,000
Epinephrine, 1:10,000
Frog Ringer's solution in dropping bottles
Frog, double pithed

Assignment 4

1. Secure a double-pithed grass frog on its back to the bottom of a dissecting pan by placing pins through the feet.
2. Use scissors to make an incision through the body wall just to one side of the midline to avoid the large abdominal vein. Extend this incision anteriorly from the lower abdominal area to the sternum (breastbone). Make lateral cuts just posterior to the sternum and pin back the flaps.
3. Lift the sternum with forceps and carefully cut through its length. Avoid cutting the underlying blood vessels. Pin back the

"chest" wall laterally to expose the heart. The liver is also now exposed.
4. Keep the heart and liver wet with frog Ringer's solution throughout the experiment.
5. Record the heart rate (beats per minute). Count the beats for 15 sec and multiply by 4.
6. Fill two hypodermic syringes with 1:10,000 epinephrine and 1:10,000 acetylcholine, respectively, as directed by your instructor.
7. Slowly inject 0.05 ml of 1:10,000 epinephrine into the liver with the hypodermic syringe. To insert the needle easily, "firm up" the liver by holding it between thumb and forefinger. *Exactly 30 sec after the injection, record the heart rate.* The epinephrine is carried to the heart via the hepatic vein and posterior vena cava.
8. Wait 3 min and record the heart rate again.
9. In the same manner as before, slowly inject 0.05 ml of acetylcholine into the liver. Determine the time required for acetylcholine to take effect. Exactly 30 sec after the injection, record the heart rate. Wait 3 min and record the heart rate again.
10. Slowly inject 0.05 ml of 1:10,000 epinephrine into the liver. Determine how long it takes for an effect to be noticed. Again, record the heart rate 30 sec after the injection and 3 min later.
11. Slowly inject 0.25 ml of 1:10,000 epinephrine. How quickly does the effect take place? Record the heart rate after 30 sec and 3 min later.
12. Slowly inject 0.25 ml of 1:10,000 acetylcholine. How quickly does the effect take place? Record the heart rate after 30 sec and 3 min later.
13. **Complete items 4 and 5 on the laboratory report.**

NEURAL CONTROL

Neural coordination occurs in all but the simplest animals. For example, a nerve net coordinates movement in coelenterates, widely separated ventral nerve cords occur in flatworms, and a **nervous system** composed of (1) a fused, double, ventral nerve cord with segmental ganglia; (2) a "brain"; and (3) well-developed sensory receptors occurs in annelids and arthropods.

Vertebrates possess the most highly developed nervous system, and it may be subdivided into two major components. The **central nervous system (CNS)** is composed of a **dorsal tubular nerve cord** and a **brain**. The **peripheral nervous system (PNS)** is composed of **cranial nerves**, which emanate from the brain and extend to the head and certain internal organs,

and the **spinal nerves**, which extend from the spinal cord to all parts of the body except the head. Of course, well-developed **sensory receptors** are evident.

In this exercise, you will investigate the structure and function of the vertebrate nervous system as found in mammals. The nervous system provides rapid coordination and control of body functions by transmitting **impulses** over neuron processes. The impulses may originate in either the brain or **receptors**, and they are carried to **effectors** (muscles and glands) where the action occurs.

NEURONS

Neural tissue is composed of two basic types of cells. **Neurons**, or nerve cells, transmit impulses. Neuroglial cells provide structural support and prevent contact of neurons except at certain sites. Although neurons may be specialized in structure and function in various parts of the nervous system, they have many features in common. Figure 23.1 shows the basic structure of neurons, and Table 23.1 identifies the characteristics of the three functional types of neurons.

The **cell body** is an enlarged portion of the neuron that contains the nucleus. Two types of neuron processes or fibers extend from the cell body. **Dendrites** receive impulses from receptors or other neurons and carry impulses *toward* the cell body. **Axons** carry impulses *away from* the cell body. A neuron may have many dendrites, but only one axon is present.

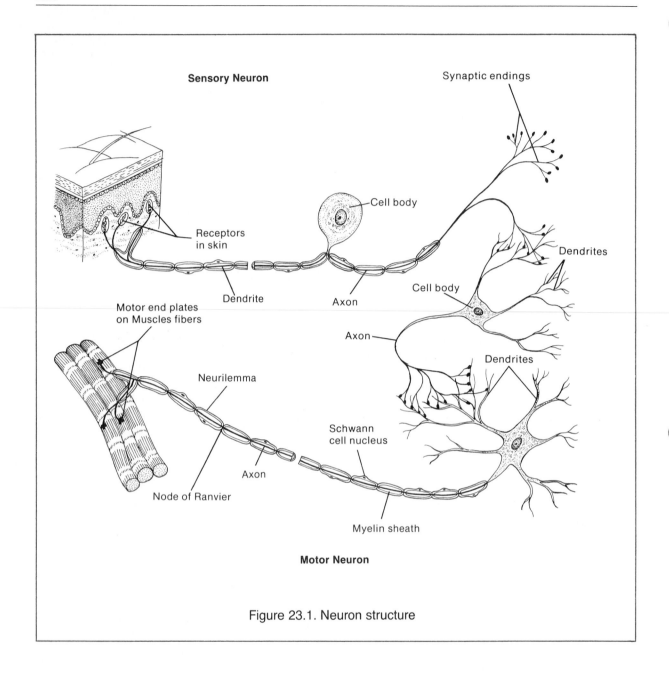

Sensory Neuron

Synaptic endings

Cell body

Receptors
in skin

Dendrite

Axon

Dendrites

Cell body

Axon

Dendrites

Motor end plates
on Muscles fibers

Neurilemma

Schwann
cell nucleus

Axon

Node of Ranvier

Myelin sheath

Motor Neuron

Figure 23.1. Neuron structure

TABLE 23.1
Functional Types of Neurons

Neuron Type	Structure	Function
Sensory	Long dendrite, short axon	Carry impulses from receptors to the CNS
Interneuron	Short dendrites, short or long axon	Carry impulses within the CNS
Motor	Short dendrites, long axon	Carry impulses from the CNS to effectors

Neuron processes of the peripheral nervous system are enclosed by a covering of **Schwann cells**. In larger processes, the multiple wrappings of Schwann cells form an inner **myelin sheath**, a fatty, insulating material, while the outer layer constitutes the **neurilemma**. The minute spaces between Schwann cells, where the neuron process is exposed, are called **nodes of Ranvier**. Impulses are transmitted more rapidly by myelinated fibers than by unmyelinated fibers. Schwann cells provide a pathway for the regeneration of neuron processes and are essential for regrowth. Schwann cells are absent in the central nervous system, but certain neuroglial cells form the myelin sheath of CNS neurons.

The junction of an axon tip of one neuron and a dendrite or cell body of another neuron is called a **synapse**. See Figure 23.2. Impulses passing along the axon cause the release of a **neurotransmitter** from the axon tip, or **synaptic knob**, into the **synaptic cleft**, the minute space between the neurons. The neurotransmitter binds with receptor sites on the postsynaptic membrane, producing either stimulation or inhibition of impulse formation, depending on the type of neurotransmitter involved. Immediately thereafter, an enzyme breaks down or inactivates the neurotransmitter, which prevents continuous stimulation or inhibition of the postsynaptic membrane. Transmission of neural impulses across synapses is always in one direction, axon to dendrite, because only axon tips can release neurotransmitters to activate the adjacent neuron.

Numerous substances are either known or suspected to be neurotransmitters. Among known neurotransmitters, **acetylcholine** and **norepinephrine** are stimulatory neurotransmitters, while **glycine** and **gamma aminobutyric acid (GABA)** are inhibitory neurotransmitters.

THE BRAIN

The **brain** is the control center of the nervous system. It is enclosed within the cranium and covered by the **meninges**, three layers of protective membranes. **Cerebrospinal fluid** within the meninges provides an additional cushion to absorb shocks. Twelve pairs of cranial nerves are attached to the brain. All but one pair innervate structures of the head and neck; vagus nerves innervate the internal organs. Refer to Figures 23.3 to 23.7 as you study the major parts of the brain.

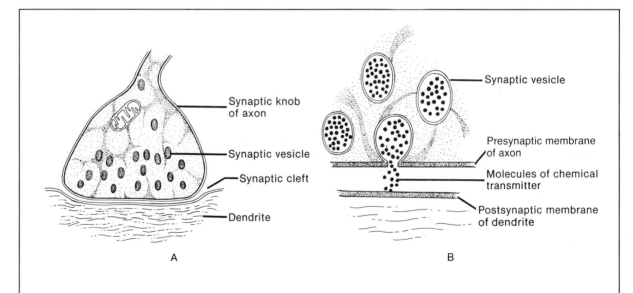

Figure 23.2. The synaptic junction.
A. Relationship of the synaptic knob of an axon to a dendrite. Synaptic vesicles migrate from the cell body to the synaptic knob. B. Synaptic transmission. An impulse moving down the axon causes a synaptic vesicle to release a chemical transmitter that either stimulates or inhibits the formation of an impulse in the dendrite.

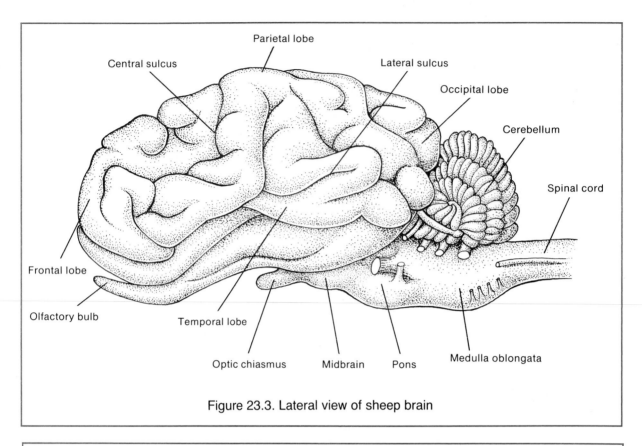

Parietal lobe

Central sulcus

Lateral sulcus

Occipital lobe

Cerebellum

Spinal cord

Frontal lobe

Olfactory bulb

Temporal lobe

Optic chiasmus

Midbrain

Pons

Medulla oblongata

Figure 23.3. Lateral view of sheep brain

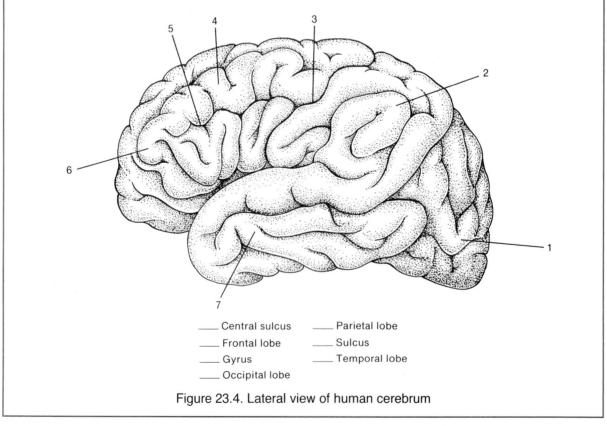

5 4 3

2

6

1

7

_____ Central sulcus   _____ Parietal lobe

_____ Frontal lobe   _____ Sulcus

_____ Gyrus   _____ Temporal lobe

_____ Occipital lobe

Figure 23.4. Lateral view of human cerebrum

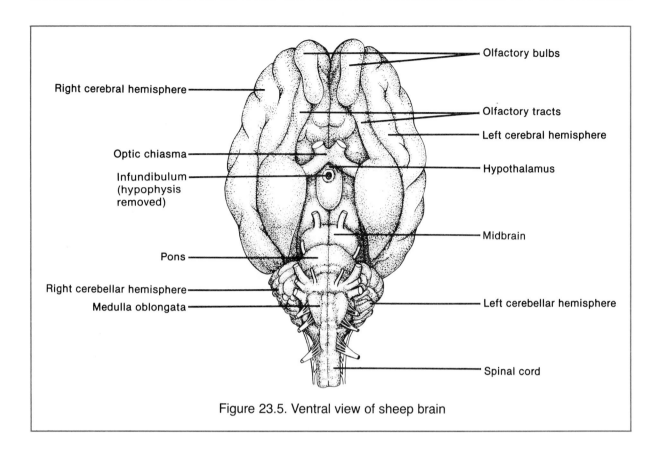

Figure 23.5. Ventral view of sheep brain

Cerebrum

The **cerebrum** is the largest part of the brain. It consists of right and left **cerebral hemispheres** that are separated by a median **longitudinal fissure**. A mass of neuron fibers, the **corpus callosum**, enables impulses to pass between the two hemispheres. The outer portion of the cerebrum, the **cerebral cortex** or **gray matter**, is composed of neuron cell bodies and unmyelinated neuron processes. Its surface area is increased by numerous **gyri** (ridges) and **sulci** (grooves). The cerebrum initiates voluntary actions and interprets sensations. In humans, it is the seat of will, memory, and intelligence.

Each hemisphere is divided into four lobes by fissures (deep grooves). Table 23.2 indicates the location and major functions of these lobes.

Cerebellum

The **cerebellum** lies just below and posterior to the occipital lobe of the cerebrum. It is divided into left and right hemispheres by a shallow fis-

sure. Muscle tone and muscular coordination are subconsciously controlled by the cerebellum.

Brain Stem

The **brain stem** is composed of a number of structures located between the cerebrum and the spinal cord. A midsagittal section is required to locate the various parts. A major function is the linking of higher and lower brain areas, but the individual portions also have some specific functions.

The **thalamus**, which is located at the upper end of the brain stem, consists of two lateral globular masses joined by an isthmus of tissue called the *intermediate mass*. It provides an uncritical awareness of sensations such as pain and pleasure and is a relay station between lower brain centers and the cerebrum.

The **hypothalamus** is located just below the thalamus. It plays a major role in homeostasis of the body by controlling things such as appetite, sleep, body temperature, and water balance. A major endocrine gland, the **hypophysis** or **pituitary gland**, is attached to the ventral

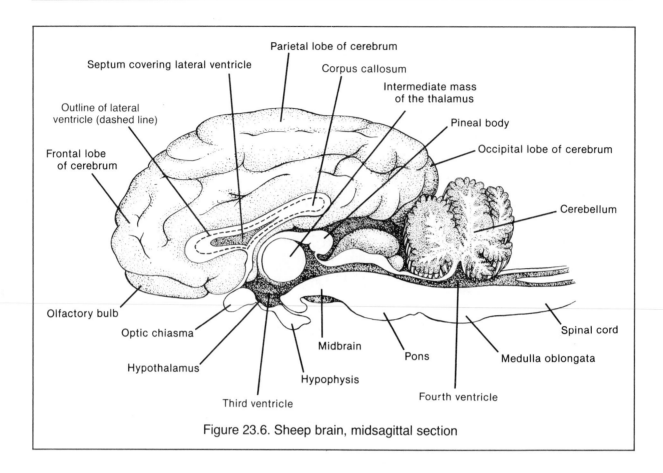

Figure 23.6. Sheep brain, midsagittal section

wall of the hypothalamus by a short stalk, the **infundibulum**.

The **midbrain** is a small area between the thalamus and pons that is associated with certain visual reflexes. It contains the **pineal body**.

The **pons** is a rounded bulge on the ventral side of the brain stem. It is an important connecting pathway for higher and lower brain centers.

The **medulla oblongata** lies between the pons and the spinal cord. It is the lowest part of the brain, and all impulses passing between the brain and spinal cord must pass through it. In addition, it controls heart rate, blood pressure, and breathing.

Materials

Colored pencils
Dissecting instruments and pans
Models of sheep and human brains
Sheep brains, preserved

Prepared slides of:
    giant multipolar neurons
    nerve, x.s.

Assignment 1

1. Add arrows to Figure 23.1 to indicate the direction of impulse transmission in motor and sensory neurons. Color-code the three types of neurons.
2. ***Complete items 1a and 1b on Laboratory Report 23 that begins on page 431.***
3. Examine a slide of giant multipolar neurons. Draw a neuron and label the cell body, nucleus, nucleolus, and neuron processes.
4. Examine a slide of nerve, x.s., and make a drawing of your observations. Note how the axons are arranged in bundles separated by connective tissue. Label an axon, myelin sheath, and connective tissue.
5. ***Complete item 1 on the laboratory report.***

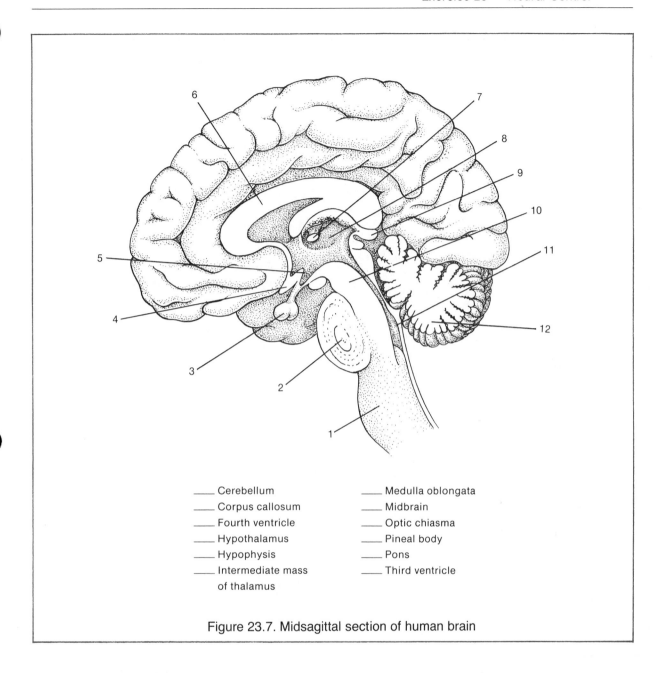

Figure 23.7. Midsagittal section of human brain

_____ Cerebellum

_____ Corpus callosum

_____ Fourth ventricle

_____ Hypothalamus

_____ Hypophysis

_____ Intermediate mass
of thalamus

_____ Medulla oblongata

_____ Midbrain

_____ Optic chiasma

_____ Pineal body

_____ Pons

_____ Third ventricle

TABLE 23.2
Location and Function of the Cerebral Lobes

Lobe	Location	Function
Frontal	Anterior to the central sulcus	Voluntary muscular movements; intellectual processes
Parietal	Between frontal and occipital lobes	Interprets sensations from skin; speech interpretation
Temporal	Inferior to frontal and parietal lobes	Hearing; interprets auditory sensations
Occipital	Posterior part of cerebrum	Vision; interprets visual sensations

Assignment 2

1. Study and color-code Figures 23.3, 23.5, and 23.6 until you are familiar with the structures of the sheep brain.
2. Label Figures 23.4 and 23.7 and locate these parts on a model of a human brain. Color-code the cerebral lobes in Figure 23.4 and the parts of the brain in Figure 23.7.
3. Obtain an entire sheep brain for study. Note the meninges, if present, and remove them with scissors. Locate the major parts of the brain shown in Figures 23.3 and 23.5. Observe the ridges and furrows that increase the surface area of the cerebrum. Is there an advantage in this?
4. Obtain half of a sheep brain that has been sectioned along the longitudinal fissure. Locate the structures shown in Figure 23.6. Note that the hypothalamus forms the floor of the third ventricle and that the hypophysis (pituitary gland) projects ventrally from it. The **pineal body** is a remnant of a third eye found in primitive reptiles. Its function in some mammals may be to control seasonal gonadal activity based on photoperiod variation.
5. Examine the demonstration coronal sections of a sheep brain and locate the two **lateral ventricles**. Cerebrospinal fluid is secreted from blood vessels in each of the four ventricles of the brain. It circulates through the ventricles and within the meninges covering the brain and spinal cord, and it is subsequently reabsorbed back into the blood.
6. **Complete item 2 on the laboratory report**.

THE SPINAL CORD

The **spinal cord** is located in the vertebral canal and is covered by the meninges like the brain. It serves as a pathway for impulses between the brain, different levels of the spinal cord, and the **spinal nerves**. In humans, there are 31 pairs of spinal nerves that lead to all parts of the body except the head. Each spinal nerve joins the spinal cord to form ventral and dorsal roots. The **ventral root** contains fibers of motor nerves only. The **dorsal root** contains fibers of sensory nerves only, and the cell bodies of the sensory neurons are located in the **dorsal root ganglion**. See Figure 23.8. Unlike the cerebrum, the gray matter of the spinal cord is located interiorly, while the white matter is found exteriorly. What causes the difference in appearance of gray and white matter?

Materials

Chart or model of the spinal cord
Prepared slide of cat spinal cord, x.s.

Assignment 3

1. Label Figure 23.8.
2. Examine a prepared slide of cat spinal cord, x.s., at $40\times$ and observe the gross features. Compare your slide with Figure 23.8. Use $100\times$ to locate the cell bodies of neurons in the gray matter, and examine them at $400\times$. What do the neuron fibers look like in the white matter?
3. Add arrows to Figure 23.8 to show the path of impulses.
4. **Complete item 3 on the laboratory report**.

Reflexes

Reflexes are involuntary responses (require no conscious act) to specific stimuli, and they are important mechanisms for maintaining the well-being of the individual. For example, coughing and sneezing are respiratory reflexes controlled by the brain. Spinal reflexes do not involve the brain.

Simple reflexes require no more than three neurons to produce an action to a stimulus, and this is accomplished through a **reflex arc**. See Figure 23.8. A reflex arc consists of (1) a receptor that forms impulses on stimulation, (2) a sensory neuron that carries the impulses to the brain or spinal cord, (3) an interneuron that receives the impulses and transmits them to a motor neuron, (4) a motor neuron that carries impulses to an effector, and (5) an effector that performs the action.

Materials

Reflex hammer
Laboratory lamp
Penlight
Dissecting instruments and pan

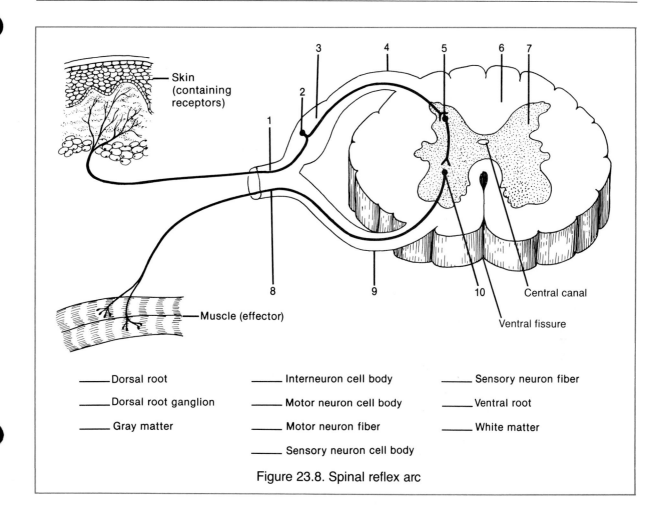

Figure 23.8. Spinal reflex arc

_____ Dorsal root

_____ Dorsal root ganglion

_____ Gray matter

_____ Interneuron cell body

_____ Motor neuron cell body

_____ Motor neuron fiber

_____ Sensory neuron cell body

_____ Sensory neuron fiber

_____ Ventral root

_____ White matter

Filter paper
Wire, thin
Acetic acid, 2%
Frog, single pithed

Patellar Reflex

Physicians use reflex tests to assess the condition of the nervous system. The **patellar reflex** is one commonly used. When the patellar tendon is struck just below the kneecap with a reflex hammer, the reflex action is a slight, instantaneous contraction of the large muscle (quadriceps femoris) on the front of the thigh that extends the lower leg. Striking the patellar tendon causes the muscle to be slightly stretched for an instant. This stretching causes impulses to be formed and carried along a sensory neuron to the spinal cord where they are passed to a motor neuron that carries the impulses to the muscle, thereby causing a weak,

brief contraction. Work in pairs to perform the reflex test.

Photopupil Reflex

This reflex enables a rapid adjustment of the size of the pupil to the existing light intensity, and it is coordinated by the brain. It is most easily observed in persons with light-colored eyes. When a bright light stimulates the retina of the eye, impulses are carried to the brain by sensory neurons. In the brain, the impulses are transmitted to interneurons and on to motor neurons that carry impulses to the muscles of the iris, causing them to contract. Contraction of the iris muscles decreases the size of the pupil and controls the amount of light entering the eye.

Assignment 4

1. Perform the patellar reflex as follows.
   a. The subject should sit on the edge of a

table with the legs hanging over the edge but not touching the floor.

b. Strike the patellar tendon (see Figure 23.9) with the small end of the reflex hammer and observe the response.

c. Divert the subject's attention by having the subject interlock the fingers of both hands and pull the hands against each other while you strike the patellar tendon again. Is the response the same as before? If not, how do you explain it?

d. Test both legs and record the results.

e. Add arrows to Figure 23.9 to show the path of impulses.

f. ***Complete items 4a and 4b on the laboratory report.***

2. Perform the photopupil reflex as follows.

a. Have the subject sit with eyes closed facing a darkened part of the room for 1 to 2 min. When the subject opens his or her eyes, note any change in pupil size.

b. While the subject is looking into the darkened area, shine a desk lamp in his or her eyes (from about 3 ft away). Note any change in pupil size.

c. To observe the effect of unilateral stimulation, have the subject look into the darkened area and, while observing the right eye, shine a penlight in the left eye. Note any response. Then observe the left eye while stimulating the right eye.

d. ***Complete item 4 on the laboratory report***.

Spinal Reflex in a Frog

Observe a spinal reflex in a demonstration performed by your instructor using a pithed frog, a frog whose brain has been destroyed so that any reaction to stimuli involves only the spinal cord. The frog is brain dead. The demonstration is performed as follows.

Assignment 5

1. Insert a wire through the anterior region of both jaws and suspend the frog by the wire from a ring on a ring stand. This allows the legs to hang free.

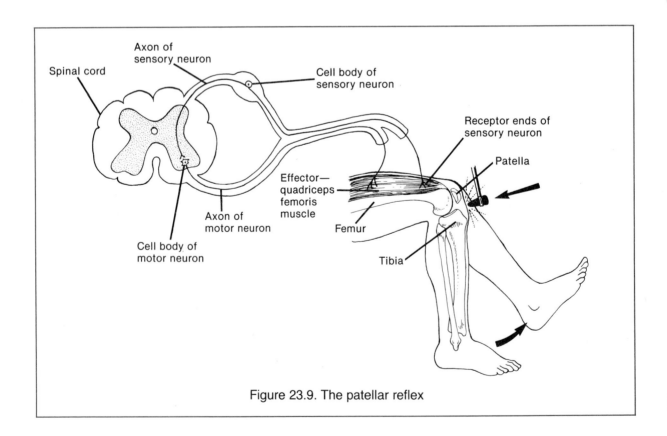

Figure 23.9. The patellar reflex

2. Dip a small piece of filter paper into 2% acetic acid, a mild irritant, and hold it against the calf of the right leg. Note the response. Immediately rinse the leg with water. Which leg responded? Record your observations in item 5a on the laboratory report.

3. Place the filter paper with acetic acid on the calf of the left leg. Immediately rinse the leg with water. Record your observations.

4. Remove the frog from the support and place it, dorsal side up, in a wax-bottomed dissecting pan. Place a pin through each foot. Slit open the skin on the dorsal surface of the right thigh; separate the muscles to locate the **sciatic nerve**, the large nerve innervating the leg. Be careful not to damage adjacent blood vessels. See Figure 23.10. Cut the nerve and resuspend the frog.

5. Apply the acetic acid as before to the lower right leg. What happens? Rinse the leg and repeat on the lower left leg. Rinse the left

Figure 23.10. Separate the thigh muscles to expose the sciatic nerve

leg. Record your observations in item 5 on the laboratory report.

6. *Complete item 5 on the laboratory report*.

24

SENSORY PERCEPTION
IN HUMANS

Sensations result from the interaction of three
components of the nervous system.
1. **Sensory receptors** generate impulses
   upon stimulation.
2. **Sensory neurons** carry the impulses to
   the brain or spinal cord, and **interneurons** carry the impulses to the sensory in-
   terpretive centers in the brain.
3. The **brain** interprets the impulses as sensations.
The type of sensation is determined by the
part of the brain receiving the impulses rather
than by the type of receptors being stimulated.
For example, the auditory center interprets all
impulses it receives as sound stimuli regardless
of their origin.

In this exercise, you will study the structure
and function of the eye and the ear and perform
tests that will demonstrate certain characteristics of sensory perception.

THE EYE

The eyes contain the receptors for light stimuli, and they are well protected by the surrounding skull bones and the eyelids. The eyelids and
the anterior surface of the eye are covered by a
mucous membrane, the **conjunctiva**, which
contains many blood vessels and pain receptors
except where it covers the cornea. The **lacrimal
gland**, located over the upper, lateral portion of
the eye, produces tears that cleanse the surface
of the conjunctiva and keep it moist.

Structure of the Eye

Refer to Figure 24.1 as you study this section.
The eye is a hollow ball, roughly spherical in
shape. Its wall is composed of three distinct
layers.

The outer layer is composed of (1) the fibrous
sclera, which forms the white portion of the eye,
and (2) the transparent **cornea**, which forms
the anterior bulge where light enters the eye.

The middle layer consists of three parts. The
choroid coat, a black layer, absorbs excess light
that passes through the retina and contains the
blood vessels that nourish the eye. Anteriorly,
the **ciliary body** forms a ring of muscle that
controls the shape of the **lens**, which is suspended from the ciliary body by the **suspensory
ligaments**. The **iris** controls the amount of light
entering the eye by controlling the size of the
pupil, the opening in the center of the iris.

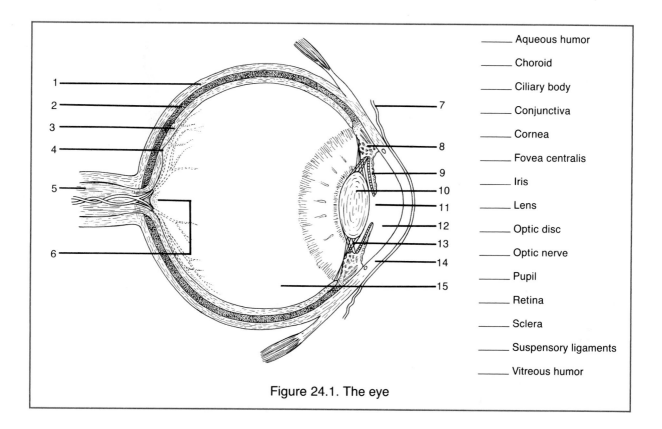

Figure 24.1. The eye

The **retina** composes the inner layer of the wall of the eye. It contains the photoreceptors: **rods** for black and white vision and **cones** for color vision. The neuron fibers coalesce at the **optic disc** where they enter the **optic nerve**, which carries impulses to the brain. The optic disc is known as the blind spot since it has no receptors. A tiny depression just lateral to the optic disc, the **fovea centralis**, contains densely packed cones for sharp, direct vision.

The interior of the eye behind the lens is filled with **vitreous humor**, a transparent, jellylike substance that helps hold the retina in place and gives shape to the eye. A watery fluid, the **aqueous humor**, fills the space between the lens and cornea.

Assignment 1

1. Label Figure 24.1 and color-code the three layers of the eye.
2. Locate the parts of the eye labeled in Figure 24.1 on an eye model or chart.
3. ***Complete item 1 on Laboratory Report 24 that begins on page 435.***

Function of the Eye

Light waves reflecting from objects are bent as they pass through the cornea. They continue through the pupil and lens, which focuses the light rays on the retina and accommodates for near and distance vision as its shape is changed by muscles in the ciliary body. Photoreceptors in the retina form impulses that are transmitted via the optic nerve to the visual center in the brain where they are interpreted as visual images.

Materials

Meterstick
Color-blindness test plates (Ishihara)
Astigmatism chart
Snellen eye chart

Assignment 2

Perform the following visual tests to learn more about visual sensations. Work in pairs. If you wear corrective lenses, perform the tests with and without them and note any differences in your results.

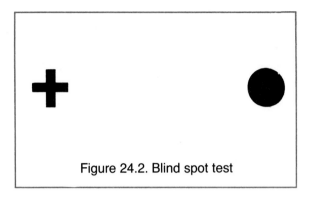

Figure 24.2. Blind spot test

TABLE 24.1
Age and Accommodation

| Age | Near Point | |
	Inches	Centimeters
10	3.0	7.5
20	3.5	8.8
30	4.5	11.3
40	6.8	17.0
50	20.7	51.8
60	33.0	82.5

Blind Spot. No rods or cones are located at the junction of the retina and optic nerve. This site is known as the optic disc or blind spot. It can be located by using the following procedure.

1. Hold Figure 24.2 about 50 cm (20 in.) in front of your eyes.
2. Cover your left eye and focus with the right eye on the cross. You will be able to see the dot as well.
3. Slowly move the figure toward your eyes while focusing on the cross, until the dot disappears.
4. Have your partner measure and record the distance from your eye to the figure at the point where the dot disappears.
5. Test the left eye in a similar manner but focus on the dot and watch for the cross to disappear.
6. ***Complete item 2a on the laboratory report.***

Near Point. The shortest distance from your eye that an object is in sharp focus is called the near point. The closer the distance, the more elastic the lens and the greater the eye's ability to accommodate for changes in distance. Elasticity of the lens is greatest in infants, and it gradually decreases with age. See Table 24.1. Accommodation is minimal after 60 yr of age, a condition called presbyopia. How does this relate to the common usage of bifocal lenses by older persons?

1. Hold this page in front of you at arm's length. Close one eye, focus on a word in this sentence, and slowly move the page toward your face until the image is blurred. Then move the page away until the image is sharp. Have your partner measure the distance between your eye and the page.

2. Test the other eye in the same manner.
3. ***Complete item 2b on the laboratory report.***

Astigmatism. This condition results from an unequal curvature of either the cornea or the lens, which prevents the light rays from being focused sharply on the retina.

1. Cover one eye and focus on the circle in the center of Figure 24.3. If the radiating lines appear equally dark and in sharp focus, no astigmatism exists. If astigmatism exists, record the number of the lines that appear lighter in color or blurred.
2. Test the other eye in the same manner.
3. ***Complete item 2c on the laboratory report.***

Acuity. Visual acuity refers to the ability to distinguish objects in accordance to a standardized scale. It may be measured using a Snellen eye chart. If you can read the letters that are designated to be read at 20 ft at a distance of 20 ft, you have 20/20 vision. If the smallest letters that can be read at 20 ft are those designated to be read at 30 ft, you have 20/30 vision.

1. Stand 20 ft from the Snellen eye chart on the wall while your partner stands next to the chart and points out the lines to read.
2. Cover one eye and read the lines as requested. Record the rating of the smallest letters read correctly.
3. Test the other eye in the same manner.
4. ***Complete item 2d on the laboratory report.***

Color Blindness. The sensation of color vision depends on the degree to which impulses are formed by the three types of cones (receptors

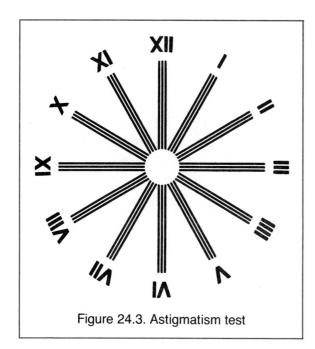

Figure 24.3. Astigmatism test

for red, green, and blue light) in the retina. The most common type of color blindness is red-green color blindness, which is caused by a deficit in cones stimulated by either red or green light. People with such a deficit have difficulty distinguishing reds and greens. A totally color-blind person sees everything as shades of gray.

1. Your partner holds the color-blindness test plates about 30 in. from your eyes in good light and allows you 5 sec to view each plate and give your response. Your partner records your responses in item 2e on your laboratory report.

2. *Complete item 2e on the laboratory report*.

THE EAR

The ear contains not only receptors for sound stimuli but also receptors involved in maintaining equilibrium. For ease of study, the ear may be subdivided into the external ear, middle ear, and internal ear. Refer to Figure 24.4 as you study this section.

External Ear

The **external ear** includes the **auricle** or **pinna**, the flap of cartilage and skin commonly

called the "ear," and the **auditory canal** that leads inward through the temporal bone to the **tympanic membrane**, or eardrum.

Middle Ear

The **middle ear**, a small cavity in the temporal bone, is connected to the pharynx by the **eustachian tube**. It is filled with air that enters or leaves via the eustachian tube depending on the air pressure at each end of the tube. Three small bones, the **ear ossicles**, form a lever system that extends from the eardrum to the inner ear. In sequence, they are (1) the **malleus** (hammer), which is attached to the ear drum, (2) the **incus** (anvil), and (3) the **stapes** (stirrup), which is inserted into the **oval window** of the inner ear.

Inner Ear

The **inner ear** consists of a complex of interconnecting tubes and chambers that are embedded in the temporal bone and that are filled with fluid. It is subdivided into three major parts.

1. The **cochlea** is coiled like a snail shell and contains the receptors for sound stimuli.
2. The **vestibule** is the enlarged portion at the base of the cochlea. The stirrup is inserted into the oval window of the vestibule, and the **round window** is covered by a thin membrane. These two windows are involved in the transmission of sound stimuli. In addition, the vestibule contains receptors for static equilibrium that inform the brain of the position of the head.
3. The three **semicircular canals** contain receptors for dynamic equilibrium that inform the brain when the head is turned or when the entire body is rotated.

The Hearing Process

The auricle channels sound waves into the auditory canal, and as they strike the eardrum, it vibrates at their frequency. This vibration causes a comparable movement of the ear ossicles (hammer, anvil, and stirrup), which, in turn, transfer the movements to the fluid in the vestibule and cochlea of the inner ear. The movement of the fluid is enabled by the thin membrane of the round window, which moves *out and in* in synchrony with the *in and out* movemen'

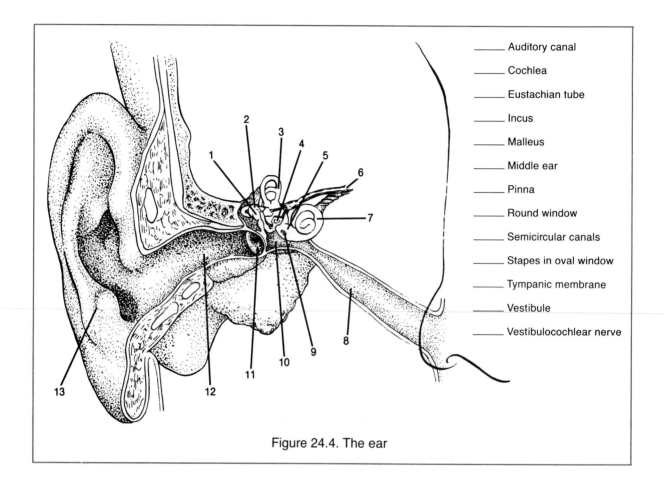

_____ Auditory canal

_____ Cochlea

_____ Eustachian tube

_____ Incus

_____ Malleus

_____ Middle ear

_____ Pinna

_____ Round window

_____ Semicircular canals

_____ Stapes in oval window

_____ Tympanic membrane

_____ Vestibule

_____ Vestibulocochlear nerve

Figure 24.4. The ear

of the stirrup in the oval window since the fluid cannot be compressed. The movement of the fluid in the cochlea stimulates the sound receptors, which send impulses to the auditory center in the brain where they are interpreted as sound sensations. Tone or pitch is determined by which receptors are stimulated and which part of the brain receives the impulses. Loudness is determined by the frequency of impulses formed and transmitted, which, in turn, depends on the intensity of the vibrations.

There are two kinds of hearing loss. **Conduction deafness** results from damage that prevents sound vibrations from reaching the inner ear. It is usually correctable by surgery or hearing aids. **Nerve deafness** is caused by damage to the sound receptors or neurons that transmit impulses to the brain. Nerve deafness usually results from exposure to loud sounds and is not correctable.

Assignment 3

1. Label and color-code Figure 24.4.
2. Locate the structures shown in Figure 24.4 on the ear model or chart.
3. ***Complete item 3 on the laboratory report***.

Materials

Pocket watch, spring wound
Cotton
Meterstick
Tuning fork

Assignment 4

Watch-Tick Test. This is a simple test to detect hearing loss at a single sound frequency. It requires a quiet area. Work in teams of three students. One student is the subject, another moves the watch, and the third measures and records distances.

1. Have the subject sit in a chair and plug one ear with cotton. The subject is to look straight ahead and indicate by hand signals when the first and last ticks are heard.
2. Start with the watch about 3 ft laterally from the ear being tested and move it *slowly* toward the ear until the subject indicates the first tick is heard. Measure and record the distance from ear to watch.
3. Start with the watch close to the ear and *slowly* move it away from the ear until the last tick is heard. Measure and record the distance from ear to watch.
4. Calculate the average of the two measurements.
5. Test the hearing of the other ear in the same manner.
6. ***Complete item 4a on the laboratory report.***

Rinne Test. A tuning fork is used in this test, which distinguishes between nerve and conduction deafness *where some hearing loss exists*. Work in pairs.

1. Have the subject sit in a chair and plug one ear with cotton. The subject is to indicate by hand signals when the sound is heard or not heard.
2. Strike the tuning fork against the heel of your hand to set it in motion. *Never strike it against a hard object.*
3. Hold the tuning fork 6 to 9 in. away from the ear being tested with the edge of the tuning fork toward the ear as shown in Figure 24.5.
4. The sound will be heard initially by persons with normal hearing and those with minimal hearing loss. As the sound fades, a point will be reached at which it will no longer be heard. (The subject is to indicate this point by a hand signal.) When this occurs, place the end of the tuning fork against the temporal bone behind the ear. See Figure 24.5. *Where a slight hearing loss exists* and the sound reappears, some conduction deafness is present.
5. Persons with a severe hearing loss will not hear the sound or will hear it only briefly.

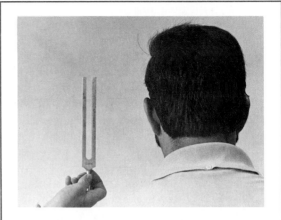

1. Hold a vibrating tuning fork 6 to 9 in. from the ear with the edge of the tuning fork toward the ear.

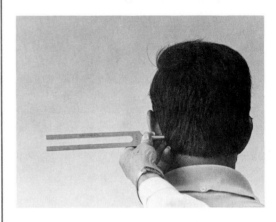

2. When the sound is no longer heard, place the end of the tuning fork against the temporal bone behind the ear.

Figure 24.5. Rinne test

If the sound reappears when the end of the tuning fork is placed against the temporal bone, conduction deafness is evident. If it does not reappear, nerve deafness exists.

6. ***Complete item 4b on the laboratory report.***

Static Balance. The inner ear is not the only receptor involved in the maintenance of equilibrium. Pressure and touch receptors in the skin, stretch receptors in the muscles, and the eyes also are involved. The brain constantly receives impulses from these receptors and subconsciously initiates any necessary corrective motor actions. The following test is a simple way to observe how this interaction functions. Work in groups of two to four students.

1. Use a meterstick to draw a series of vertical lines on the chalkboard about 5 cm (2 in.) apart. Cover an area about 1 m wide. This will help you detect body movements.
2. Have the subject remove his or her shoes and stand in front of the lined area facing you.
3. With feet together and arms at the side, the subject is to try to stand perfectly still for 30 sec while you watch for any swaying motion. Record the degree of swaying motion as slight, moderate, or great.
4. Repeat the test with the subject's eyes closed. Record the degree of movement. Try it again to see if extending the arms laterally helps the subject to maintain balance. What happens when the subject stands on only one foot with the hands at the side and with the eyes closed?
5. ***Complete item 4 on the laboratory report***.

SKIN RECEPTORS

Human skin contains receptors for touch, pain, pressure, hot, and cold stimuli. The following tests will enable you to detect certain characteristics of sensations involving some of these receptors. Work in pairs to perform these experiments.

Materials

Beakers, 400 ml
Coins
Hot plate
Metric ruler
Celsius thermometer
Crushed ice
Dividers or scissors, pointed blades
Clock with second hand

Assignment 5

Distribution of Touch Receptors. For you to perceive two simultaneous stimuli as two touch sensations, the stimuli must be far enough apart to stimulate two touch receptors that are

separated by at least one unstimulated touch receptor. This characteristic can be used to determine the density of touch receptors in the skin.

1. With the subject's eyes closed, touch his or her skin with one or two points of the dividers. The subject reports the sensation as either "one" or "two." Start with the points of the dividers close together and gradually increase the distance between the points until the subject reports a two-point stimulus as a two-touch sensation about 75% of the time. Measure and record the distance between the tips of the dividers as the minimum distance evoking a two-point sensation.
2. Use the procedure in step 1 to determine the minimum distance giving a two-point sensation on the (1) inside of the forearm, (2) back of the neck, (3) palm of the hand, and (4) tip of the index finger.
3. ***Complete item 5a on the laboratory report***.

Adaptation to Stimuli. Your nervous system has the ability to ignore stimuli or impulses so that you are not constantly bombarded with insignificant sensations.

1. Have the subject rest a forearm on the top of the table with the palm of the hand up.
2. With the subject's eyes closed, place a coin on the inner surface of the forearm. The subject is to indicate awareness of the presence of the coin and also the instant the sensation disappears. Record the time between these two events as the adaptation time.
3. Repeat the test using several coins stacked up to make a heavier object that increases the intensity of the stimulus. Determine the adaptation time.
4. ***Complete item 5b on the laboratory report***.

Intensity of Sensations. The intensity of a sensation usually is proportional to the intensity of the stimulus. This occurs because the receptors form more impulses when the strength of the stimulus is increased, and the brain interprets the arrival of more impulses as a greater sensation.

Experiment 1

1. Fill three 400-ml beakers with ice water, tap water, and warm water (about 50° C), respectively.
2. Place your index finger in the water in each beaker, in sequence, and note the sensations. Can you recognize the temperature differences?
3. Use some small beakers to prepare water with slight differences in temperature to determine the smallest difference that is detectable. Record this differential.
4. Now return to the original three beakers. Place your index finger in the ice water and note the sensation. Then immerse your whole hand and note the sensation. Repeat this procedure with the warm water.

5. *Complete items 5c and 5d on the laboratory report.*

Experiment 2

1. Use the three beakers of ice water, tap water, and warm water as in the previous experiment. Be sure that the warm water has not cooled.
2. Place one hand in ice water and the other in warm water. Note the sensation. Does it change with time? Explain.
3. After 2 min, place both hands in the beaker of tap water. Note the sensation. Does it change with time?
4. *Complete item 5 on the laboratory report.*

25

THE SKELETAL SYSTEM

Two basic types of rigid skeletons are found in
animals. An **exoskeleton** is formed exterior to
the body surface; it is the most common type of
skeleton in invertebrates. An **endoskeleton** is
formed within the body of the animal; it occurs almost exclusively in vertebrates. Some
invertebrates, such as earthworms, have a
hydraulic skeleton that gives shape to the organism and is involved in body movements. It
consists of the interaction of noncompressible
body fluids and muscles in the body wall. This
exercise will focus on rigid skeletons, especially
the endoskeletons of vertebrates.

EXOSKELETONS AND ENDOSKELETONS

Exoskeletons of some invertebrates provide
only support and protection as in the calcareous
skeletons of corals and many mollusks. The chitinous exoskeleton of arthropods performs these
functions and also provides sites for muscle attachment that enables locomotion and gives
shape to the animal. Terrestrial arthropods secrete a waxy coating over the exoskeleton that
retards evaporative water loss from internal
tissues.

The arthropod exoskeleton is somewhat like a
coat of armor. It is flexible at the joints to allow
movement but rigid and hard between joints due
to the accumulation of mineral salts. The exoskeleton must be shed periodically by molting to
allow growth of the internal tissues.

With the exception of echinoderms, all endoskeletons occur in vertebrates. Echinoderms
form calcareous plates and spines in the body
wall that are covered by a thin epidermis. The
endoskeleton of vertebrates is usually formed of
bone, but lampreys, sharks, and rays have cartilaginous endoskeletons, a condition that is
usually considered to be degenerate. The bony
vertebrate skeleton provides (1) protection for
vital organs, (2) support for the body, (3) sites for
muscle attachment, (4) a storehouse for certain
minerals, and (5) formation of blood cells. The
motility of vertebrates depends on the interaction of the skeleton and skeletal muscles.

A vertebrate skeleton usually consists of (1)
numerous **bones**, (2) **ligaments** that hold the
bones together, and (3) associated **cartilages**.

Materials

Endoskeletons of echinoderms and vertebrate
classes
Exoskeletons of sponges, corals, mollusks, and
arthropods
Crayfish, preserved
Dissecting instruments
Dissecting pan

Assignment 1

1. Examine the exoskeletons of representative invertebrates. Note the general structure and function provided. In bivalve
mollusks, locate the scars at the point of attachment of the adductor muscles used to
close the shells. Do molluscan shells grow
at the outer edge?
2. Examine the exoskeleton of a preserved
crayfish. Note the flexibility at the joints
and the hardness elsewhere. Make a transverse cut between the segments of the abdomen. Examine the cut surface and note
how tightly the exoskeleton envelops the
body. Can you locate the epidermis just inside the exoskeleton? The epidermis secretes the exoskeleton.
3. Cut off a cheliped and use scissors to make
lengthwise cuts through the exoskeleton
extending across one or two joints. Spread
the exoskeleton and observe the way muscles are attached to the interior of the
skeleton.
4. Examine the demonstration endoskeletons. Note their general differences and
similarities, especially among the vertebrate classes.
5. ***Complete item 1 on Laboratory Report
25 that begins on page 439.***

MACROSCOPIC BONE STRUCTURE

Figure 25.1 depicts the structure of a typical
long bone, a human humerus. A long bone is
characterized by a shaft of bone, the **diaphysis**,
which extends between two enlarged portions
forming the ends of the bone, the **epiphyses**.
The articular surface of each epiphysis is covered by an **articular cartilage** (hyaline cartilage), which reduces friction in the joints and
protects the ends of the bone. The rest of the
bone is covered by the **periosteum**, a tough,
tightly adhering membrane containing tiny
blood vessels that penetrate into the bone. Bone
deposition by the periosteum contributes to the
growth in diameter of the bone. Larger blood
vessels and nerves enter the bone through a
channel called a **foramen**.

A longitudinal section of the bone reveals that
the epiphyses consist of **cancellous** (spongy)
bone covered by a thin layer of **compact bone**,
while the diaphysis is formed of heavy, compact
bone. **Red marrow** fills the spaces in the
spongy bone. It forms red and white blood cells.
The **medullary cavity** is filled with the fatty
yellow marrow. In immature bones, an **epiphyseal disc** of cartilage is located between the
diaphysis and each epiphysis; this is the site of
linear growth. A mature bone lacks this cartilage since it has been replaced by bone, and only
an **epiphyseal line** of fusion remains.

Materials

Colored pencils
Beef femur or tibia, fresh and split
Human femur, split
Dissecting instruments
Dissecting tray

Assignment 2

1. Label and color-code Figure 25.1.
2. Examine a fresh beef bone that has been
split. Locate the parts labeled in Figure
25.1. Feel the surface of the articular cartilage. How would you describe it? Is the
bone supplied with blood? Locate the stubs
of ligaments near the joint. Are they elastic, pliable, and easy to cut?
3. ***Complete item 2 on the laboratory report.***

MICROSCOPIC STUDY

Bones, ligaments, articular cartilages, and intervertebral discs are composed of **connective
tissues**. Such tissues are characterized by an
abundant intercellular material called **matrix**
and relatively few scattered cells. See Figures
25.2 through 25.5.

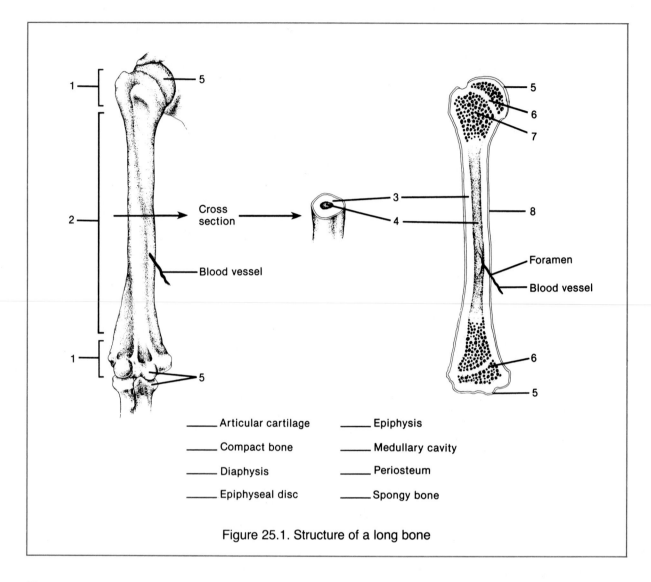

_____ Articular cartilage   _____ Epiphysis

_____ Compact bone   _____ Medullary cavity

_____ Diaphysis   _____ Periosteum

_____ Epiphyseal disc   _____ Spongy bone

Figure 25.1. Structure of a long bone

Bone

In a living bone, the **osteocytes** (bone cells) are trapped in **lacunae** (tiny spaces) in the **matrix** due to the deposition of calcium salts around them. The cells and lacunae are arranged in concentric circles around **Haversian canals** containing blood vessels and nerves. Concentric rings of bone called **lamellae** are formed between the rings of bone cells. Tiny channels between the lacunae, the **canaliculi**, provide passageways for materials to move between the cells.

Hyaline Cartilage

The articular cartilages are formed of hyaline cartilage. It is characterized by the absence of

fibers and the clear, homogeneous nature of the matrix that gives a smooth, glassy appearance to articular cartilages.

Fibrocartilage

The intervertebral discs are formed of fibrocartilage, which contains many nonelastic white fibers in the matrix that account for its characteristic toughness and resiliency.

Dense Fibrous Connective Tissue

Ligaments and tendons are formed by nonelastic, pliable, and tough connective tissue. The many white fibers give its characteristic coloration.

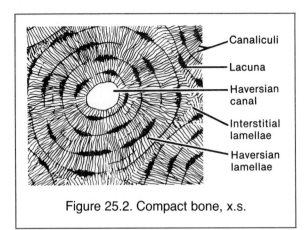

Figure 25.2. Compact bone, x.s.

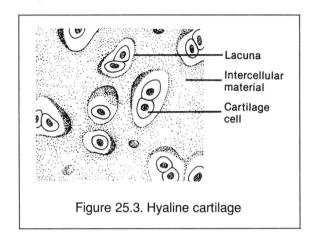

Figure 25.3. Hyaline cartilage

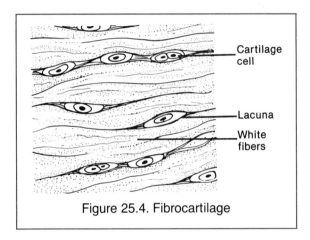

Figure 25.4. Fibrocartilage

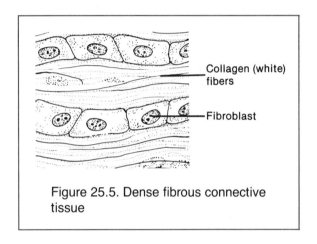

Figure 25.5. Dense fibrous connective tissue

Materials

Prepared slides of:
   compact bone, x.s., ground section
   hyaline cartilage
   fibrocartilage
   dense fibrous connective tissue

Assignment 3

1. Examine prepared slides of compact bone, hyaline cartilage, fibrocartilage, and dense fibrous connective tissue. Locate the structures shown in Figures 25.2 through 25.5.
2. ***Complete item 3 on the laboratory report***.

THE HUMAN SKELETON

The skeleton consists of two major subdivisions. The **axial skeleton** is composed of the skull, vertebral column, ribs, and sternum. The **appendicular skeleton** consists of the bones of the upper extremities and the pectoral girdle and the lower extremities and the pelvic girdle. See Figure 25.6.

Axial Skeleton

The **skull** is composed of the **cranium** (8 fused bones encasing the brain), 13 fused **facial bones**, and the movable **mandible** (lower jaw).

The **vertebral column** consists of vertebrae separated by **intervertebral discs** composed of fibrocartilage. Vertebrae are subdivided as follows:

Cervical vertebrae: 7 vertebrae of the neck
Thoracic vertebrae: 12 vertebrae of the thorax to which ribs are attached
Lumbar vertebrae: 5 large vertebrae of the lower back
Sacrum: bone formed of 5 fused vertebrae

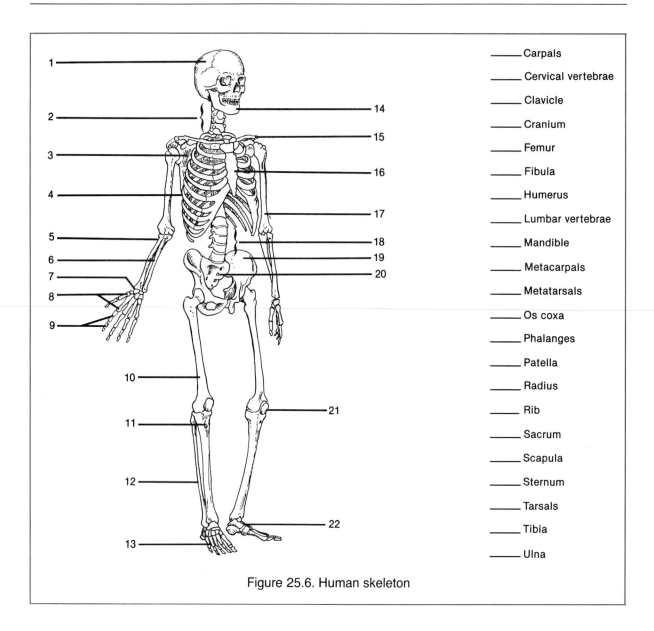

Figure 25.6. Human skeleton

1 _____ Carpals
2 _____ Cervical vertebrae
3 _____ Clavicle
4 _____ Cranium
5 _____ Femur
6 _____ Fibula
7 _____ Humerus
8 _____ Lumbar vertebrae
9 _____ Mandible
10 _____ Metacarpals
11 _____ Metatarsals
12 _____ Os coxa
13 _____ Phalanges
14 _____ Patella
15 _____ Radius
16 _____ Rib
17 _____ Sacrum
18 _____ Scapula
19 _____ Sternum
20 _____ Tarsals
21 _____ Tibia
22 _____ Ulna

Coccyx: tailbone formed of 3 to 5 fused, rudimentary vertebrae

There are 12 pairs of **ribs**. The first 10 pairs are joined to the **sternum** (breastbone) by costal cartilages to form the thoracic cage. The last 2 pairs are short and unattached anteriorly. They are called floating ribs.

Appendicular Skeleton

The **pectoral girdle** supports the upper extremities. It consists of a **clavicle** (collarbone) and **scapula** (shoulder blade) on each side of the body. The clavicle is attached to the sternum on one end and the scapula on the other. The scapula is supported by muscles that allow mobility for the shoulder.

The **humerus** (upper arm bone) articulates with the scapula at the shoulder and the **ulna** and **radius** at the elbow. The wrist is composed of eight **carpal bones** that lie between the (1) ulna and radius and (2) **metacarpals**, the bones of the hand. The **phalanges** are the bones of the fingers and thumb.

The **pelvic girdle** consists of two **os coxae** (hipbones) that join together anteriorly at the pubic symphysis and are fused posteriorly to the sacrum. This provides a sturdy support for the lower extremities. Each os coxa consists of three fused bones: **ilium**, **ischium**, and **pubis**.

The **femur** (thighbone) articulates with the os coxa at the hip and with the **tibia** (shinbone) at the knee. The **patella** (kneecap) is embedded in the tendon anterior to the knee joint. The smaller bone of the lower leg is the **fibula**. Both tibia and fibula articulate with the **tarsal bones** forming the ankle and posterior part of the foot. The anterior foot bones are five **metatarsals**, and the toe bones are the **phalanges**.

Articulations

The bones of the skeleton are attached to each other by ligaments in such a way that varying degrees of movement occur at the joints. Articulations are categorized according to the degree of movement that is possible.
1. **Immovable joints** are rigid, such as those that occur between the skull bones.
2. **Slightly movable joints** allow a little movement such as those between vertebrae and at the pubic symphysis.
3. **Freely movable joints** are the most common and allow the broad range of movement noted in the arms and legs. There are several types:
   a. **Hinge joints** allow movement in one direction only.
   b. **Ball-and-socket joints** allow angular movement in all directions plus rotation.
   c. **Gliding joints** occur where bones slide over each other.
   d. **Pivot joints** enable rotation in only one axis.

Materials

Colored pencils
Articulated human skeleton
Male and female pelvic girdles
Calipers or dividers
Human fetal skeleton
Metric ruler

Assignment 4

1. Label Figure 25.6 and color-code the axial skeleton.
2. Identify the bones of an articulated human skeleton.
3. ***Complete item 4a and 4b on the laboratory report***.
4. Locate the various types of joints on the articulated skeleton.
5. ***Complete item 4 on the laboratory report***.

SEXUAL DIFFERENCES OF THE PELVIS

The structure of the pelvic girdle is different in males and females, primarily because the female pelvis is adapted for childbirth. Table 25.1 lists some of the major differences between male and female pelvic girdles. Compare these characteristics with Figure 25.7. Since the characteristics of a pelvis may have considerable variations from the ideal, sex determination of a pelvis is based on a combination of characteristics rather than on a single characteristic.

TABLE 25.1
Comparison of the Male and Female Pelvis

Characteristic	Male	Female
General structure	Not tilted forward; narrower and longer; heavier bones	Tilted forward; broader and shorter; lighter bones
Acetabula	Larger and closer together	Smaller and farther apart
Pubic angle	Less than 90°	Greater than 90°
Sacrum	Longer and narrower	Shorter and wider
Coccyx	More curved; less movable	Straighter; more movable
Pelvic brim	Narrower and heart shaped	Wider and oval shaped
Ischial spines	Longer, sharper, closer together, and project more medially	Shorter, blunt, farther apart and project more posteriorly

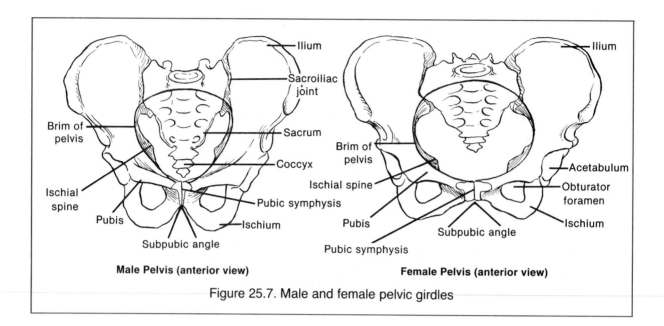

Figure 25.7. Male and female pelvic girdles

Calculation of a **pelvic ratio** is helpful in determining the sex of a pelvic girdle. Two measurements are required. The ratio is calculated by dividing the first measurement by the second.

1. The distance between the tips of the ischial spines
2. The distance between the inner surface of the pubic symphysis and the upper, inner surface of the sacrum

$$\text{Pelvic ratio} = \frac{\text{first measurement}}{\text{second measurement}}$$

Females usually have a ratio of 1.0 or more, while males usually have a ratio of 0.8 or less.

Assignment 5

1. Examine the male and female pelvic girdles provided. Verify the characteristics in Table 25.1 as you compare the male and female pelvic girdles.
2. Use calipers or dividers and metric rulers to make the two pelvic measurements. Then calculate the pelvic ratio for the male and female pelvic girdles.
3. Determine the pelvic ratio for the pelvis of the articulated skeleton available. Use the characteristics in Table 25.1 and the pelvic

ratio to determine the sex of the articulated skeleton.
4. ***Complete item 5 on the laboratory report***.

FETAL SKELETON

The skeleton of a newborn baby is composed of a large amount of cartilage. The cartilage is gradually replaced by bone as the individual grows. Skeletal growth is complete by age 25.

Assignment 6

1. Examine a human fetal skeleton. Note the cartilage, bones, and the relative size of head and body.
2. ***Complete item 6 on the laboratory report***.

SKELETAL ADAPTATIONS FOR LOCOMOTION

The skeletons of vertebrates are adapted for specific types of locomotion. In this section, you will examine the skeletons provided to correlate major skeletal features with the type of locomo-

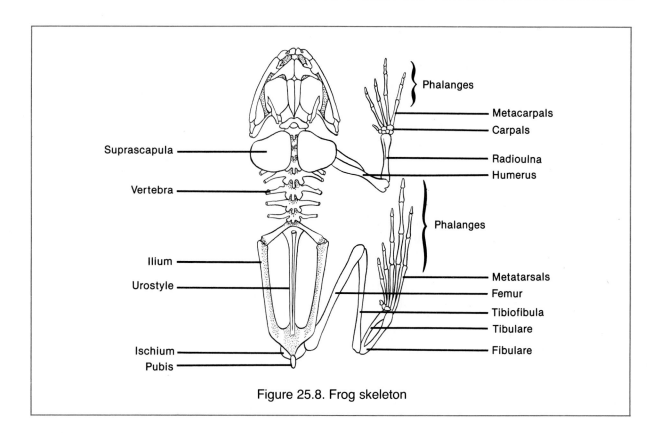

Figure 25.8. Frog skeleton

tion used by the animals. To assist you in getting started the skeletons of a frog and a chicken are shown in Figures 25.8 and 25.9, respectively. Correlate the brief discussions that follow with these figures before examining the skeletons.

The hind legs of a frog are adapted to provide extra leverage for jumping. The tibia and fibula have fused to form a single bone, while two "anklebones," the **fibulare** and **tibulare**, have elongated.

The bird skeleton shows a number of adaptations for flight. Bones are hollow and light in weight. The large sternum provides attachment for flight muscles. The "wrist" and "hand" bones have been reduced in the wing. The tibia and proximal tarsal bones have fused to form the **tibiotarsus**, and the distal tarsals and metatarsal bones have fused and elongated to form the **tarsometatarsus**, the lowest part of the leg. This extension provides additional leverage for springing into the air and absorbing shocks upon landing. All vertebrae except those in the

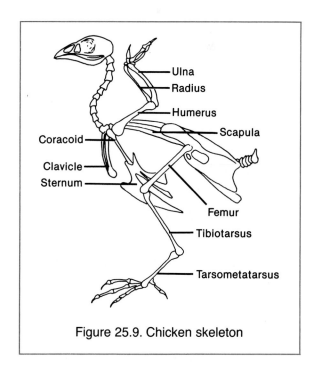

Figure 25.9. Chicken skeleton

neck and tail are fused with the ribs and pelvic girdle and provide a rigid framework.

A cat skeleton is provided as an example of a vertebrate adapted for running. Cats are fast runners for short distances. Note how the legs are located under the body rather than extending laterally as in the frog or lizards. Observe the orientation of the bones of the front and hind legs and their attachment to the axial skeleton to provide good leverage and freedom of movement. Note the long tail. What is its function?

The human skeleton shows adaptations for an erect, bipedal mode of locomotion. The major adaptations include (1) a broad pelvis that provides support for the erect body, (2) an almost central inferior attachment of the skull to the vertebral column and the vertebral curvatures that are a result of erect posture, (3) hands that are specialized for handling and grasping objects while the feet have lost this capability in favor of efficient bipedal walking and running, and (4) eye orbits that are forward facing, enabling stereoscopic vision.

Materials

Articulated skeletons of:
  cat
  chicken
  frog
  human
  monkey

Assignment 7

1. Examine the skeletons of the frog, chicken, and cat noting the adaptations for their particular modes of locomotion.
2. Examine the articulated human skeleton and note the major adaptations to erect posture and bipedal locomotion.
3. Compare the human and monkey skeletons and note the differences in skeletal structure between an erect, bipedal human and a monkey that lives primarily in trees.
4. ***Complete item 7 on the laboratory report***.

26

MUSCLES AND MOVEMENT

Minute **contractile fibrils** are present in many of the unicellular organisms. They produce a variety of movements including whip-like beating of flagella and cilia. In plants, such fibrils are found only in the flagella of motile cells, but they have been exploited to a remarkable degree by advanced animals.

In animals, contractile fibrils are incorporated into contractile cells. Sponges, which lack neurons, possess primitive contractile cells that react to direct stimulation by the environment. In all higher animals, contractile cells are under **neural control**. Other evolutionary trends have been (1) the organization of contractile cells into large groups called anatomical muscles and (2) functional and structural specialization of contractile cells.

VERTEBRATE MUSCLE TISSUES

Muscle tissues of vertebrates can be divided into three distinct groups on the basis of structure, location, and function.

Smooth Muscle Tissue

This type of muscle tissue occurs in the walls of hollow internal organs such as the digestive tract and blood vessels. Each cell is spindle shaped with a centrally located nucleus. Striations are absent. The cells are not enveloped by a sarcolemma as in skeletal muscle. Contractions are slow and untiring. Since smooth muscle is controlled by the autonomic nervous system and is not under conscious control, contractions are said to be *involuntary*. See Figure 26.1.

Skeletal Muscle Tissue

The muscles attached to bones or other muscles are composed of a large number of skeletal muscle fibers bound together by connective tissue. A skeletal muscle fiber may not be a single cell but an aggregation of cells that lack delimiting membranes. The fact that each fiber has several nuclei located on the periphery of the fiber supports this view. A thin membrane, the **sarcolemma**, envelops each fiber but is difficult to see unless it has been accidentally separated from the fiber. The contractile elements are the **myofibrils** that extend the length of the fiber. They exhibit alternating dark and light **striations** (cross-banding), which is the basis for an alternate name for this tissue, striated muscle. Skeletal muscle is controlled by the somatic nervous system, which enables its functions to be *voluntary* (under conscious control), although it contracts involuntarily in reflexes. Contractions are rapid, but the muscles are easily fatigued. See Figure 26.2.

Cardiac Muscle Tissue

The muscle of the heart is cardiac muscle tissue. It is striated like skeletal muscle, but the fibers form an interwoven network. The boundaries of cells are denoted by **intercalated discs**, and each cell has a single centrally located nucleus. A sarcolemma envelops each cell. Contractions are rhythmic, untiring, and *involuntary*. Control is by the autonomic nervous system. See Figure 26.3.

SKELETAL MUSCLE FUNCTION

Anatomical muscles and bones of vertebrates are arranged to form a system of levers that work together to produce a variety of movements. Further, the muscles are arranged in **antagonistic groups** so that opposing muscles move a body part in opposite directions. This is necessary because muscles can contract only upon stimulation. Examples of antagonistic actions are the following:

Flexors and Extensors. Flexors decrease the angle of bones forming a joint, while extensors increase the angle.

Abductors and Adductors. Abductors move a body part away from the midline of the body, while adductors move it toward the midline.

In humans, the biceps flexes the forearm while the triceps extends it. See Figure 26.4. When a muscle is stimulated to contract by neural impulses, the antagonist is automatically inhibited from contracting.

Muscles are attached at each end to bones or other muscles by **tendons**. The end that is stationary (does not move during contraction) is the

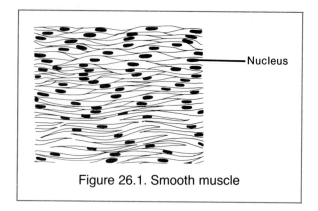

Figure 26.1. Smooth muscle

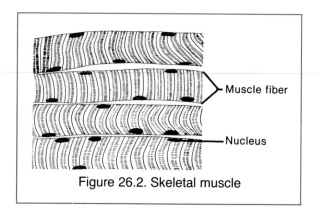

Figure 26.2. Skeletal muscle

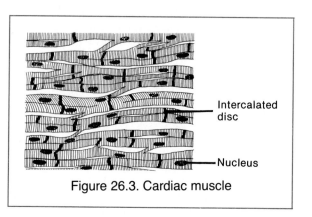

Figure 26.3. Cardiac muscle

origin. The end where the movement occurs is the **insertion**.

Levers

A lever consists of a rigid rod (a bone) that moves about a fixed point (a joint) called a **fulcrum (F)**. Two opposing forces act on a lever. The **resistance (R)** is the weight to be moved.

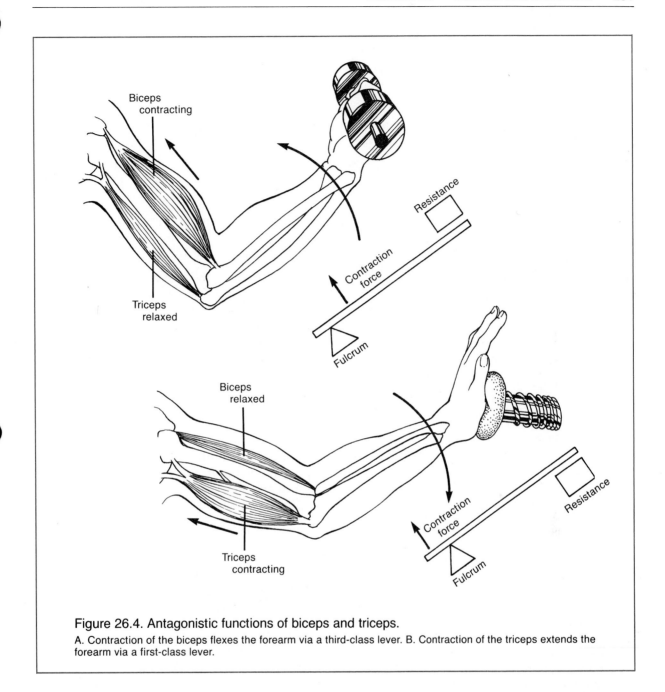

Figure 26.4. Antagonistic functions of biceps and triceps.
A. Contraction of the biceps flexes the forearm via a third-class lever. B. Contraction of the triceps extends the forearm via a first-class lever.

The **contraction force (CF)** is the force applied by a contracting muscle at the point of its insertion.

The types of levers differ from each other by the relative positions of the resistance, fulcrum, and contraction force. Figure 26.5 shows the three types of levers. Be sure that you know their characteristics before proceeding.

Study Figure 26.4 and note the two types of levers involved in the flexion and extension of the forearm. The fulcrum is always at a joint, and the contraction force is always applied at the site of the muscle insertion. The farther the point of insertion is from the fulcrum, the greater is the mechanical advantage and the greater the resistance that can be moved.

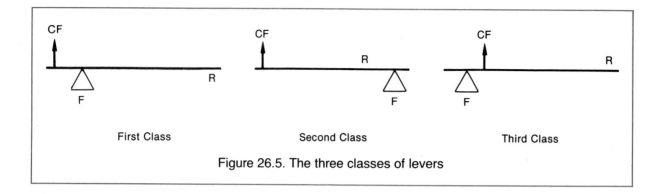

Figure 26.5. The three classes of levers

The nearer the insertion is to the fulcrum, the greater is the freedom and speed of movement of the body part.

Materials

Colored pencils
Dissecting instruments and pan
Frog, preserved or freshly killed
Prepared slides of:
  smooth muscle, teased
  skeletal muscle, teased
  cardiac muscle, teased

Assignment 1

1. Examine prepared slides of smooth, skeletal, and cardiac muscle tissue, and compare your observations with Figures 26.1, 26.2, and 26.3.
2. *Complete item 1 on Laboratory Report 26 that begins on page 443.*

Assignment 2

1. Determine the types of levers and movements involved in each of the actions shown in Figure 26.6. Indicate the location of the fulcrum (F), resistance (R), and contraction force (CF) by placing their symbols on the diagrams.
2. *Complete item 2 on the laboratory report.*

Study of Frog Muscles

The hind legs of a frog are suitable subjects for studying the way muscles are arranged to enable movement.

Assignment 3

1. Using Figures 26.7 and 26.8 as guides, skin a hind leg of a frog.
2. Examine the muscles of the leg. If the frog is freshly killed, note the color and texture of the muscles and connective tissue.
3. Refer to Figure 26.9 to locate the following muscles, and separate them from adjacent muscles with a blunt probe. Color-code each muscle in Figure 26.9.

Dorsal Muscles
Triceps femoris
Semimembranous
Gastrocnemius
Peroneus

Ventral Muscles
Sartorius
Gracilis major
Gastrocnemius
Extensor cruris

4. Note the origins and insertions of the muscles in Table 26.1, and locate their general positions in the frog. Determine and record the action of each muscle by pulling the body of the muscle toward the origin with forceps while holding the origin in place with your other hand.
5. If time permits, determine the action of other leg muscles in a similar manner.
6. *Complete item 3 on the laboratory report.*

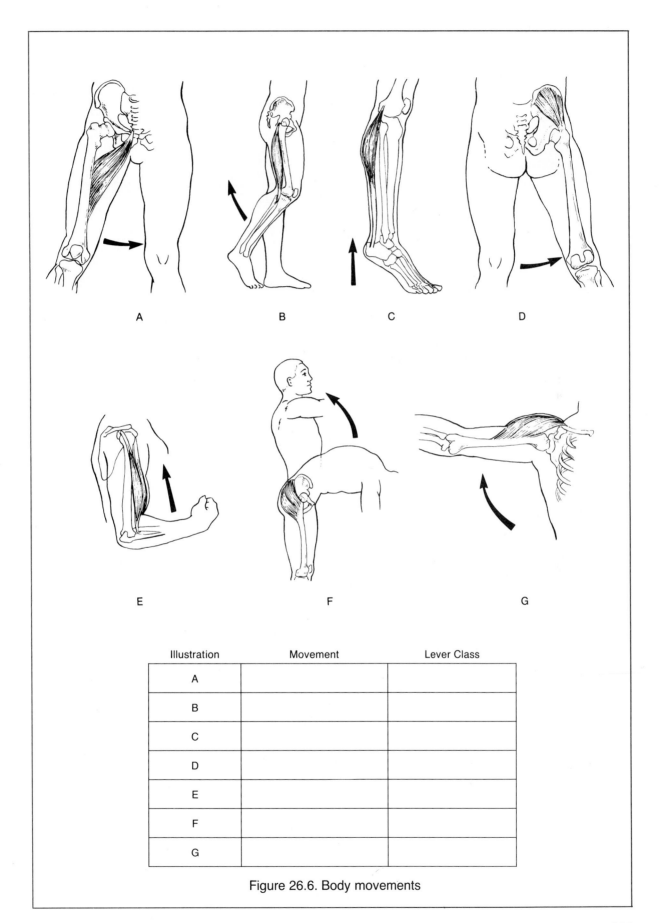

Illustration	Movement	Lever Class
A		
B		
C		
D		
E		
F		
G		

Figure 26.6. Body movements

Figure 26.7. Cut the skin around the uppermost part of the thigh.

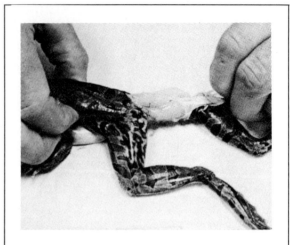

Figure 26.8. Strip the skin off the entire leg.

TABLE 26.1
Selected Muscles of a Frog's Leg

Muscle	Origin	Insertion	Action
Triceps femoris	Ilium	Tibiofibula, femur	
Semimembranosus	Ischium, pubis	Tibiofibula	
Gastrocnemius	Femur	Tarsals	
Peroneus	Femur	Tibiofibula	
Sartorius	Pubis	Tibiofibula	
Gracilis major	Pubis	Tibiofibula	
Extensor cruris	Femur	Tibiofibula	

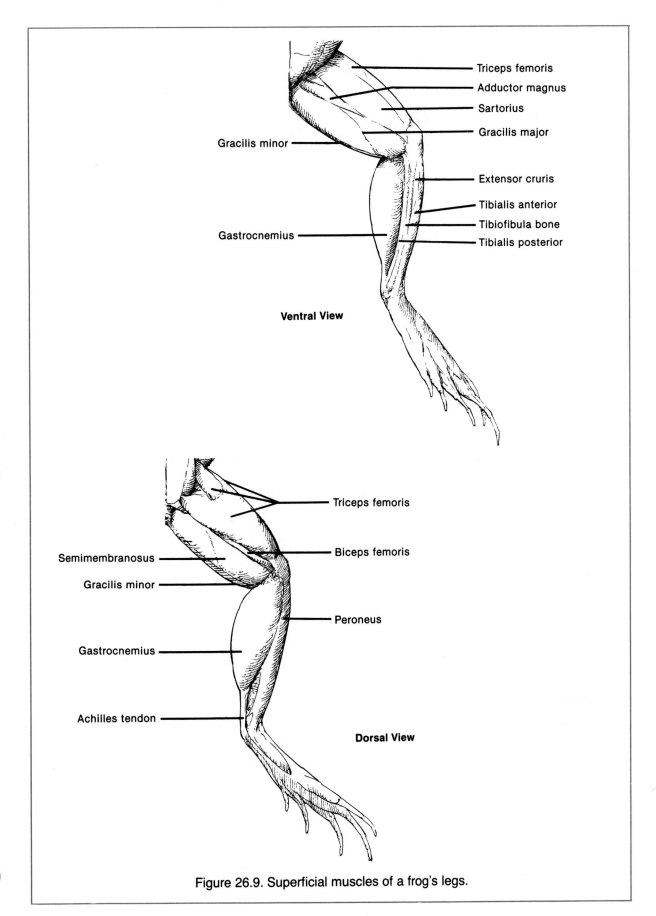

Triceps femoris

Adductor magnus

Sartorius

Gracilis major

Gracilis minor

Extensor cruris

Tibialis anterior

Tibiofibula bone

Gastrocnemius

Tibialis posterior

Ventral View

Triceps femoris

Biceps femoris

Semimembranosus

Gracilis minor

Peroneus

Gastrocnemius

Achilles tendon

Dorsal View

Figure 26.9. Superficial muscles of a frog's legs.

MYOFIBRIL ULTRASTRUCTURE AND CONTRACTION

The striations of a skeletal muscle fiber result from the arrangement of actin and myosin **myofilaments** within each myofibril. See Figure 26.10. The thicker **myosin** myofilaments are the main components of the **A band**. A light-colored **H zone** occurs in the center of the A band where only myosin is present. Thinner **actin** myofilaments extend into the A band from the **Z lines**. The Z Lines form the boundaries of a **sarcomere**, the contractile unit of a myofibril. A light-colored **I band** lies between the A band and a Z line.

Mechanics of Contraction

Contraction of a muscle fiber results from the interaction of actin and myosin in the presence of calcium and magnesium ions. ATP supplies the required energy.

When a muscle fiber is activated by a neural impulse, the cross-bridges of the myosin myofilaments attach to active sites on the actin myofilament and bend to exert a power stroke that pulls the actin filaments toward the center of the A band. After the power stroke, the cross-bridges separate from the first actin active sites, attach to the next active sites, and produce another power stroke. This process is repeated until maximum contraction is attained.

Experimental Muscle Contraction

In this section, you will induce muscle fibers to contract using ATP and magnesium and potassium ions. Your instructor has prepared 2-cm segments from a glycerinated rabbit psoas muscle. The segments have been placed in a petri dish of glycerol and teased apart to yield thin strands consisting of very few muscle fibers. The strands should be no more than 0.2 mm thick.

Materials

Microscopes, compound and dissecting
Dissecting instruments
Microscope slides and cover glasses
Plastic ruler, clear and flat
Glycerinated rabbit psoas muscle
Dropping bottles of:
 ATP, 0.25%, in triple-distilled water
 glycerol
 magnesium chloride, 0.001 M
 potassium chloride, 0.05 M

Assignment 4

1. Place a few muscle strands in a small drop of glycerol on a microscope slide and add a cover glass. Observe them with the high-dry or oil-immersion objective. *Draw the pattern of striations of the relaxed muscle fibers in item 4c of the laboratory report*.
2. Place three to five of the thinnest strands on another slide in just enough glycerol to moisten them. Arrange the strands straight, parallel, and close together.
3. Use a dissecting microscope to measure the length of the relaxed strands by placing the slide on a clear plastic ruler. *Record the length on the laboratory report*. Note the width of the strands.
4. While observing through the dissecting microscope, add one drop from each of the solutions: ATP, magnesium chloride, and potassium chloride. Note any changes in length or width of the strands.
5. Remeasure and record the length of the contracted strands.
6. Place a few contracted strands in a small drop of glycerol on another slide and add a cover glass. Observe the strands with the high-power or oil-immersion objective. *Draw the pattern of striations in item 4c on the laboratory report*.
7. *Complete item 4 on the laboratory report*.

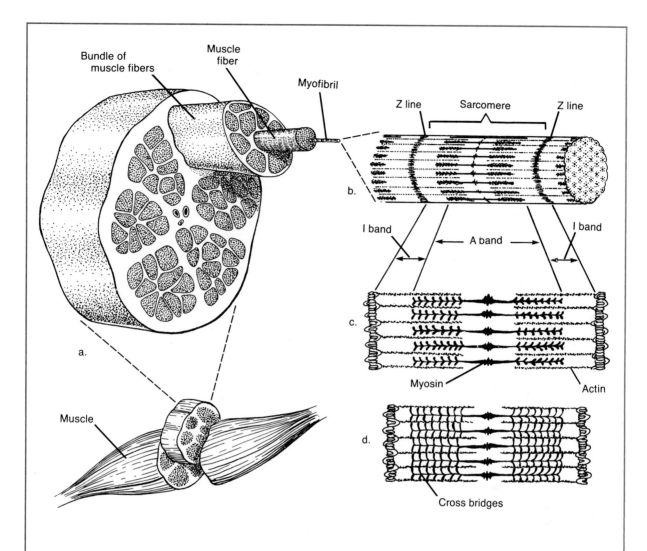

Figure 26.10. Structure of skeletal muscle.

A. The arrangement of muscle fibers within a muscle. B. The ultrastructure of a myofibril showing the relationship of actin and myosin filaments. Note how the arrangement of actin and myosin differs in (C) the relaxed state and (D) the contracted state.

PART IV

CONTINUITY OF LIFE

27

CELL DIVISION

All new cells are formed by the division of preexisting cells. In **prokaryotic cells**, cell division is relatively simple. The process is known as **binary fission**, and it occurs by (1) replication and separation of the circular DNA molecules and (2) the formation of additional cell membrane and cell wall material to separate the original cell into two new cells. Figure 27.1 depicts the process of binary fission. The cells are too small for you to observe this process in the laboratory, however.

In **eukaryotic cells**, two processes of cell division produce distinctly different types of cells. Study Table 27.1. Cells formed by **mitotic cell division** contain the same number and composition of chromosomes as the mother cell. In contrast, cells formed by **meiotic cell division** have only one half the number of chromosomes as the mother cell. Thus, these two types of cell

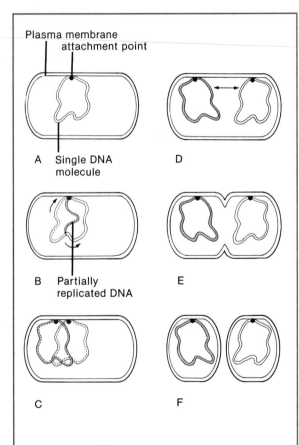

Figure 27.1. Cell division in a prokaryotic cell.

A. Cell before replication of DNA. Note attachment of DNA to cell membrane. B. Replication of DNA moving in both directions from starting point. C. DNA replication completed. D. Growth of cell membrane and cell wall occurs between the points of DNA attachment. E. Cleavage furrow of cell membrane and cell wall begins. F. New cells formed.

TABLE 27.1
Significant Differences in Mitotic and Meiotic Cell Divisions

Mitotic Cell Division	Meiotic Cell Division
1. Occurs in both haploid (n) and diploid (2n) cells.	1. Occurs in diploid (2n) cells, but not in haploid (n) cells.
2. Completed when one cell divides to form two cells.	2. Requires two successive cell divisions to produce four cells from the single parent cell.
3. Duplicated chromosomes do not align themselves in homologous pairs during division.	3. Duplicated chromosomes arrange themselves in homologous pairs during the first cell division.
4. The two daughter cells contain (a) the same genetic composition as the parent cell and (b) the same chromosome number as the parent cell.	4. The four daughter cells contain (a) different genetic compositions and (b) one half the chromosome number of the parent cell.

division differ in the way the chromosomes are dispersed to the new cells that are formed. The terms **mitosis** and **meiosis** refer to the orderly process of separating and distributing the replicated chromosomes to the new cells. **Cytokinesis** (division of the cytoplasm) is the process of actually forming the **daughter cells**.

Each organism has a characteristic number of chromosomes in the nuclei of its cells. If a single set of chromosomes is present, the cell is **haploid (n)**, and it contains only one chromosome of each chromosome pair. If two sets are present, the cell is **diploid (2n)**, and each chromosome pair is composed of **homologous chromosomes**.

The body cells of animals and most higher plants are diploid. For example, fruit flies have 8 chromosomes (4 pairs), onions have 16 (8 pairs), and humans have 46 (23 pairs). Gametes of these organisms are always haploid and contain 4, 8, and 23 chromosomes, respectively. The body cells of most simple organisms are haploid.

MITOTIC CELL DIVISION

In unicellular and some multicellular organisms, mitotic cell division serves as a means of reproduction. In multicellular organisms, it serves as a means of growth and repair. As worn-out or damaged cells die, they are replaced by new cells formed by mitotic division in the normal maintenance and healing processes. Millions of new cells are formed in the human body each day in this manner.

Mitotic cell division is an orderly, controlled process, but it sometimes breaks out of control to form massive numbers of nonfunctional, rapidly dividing cells that constitute either a benign tumor or a cancer. Seeking the causes of uncontrolled mitotic cell division is one of the major efforts of current biomedical research.

The Cell Cycle

A cell passes through several recognizable stages during its life span. These stages constitute the **cell cycle**. There are two major stages. **Mitosis**, the M stage, accounts for only 5 to 10% of the cell cycle. The **interphase** forms the remainder. See Figure 27.2.

Interphase has three subdivisions. Immediately after mitosis is a growth period, the G_1 **stage**. Next is the **synthesis (S) stage** where chromosome and centriole (if present) replication occurs. Each replicated chromosome consists of two **sister chromatids** joined at the **centromere**. See Figure 27.3. A second growth stage, the G_2 **stage**, follows and prepares the cell for the next mitotic division. Cells that will not divide again remain in the G_1 stage and carry out their normal functions.

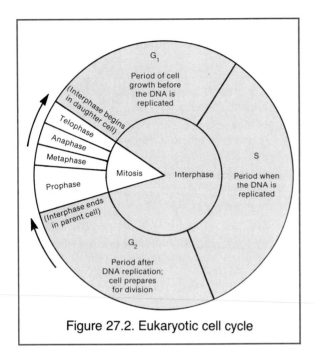

Figure 27.2. Eukaryotic cell cycle

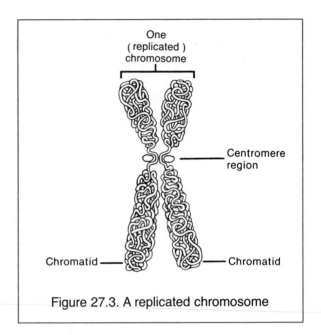

Figure 27.3. A replicated chromosome

Mitotic Phases in Animal Cells

The process of mitosis is arbitrarily divided into recognizable stages or phases to facilitate understanding, although the process is a continuous one. These phases are **prophase**, **metaphase**, **anaphase**, and **telophase**. The characteristics of each phase as observed in animal cells are noted here to aid your study. Interphase is also included for comparative purposes. Compare these descriptions with Figure 27.4.

Interphase

Cells in interphase have a distinct nucleus and two pairs of **centrioles**. The chromosomes are uncoiled and are visible only as **chromatin granules**.

Prophase

During prophase, (1) the nuclear membrane and nucleolus disappear, (2) the chromosomes coil tightly to appear as rod-shaped structures, (3) each pair of centrioles migrates to opposite ends of the cell, and (4) the **spindle** forms. Each pair of centrioles and its radiating **astral rays** constitute an **aster** at each end (pole) of the spindle.

Metaphase

This brief phase is characterized by the chromosomes lining up at the equator of the spindle. Each replicated chromosome is attached to a spindle fiber by its **centromere**.

Anaphase

Anaphase begins with the separation of the centromeres of the sister chromatids, which migrate toward opposite poles of the spindle. Once the sister chromatids separate, they are called **daughter chromosomes**. Thus, a cell in anaphase contains two complete sets of chromosomes.

Telophase

In telophase, (1) a new nuclear membrane forms around each set of chromosomes to form two new nuclei, (2) the nucleus reappears, (3) the chromosomes start to uncoil, and (4) a **cleavage furrow** forms to divide the mother cell into two **daughter cells**. Recall that the division of the mother cell is called cytokinesis.

Mitotic Division in Plant Cells

Mitotic division in plants follows the same basic pattern that occurs in animals with some

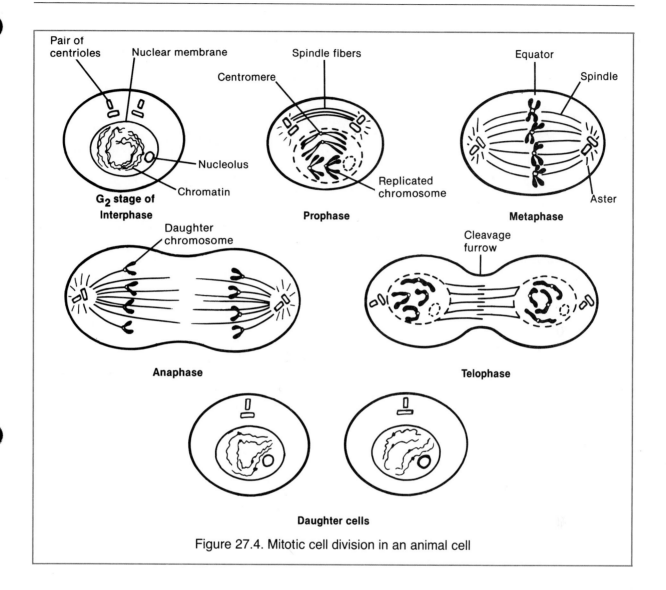

Figure 27.4. Mitotic cell division in an animal cell

notable exceptions. Most plants do not possess centrioles, although a spindle of fibers is present. The rigid cell wall prevents the formation of a cleavage furrow during cytokinesis; instead, a **cell plate** forms to separate the mother cell into two daughter cells, and a new cell wall forms along the cell plate. Cytokinesis usually, but not always, occurs during telophase. See Figure 27.5.

Microscopic Study

The rapidly dividing cells of whitefish blastula, an early fish embryo, are excellent for studying mitotic division in animals. See Figure 27.6. On the prepared slide that you will use are several thin sections of the blastula, and each

contains many cells in various stages of the cell cycle, including mitosis. A prepared slide of onion (*Allium*) root tip is used to study mitotic division in plants. Each slide usually contains three longitudinal sections of root tip. The region of cell division is near the pointed tip.

Materials

Prepared slides of:
  whitefish blastula, x.s.
  onion root tip, l.s.

Assignment 1

Complete items 1, 2a, and 2b on Laboratory Report 27 that begins on page 445.

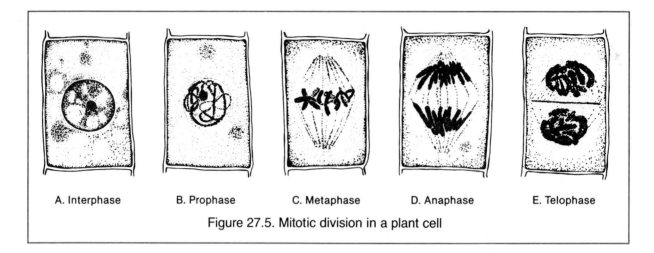

A. Interphase   B. Prophase   C. Metaphase   D. Anaphase   E. Telophase

Figure 27.5. Mitotic division in a plant cell

Assignment 2

1. Obtain a prepared slide of whitefish blastula. Locate a section for study with the 4× objective. Then switch to the 10× objective to find mitotic phases for observation with the 40× objective. ***Locate cells in each phase of mitosis, and draw them in the space for item 2c on the laboratory report***. You may have to examine all the sections on your slide, or even other slides, to observe each phase of mitosis.

2. Examine a prepared slide of onion root tip. Locate the cells in mitotic phases near the tip of the section. Observe cells in metaphase, anaphase, and telophase. Note the cell plate. Note how mitotic division in an onion root tip differs from that observed in a whitefish blastula.

3. ***Complete item 2 on the laboratory report***.

Figure 27.6. Various stages of mitotic division in whitefish blastula cells.

Courtesy of Turtox/Cambosco, Macmillan Science Co., Inc., Chicago.

Meiotic Cell Division

In contrast to mitotic cell division, meiotic cell division consists of *two* successive divisions but only *one* chromosome replication. This results in the formation of four cells that have only half the number of chromosomes of the **diploid** (2n) mother cell. Thus, the daughter cells have a **haploid** (n) number of chromosomes since they each contain only *one member of each chromosome pair*. In addition to reducing the chromosome number in the daughter cells, meiosis also reshuffles the genes, hereditary units formed of small segments of DNA, and this greatly increases the genetic variability among the daughter cells.

In humans and most animals, cells formed by meiotic division become either sperm or eggs. In plants, meiotic cell division results in the formation of meiospores that do not fuse like gametes but grow into haploid gametophytes that, in turn, produce gametes by mitotic division. In either case, the basic result of meiosis is

Meiosis I

Chromosome tetrad

A. Prophase I

B. Metaphase I

C. Anaphase I

D. Telophase I

E. Daughter Cells

Meiosis II

A. Prophase II

B. Metaphase II

C. Anaphase II

D. Telophase II

E. Daughter Cells

Figure 27.7. Meiotic cell division. For simplicity, only one of the cells formed in meiosis I is shown in meiosis II.

245

the same: haploid cells with increased genetic variation.

Meiotic Phases in Animal Cells

Study Figure 27.7 as you read the following description of meiotic cell division in an animal cell. Chromosome and centriole replication occur in the S stage of interphase prior to the start of meiosis.

Meiosis I

Prophase I exhibits the following characteristics. Each chromosome is composed of two sister chromatids joined together at the centromere. The replicated members of each chromosome pair join together in a side-by-side pairing called **synapsis**. Chromosomes in synapsis are often called **tetrads** since they consist of four chromatids. An exchange of chromosome segments (cross-over) frequently occurs between members of the tetrad and increases the genetic variability of the cells produced by meiotic division. The chromosomes coil tightly to appear as rod-shaped structures, the nuclear membrane and nucleolus disappear, and a spindle forms.

Metaphase I is characterized by the synapsed chromosomes lining up at the equatorial plane where they attach to spindle fibers by their centromeres.

Anaphase I begins with the separation of the members of each chromosome pair. The centromeres do *not* separate, so each chromosome still consists of two chromatids joined at the centromere. Members of each chromosome pair migrate to opposite poles of the spindle in the replicated state.

Telophase I proceeds to form a nuclear membrane around each set of chromosomes. The chromosomes untwist and the nucleolus reappears. Cytokinesis separates the mother cell into two daughter cells. Keep in mind that the nucleus of each daughter cell contains only *one member of each chromosome pair* in a replicated state. Thus, each daughter cell is haploid (n).

Meiosis II

Both cells formed by meiosis I divide again in meiosis II, but for discussion purposes we will follow only one of these cells in the second division. In interphase between meiosis I and II, the centrioles replicate but chromosomes do *not* replicate again. Recall that they are already replicated.

Prophase II is characterized by the usual loss of the nuclear membrane and nucleolus, spindle formation, and the appearance of rod-shaped chromosomes.

Metaphase II is characterized by the chromosomes lining up at the equator of the spindle. Each chromosome consists of two sister chromatids joined together at the centromere that is attached to a spindle fiber.

Anaphase II begins with the separation of the centromeres. The sister chromatids, now called daughter chromosomes, move toward opposite poles of the spindle.

Telophase II proceeds as usual to form the new nuclei, and cytokinesis divides the cell to form two haploid (n) daughter cells.

Since each cell entering meiosis II forms two daughter cells, a total of four haploid (n) cells are produced from the original diploid (2n) mother cell entering meiosis I. Thus, meiotic cell division may be summarized as:

$$1 \text{ cell (2n)} \xrightarrow{\text{M I}} 2 \text{ cells (n)} \xrightarrow{\text{M II}} 4 \text{ cells (n)}$$

Materials

Colored pipe cleaners or chromosome simulation kits

Assignment 3

1. Study Figure 27.7.
2. Using colored pipe cleaners to represent chromosomes or a chromosome simulation kit, simulate the replication and distribution of chromosomes in both mitosis and meiosis where 2n = 2.
3. ***Complete items 3 and 4 on the laboratory report.***

28

REPRODUCTION IN SIMPLE ORGANISMS

OBJECTIVES

On completion of the laboratory session, you should be able to:
1. Recognize and describe the patterns of reproduction.
2. Describe the life cycles and reproductive methods of the organisms studied.
3. Identify and give the function of specialized reproductive cells in the organisms studied.
4. Explain the adaptive advantage of gametic reproduction.
5. Define all terms in bold print.

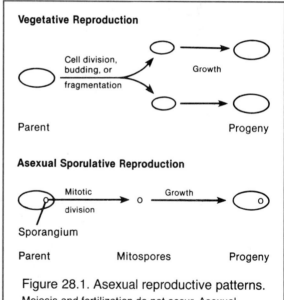

Figure 28.1. Asexual reproductive patterns.
Meiosis and fertilization do not occur. Asexual reproduction occurs in both haploid (n) and diploid (2n) organisms.

Before considering reproduction in simple organisms, you should become familiar with the basic patterns of reproduction in the biotic world. They may be grossly classified as either asexual or sexual, and a number of variations exist. Refer to Figures 28.1 and 28.2 as you study this section.

Asexual reproductive patterns may occur in either haploid (n) or diploid (2n) organisms, but meiosis and fertilization are never involved. There are two types of asexual reproduction:
1. **Vegetative reproduction** may occur by **cell division** (fission in prokaryotes, mitotic division in eukaryotes), **fragmentation**, or **budding**.
2. **Asexual sporulative reproduction** uses mitotic division to form **mitospores**, often in **sporangia**. After release, the spores

germinate to grow into a new adult organism.

Sexual reproduction occurs in organisms that have a diploid (2n) stage in their life cycle. Meiosis and fertilization are always part of the life cycle. There are two basic forms of sexual reproduction:
1. **Sexual sporulative reproduction** involves the production of **meiospores** by

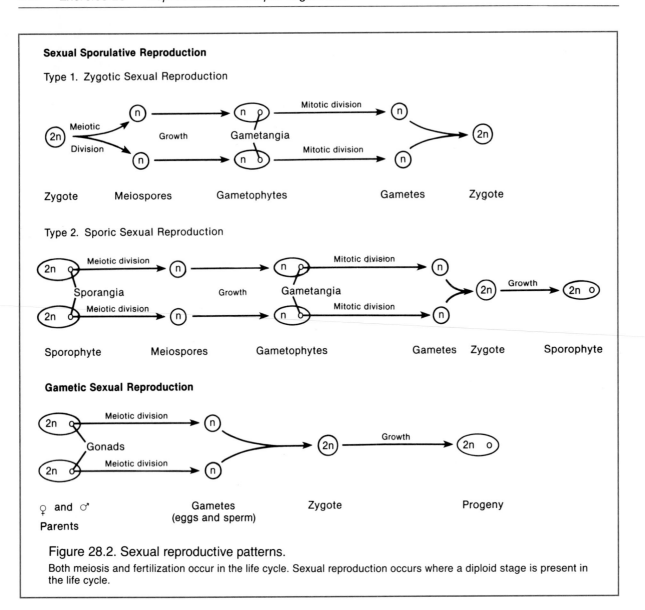

Figure 28.2. Sexual reproductive patterns.
Both meiosis and fertilization occur in the life cycle. Sexual reproduction occurs where a diploid stage is present in the life cycle.

meiotic division at one stage of the life cycle. The two types of sexual sporulative reproduction differ as to the site of meiotic division:

a. In **zygotic sexual reproduction**, only the **zygote** is diploid, and it undergoes meiotic division upon germination to form four **meiospores** (n). The meiospores germinate and grow into a **gametophyte** (n) that forms **gametes** (n) by mitotic division. Fertilization by gamete union forms a new zygote (2n) to complete the cycle.

b. **Sporic sexual reproduction** involves two adult generations that alternate in the life cycle. The sporangia of a **sporophyte** (2n) produce **meiospores** (n) by meiotic division, and these meiospores germinate and grow into **gametophytes** (n). Gametangia in a gametophyte produce **gametes** (n), eggs or sperm, by mitotic division. The fusion of gametes in fertilization yields a **zygote** (2n). The zygote grows into a new sporophyte (2n).

2. **Gametic sexual reproduction** usually, but not always, involves separate male and female parents (2n) whose gonads form ga-

metes (n), eggs or sperm, by meiotic division. Fertilization forms a **zygote** (2n) that grows into a new adult (2n).

REPRODUCTION IN MONERANS

Bacteria and cyanobacteria usually reproduce by **binary fission**, a vegetative form of reproduction. By this method, these haploid (n) organisms are able to produce prodigious numbers of haploid descendents when environmental conditions are favorable. Bacteria are able to form nonreproductive spores that can survive unfavorable environmental conditions and can germinate whenever conditions become favorable.

A sexual process, **conjugation**, also occurs in monerans and is best known in bacteria. In this process, two cells become connected by a bridge formed by the cells, and the replicate DNA of one cell passes through the bridge to the other cell. The strand of DNA is analogous to true gametes in higher forms. Subsequently, the donor DNA may combine with and replace homologous segments in the recipient DNA. After separation, the cells continue to divide by binary fission.

REPRODUCTION IN PROTISTS

Mitotic cell division is the most common form of reproduction in protists, which can produce great numbers of individuals when conditions are optimum. Two basic patterns of sexual reproduction occur depending on the diploid or haploid nature of the organism. Variations of these patterns occur in different species.

1. Haploid (n) protists may become gametes or produce gametes (n) by mitosis. Fusion of the gametes forms a zygote (2n) that divides by meiosis to yield four haploid individuals.
2. Diploid (2n) protists form four haploid (n) gametes by meiosis. Subsequently, the fusion of gametes yields a diploid zygote that divides mitotically to form additional diploid individuals. Note that only the gametes are haploid in this pattern.

Reproduction in *Paramecium*

Paramecium reproduces asexually by mitotic division and sexually by conjugation. Study the complex sexual process shown in Figure 28.3. Note that the "gametes" are only nuclei and that the mating individuals cannot be identified as either male or female.

Materials

Colored pencils
Prepared slides of:
    Paramecium, dividing
    Paramecium, conjugating
    Paramecium bursaria, cultures of mating types

Assignment 1

1. Color-code haploid stages red and diploid stages blue in Figure 28.2.
2. ***Complete item 1 on Laboratory Report 28 that begins on page 451.***

Assignment 2

1. Examine prepared slides of *Paramecium* undergoing cell division. Is the plane of division lengthwise or transverse?
2. Examine a prepared slide of *Paramecium* in conjugation. Do the cells join at a particular site or randomly? Try to find pairs corresponding to the stages shown in Figure 28.3.
3. Prepare and observe a slide of living *Paramecium* in conjugation. Are the conjugants motile? What is the site of attachment? Is the union a brief one?
4. ***Complete item 2 on the laboratory report.***

REPRODUCTION IN FUNGI

Vegetative reproduction occurs in some fungi by cell division or budding and in others by fragmentation of the mycelium. The most common method of asexual reproduction is by **mitospores**.

Fungi also reproduce by sexual reproduction. Fungi are typically haploid, and certain cells become modified to function as gametes (n). The

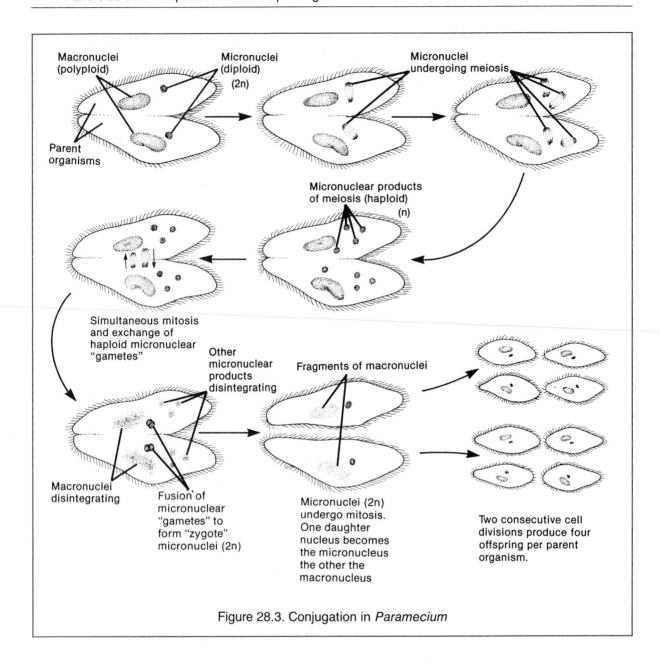

Figure 28.3. Conjugation in *Paramecium*

fusion of gametes of opposite mating types results in a zygote (2n) that divides meiotically to form either meiospores (n), as in sac and club fungi, or haploid hyphae that form sporangia, which produce mitospores (n), as in many molds. The spores later germinate to form haploid mycelia.

Reproduction in Black Bread Mold

Common bread mold, *Rhizopus,* exhibits a life cycle found in some molds. Examine Figure 28.4.

Note that the mycelium may be of either mating typing, + or −. Asexual reproduction by mitospores occurs continuously, and sexual reproduction occurs only when opposite mating types come in contact with each other. Then special cells are formed that become the gametes (n). The zygote (2n) formed by their fusion (conjugation) develops a resistant cell wall and is called a **zygospore**. Subsequently, it germinates by meiotic division to form haploid hyphae and sporangia that produce mitospores of each

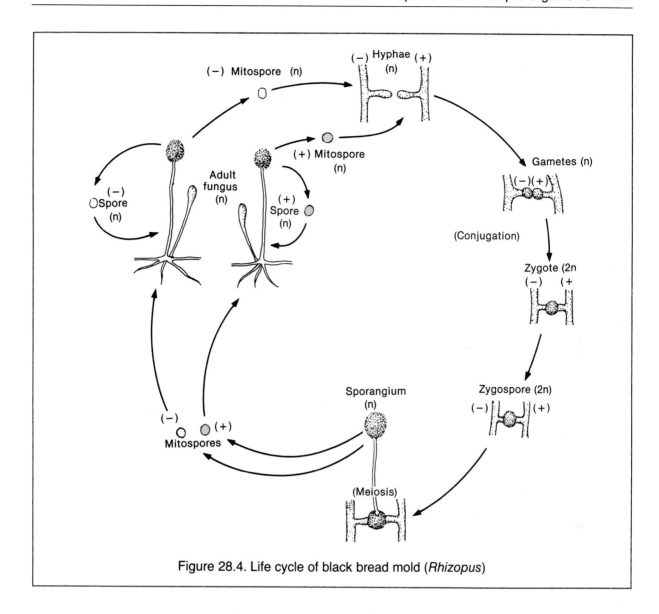

Figure 28.4. Life cycle of black bread mold (*Rhizopus*)

mating type. Germination of these spores forms new haploid mycelia.

Reproduction in a Mushroom

The life cycle of a mushroom, *Coprinus,* is more complex than that of *Rhizopus* and is representative of the club fungi. See Figure 28.5. The haploid mycelia occur as two mating types, + and −. Whenever opposite mating types come in contact with each other, cells (analogous to gametes) from one mating type grow toward cells of the opposite mating type and fuse together in pairs, but their nuclei remain separate. The cytoplasmic fusion is called **plas-**

mogamy, and the cell that is formed is a **dikaryotic cell** since it contains two nuclei, one of each mating type. An extensive mycelium is formed from the dikaryotic cell by mitotic division.

A fruiting body (basidiocarp) is formed by the dikaryotic hyphae. The spore-forming cells (basidia) on the gills are also dikaryotic. Later, the nuclei in the basidia fuse to form a single diploid nucleus (a zygote nucleus). Then, each basidium forms four **basidiospores**, two of each mating type, by meiotic division. The spores are released and carried by air currents. On germination, each spore forms a haploid mycelium by mitotic division.

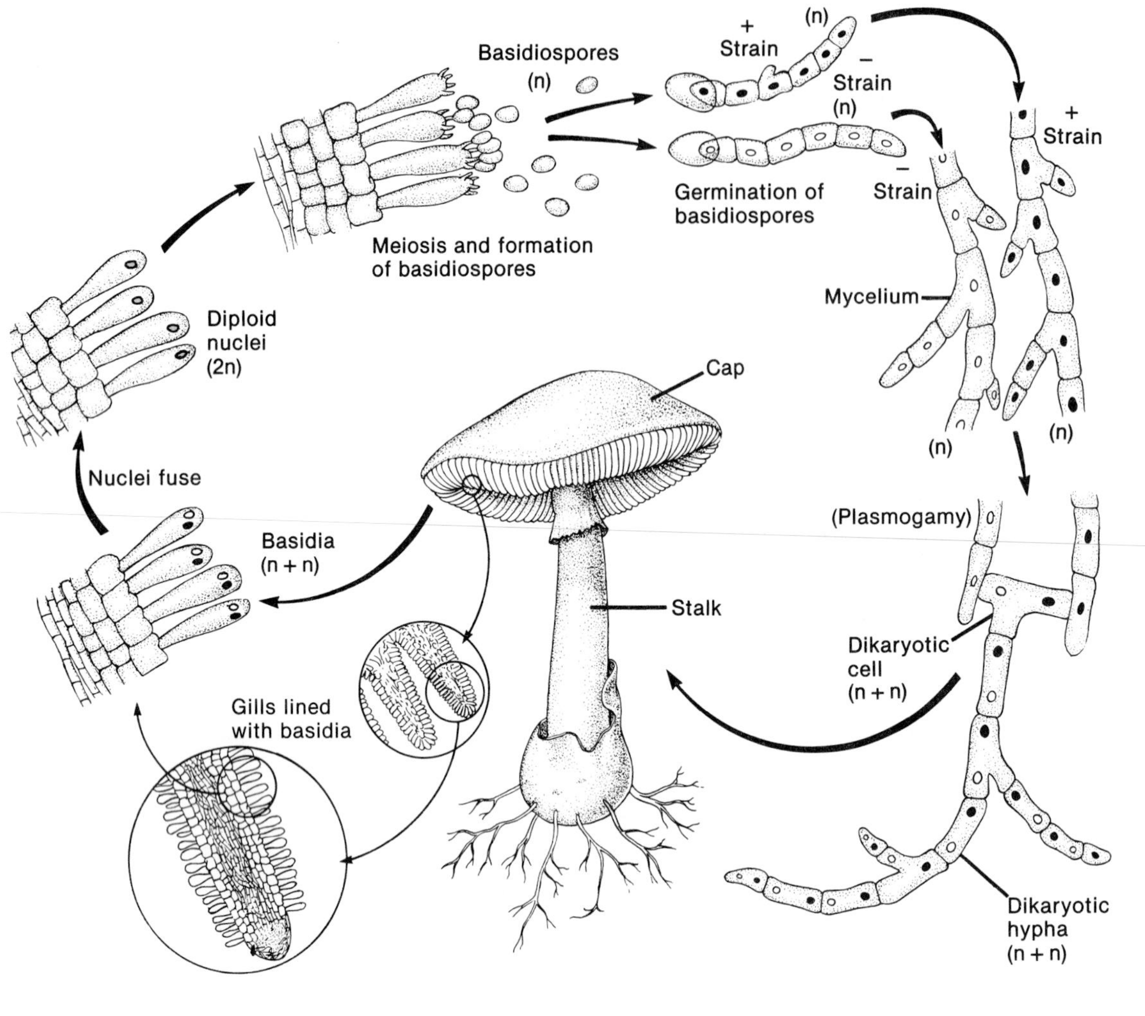

Figure 28.5. Life cycle of a mushroom (*Coprinus*)

Materials

Rhizopus culture of opposite mating types with zygospores
Yeast culture
Prepared slides of:
   Rhizopus with gametes and zygospores
   Coprinus gills, x.s., with basidia and basidiospores

Assignment 3

1. Prepare and examine a slide of yeast culture and locate the buds forming on the yeast cells. **Draw a few budding cells in item 3b on the laboratory report.**

2. Study Figures 28.4 and 28.5. Color-code haploid stages red, dikaryotic stages green, and diploid stages blue.

3. Examine the *Rhizopus* culture with zygospores set up under a demonstration dissecting microscope. Note the color and size of the gametes, zygospores, and sporangia.

4. Examine a prepared slide of gametes and zygospores in *Rhizopus*. Try to locate the stages shown in Figure 28.4. **Draw a few sporangia and zygospores in item 3b on the laboratory report.**

5. Examine a prepared slide of *Coprinus* gill, x.s., and locate the basidia, basidiospores, and dikaryote cells of the hyphae.

6. ***Complete item 3 on the laboratory report***.

REPRODUCTION IN GREEN ALGAE

Green algae reproduce vegetatively by mitotic cell division in unicellular forms and by fragmentation in colonial and multicellular forms. Most algae also reproduce by spores and gametes, and these two processes sometimes alternate in the life cycle.

Reproduction in *Spirogyra*

Figure 28.6. depicts the life cycle of *Spirogyra,* a haploid, colonial green alga. New cells are formed by mitotic cell division, and new filaments are formed by fragmentation.

Conjugation occurs at certain times of the year when opposite mating types are in contact. A bridge is formed between cells of the filaments, and the protoplasm of each cell condenses to form a gamete (n). Then the gamete of the donor cell moves through the bridge to combine with the gamete in the recipient cell to form a zygote (2n). The zygote becomes a zygospore that can withstand unfavorable conditions. Subsequently, the zygospore nucleus divides by meiosis without cell division. Three of the nuclei disintegrate, but the remaining haploid nucleus is functional and forms a new haploid filament by mitotic division.

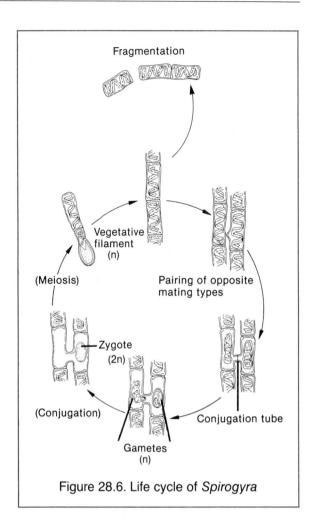

Figure 28.6. Life cycle of *Spirogyra*

Reproduction in *Oedogonium*

The life cycle of *Oedogonium* is more complex than that of *Spirogyra,* since it exhibits both asexual sporulative and zygotic sexual reproduction. The gametes are sperm and eggs. *Oedogonium* is a haploid, colonial, filamentous green alga that shows considerable specialization in the cells of a filament, a trend toward the multicellular condition. New cells are added to the filament by mitotic division, and new filaments are formed by fragmentation. Study Figure 28.7.

Asexual sporulative reproduction occurs when the zoosporangia form and release motile **zoospores** (n). On germination, each zoospore forms another haploid filament.

In sexual reproduction, the protoplast of each **antheridium** divides mitotically to produce two **sperm** (n). Some larger cells, **oogonia**, form a single *egg* (n). Fertilization occurs when a sperm swims to and unites with an egg. The resulting zygote (2n) becomes a zygospore. On germination, the zygospore undergoes meiosis to form four haploid zoospores that swim away to form new filaments by mitotic cell division.

Reproduction in *Ulva*

Ulva, a multicellular, marine green alga, exhibits an additional advancement: sporic sexual reproduction, which includes an **alternation of generations**. This is a significant advancement since this reproductive pattern is characteristic of all plants taxonomically higher than green algae.

Examine Figure 28.8. The zygote (2n) germinates to form a **sporophyte** (2n), a spore-forming plant that produces motile zoospores (n) by meiosis. On germination, zoospores form sepa-

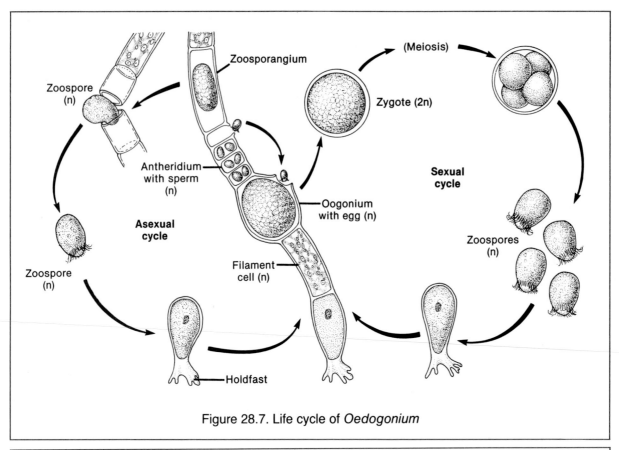

Figure 28.7. Life cycle of *Oedogonium*

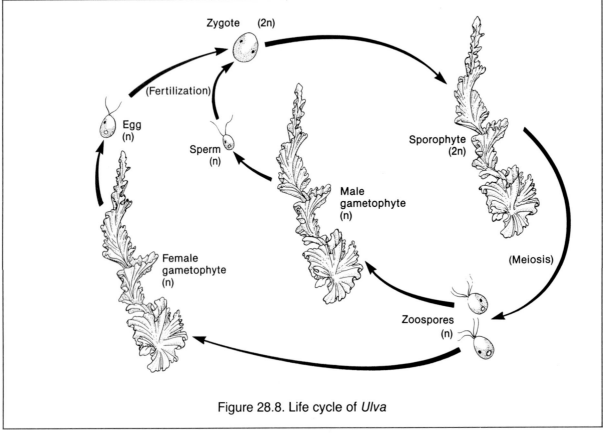

Figure 28.8. Life cycle of *Ulva*

rate male and female **gametophytes** (n), gamete-forming plants. Male gametophytes produce motile sperm (n), and female gametophytes produce motile eggs (n) by mitotic division. Fusion of sperm and egg forms a zygote (n), and the cycle continues.

Materials

Colored pencils
Prepared slides of:
    Spirogyra conjugation
    Oedogonium antheridia, oogonia, zoospores
    Spirogyra, living
    Oedogonium, living
    Ulva, fresh or preserved

Assignment 4

1. Study Figures 28.6, 28.7, and 28.8. Note the characteristics of each life cycle and the evolutionary trends. Color-code haploid stages red and diploid stages blue.
2. Prepare and observe a wet-mount slide of *Spirogyra*. Note the structure of the filament and cells.
3. Examine a prepared slide of *Spirogyra* in conjugation. Locate gametes and zygotes. Are the motile gametes on the same filament?
4. Examine prepared slides of *Oedogonium* zoospores, antheridia, and oogonia. Compare your observations with Figure 28.7.
5. Prepare and observe a wet-mount slide of *Oedogonium*. Note the structure of the filament and cells. Are antheridia, oogonia, or sporangia present? **Make drawings of your observations in item 4b on the laboratory report.**
6. Examine the specimens of *Ulva*.
7. ***Complete item 4 on the laboratory report.***

29

REPRODUCTION IN LAND PLANTS

OBJECTIVES

On completion of the laboratory session, you should be able to:
1. Describe the life cycles and reproductive methods of mosses, ferns, conifers, and flowering plants.
2. Describe the processes of pollination, fertilization, and double fertilization.
3. Identify and describe the function of the specialized reproductive cells and organs in the organisms studied.
4. Identify and describe the function of the parts of a sorus, flower, fruit, and seed.
5. Describe the adaptive advantages of the reproductive methods observed in flowering plants.
6. Define all terms in bold print.

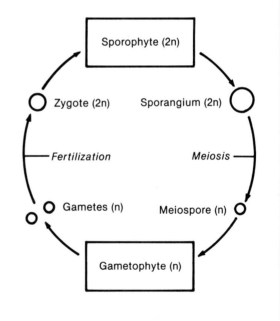

Figure 29.1. Alternation of generations in sporic sexual reproduction

Land plants exhibit vegetative asexual reproduction and sporic and gametic sexual reproduction. The degree of vegetative reproduction is variable, but the alternation of spore-forming and gamete-forming generations is a major feature. This **alternation of generations** was noted in advanced algae and is depicted in Figure 29.1. All bryophytes and vascular plants possess these distinctive characteristics:
1. Distinct **sporophyte** (diploid) and **gametophyte** (haploid) generations
2. Multicellular **sporangia** that produce spores by meiosis

3. Multicellular **sex organs**
4. Dissimilar male and female **gametes**
5. Multicellular **embryonic stage** in sporophyte development

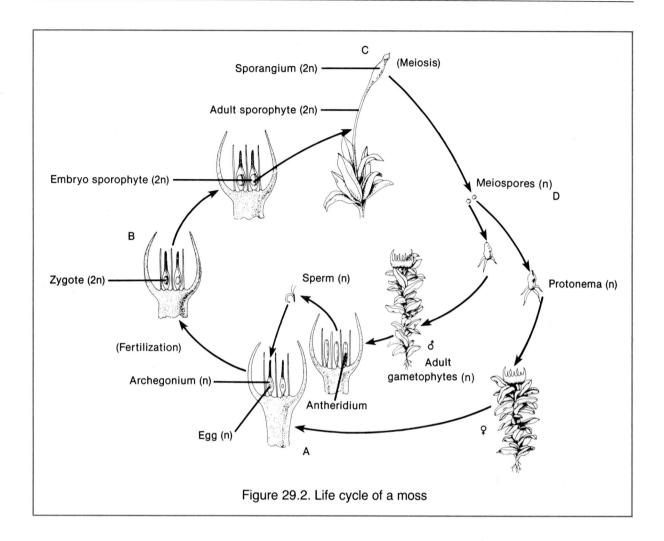

Figure 29.2. Life cycle of a moss

REPRODUCTION IN MOSSES

Moss plants are poorly adapted to most terrestrial habitats because of their lack of vascular tissue and because their sperm require free water in which to swim to the egg.

Study the moss life cycle illustrated in Figure 29.2 and described here:

A. Male and female sex organs, **antheridia** and **archegonia**, form at the tops of adult gametophyte (n) plants. When the gametes are mature and moisture is present, sperm are released from the antheridia and swim into the archegonia to fertilize the eggs.

B. Fertilization of an egg produces a zygote that divides inside the archegonium to form a multicellular **sporophyte embryo**.

C. The sporophyte embryo develops into an adult sporophyte (2n), which grows out of the archegonium of the female gametophyte. The sporophyte partially depends on the female gametophyte for nutrients.

D. Meiospores are released from the sporangium and germinate in the soil, producing **protonemas** (n) that develop into adult gametophyte plants.

Materials

Colored pencils
Moss gametophytes with sporophytes
Prepared slides of:
  moss antheridia with sperms
  moss archegonia with egg
  Moss capsule with spores

Assignment 1

1. Color-code haploid stages red and diploid stages blue in Figure 29.1.
2. **Complete item 1 on Laboratory Report 29 that begins on page 455.**

Assignment 2

1. Study Figure 29.2 and color-code haploid stages red and diploid stages blue. Examine the specimens of moss gametophytes with sporophytes.
2. Examine prepared slides of moss antheridia with sperms, archegonia with an egg, and capsule with spores.

3. **Complete item 2 on the laboratory report.**

REPRODUCTION IN FERNS

Because fern sporophytes possess vascular tissue, they are better adapted to terrestrial life than mosses, but like mosses, fern gametophytes still require free water for sperm transport.

Study the life cycle of a fern illustrated in Figure 29.3 and described here:

A. When a fern sporophyte (2n) is mature, it produces sporangia on the underside of its

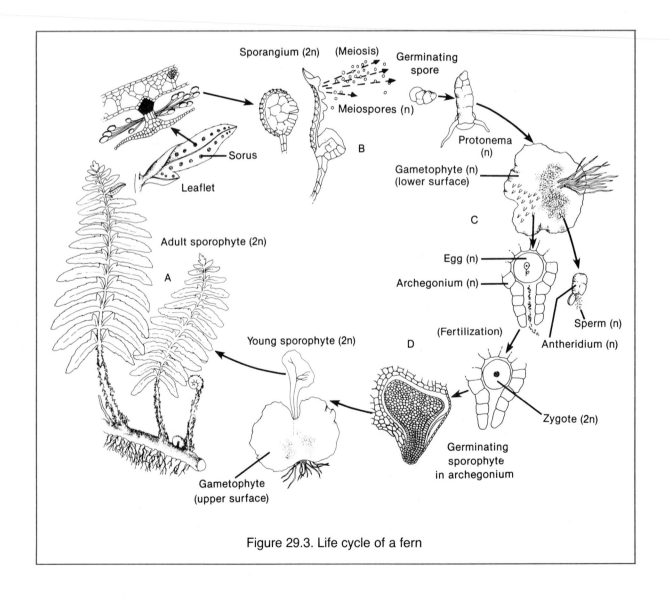

Figure 29.3. Life cycle of a fern

leaves. The sporangia usually are grouped in clusters called **sori**.

B. Meiospores formed in the sporangia are released and germinate in the soil to produce small gametophyte plants.

C. Each gametophyte (n) forms both male and female sex organs, that is, **antheridia** and **archegonia**, on its undersurface. When free water is present on the undersurface of the gametophyte plant, sperm (n) swim from the antheridia to the archegonia where fertilization of the egg (n) occurs.

D. The resultant zygote (2n) divides mitotically in the archegonium, producing a multicellular sporophyte embryo that first derives nutrients from the gametophyte and later becomes an independent sporophyte plant. The small gametophyte then dies.

Materials

Fern gametophytes, living
Fern sporophytes with sori, living
Microscope slides and cover glasses
Prepared slides of fern gametophyte

Assignment 3

1. Study Figure 29.3. Color-code haploid stages red and diploid stages blue.

2. Examine a fern sporophyte and locate the sori on the leaflets. Remove a leaflet and examine it with a dissecting microscope. Make a wet-mount slide of a sorus and examine it with a compound microscope. What composes a sorus?

3. Examine a prepared slide of fern gametophyte. Locate the archegonia near the notch and the antheridia near the rhizoids. Note the sperm and eggs.

4. Observe the living gametophytes.

5. *Complete item 3 on the laboratory report*.

REPRODUCTION IN SEED PLANTS

Cone-bearing plants and **flowering plants** exhibit several important reproductive characteristics in addition to the basic ones previously discussed for plants in general.

1. The sporophyte (2n) produces two types of meiospores: (a) **microspores** are formed by **microsporangia** and (b) **megaspores** are formed by **megasporangia**.

2. Microspores (n) develop into **male gametophytes** (pollen grains). Megaspores (n) develop into **female gametophytes**.

3. **Pollen grains** are transferred from microsporangia to megasporangia or associated structures by wind or insects—a process called **pollination**.

4. The pollen grain forms a **pollen tube** that carries the **sperm nuclei** to the egg within the female gametophyte.

5. The embryo sporophyte, female gametophyte, and associated tissues compose the **seed**.

Cone-Bearing Plants

Study the life cycle of the pine illustrated in Figure 29.4 and described here:

A. The sporophyte plant produces **staminate** (male) and **ovulate** (female) **cones** that contain many microsporangia and megasporangia, respectively.

B. Many microspores are formed by meiosis in the microsporangia, and they rapidly develop into pollen grains (male gametophytes). Pollination occurs by wind.

C. Four **megaspores** are formed by meiosis. Three degenerate; one becomes the **functional megaspore** that develops into the female gametophyte.

D. An archegonium containing one egg develops in the female gametophyte. The pollen tube carries a sperm nucleus to the egg for fertilization.

E. The zygote grows into the embryo sporophyte within the seed.

F. The seed is released from the cone and is dispersed by the wind. Upon germination, the embryo sporophyte uses nutrients stored in the seed to grow and become self-supporting.

Materials

Colored pencils
Male cones with pollen
Female cones with seeds
Pine seeds, soaked

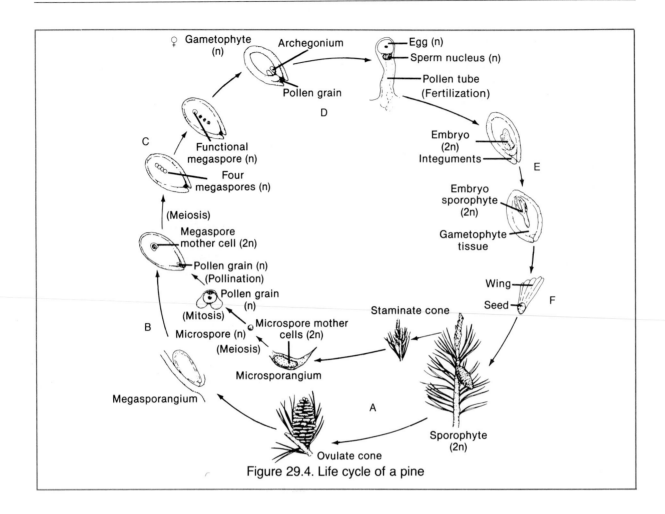

Figure 29.4. Life cycle of a pine

Prepared slides of:
  microsporangia with pollen
  megasporangium with archegonium and egg

Assignment 4

Complete item 4 on the laboratory report.

Assignment 5

1. Study Figure 29.4. Color-code haploid stages red and diploid stages blue.
2. Examine a prepared slide of microsporangia with pollen. How is the pollen adapted for wind transport?
3. Examine the male (staminate) cones. Shake a bit of pollen on a microscope slide, add 1 drop of water and a cover glass, and observe with your microscope. Are the pollen grains like those on the prepared slide?

4. Examine a prepared slide of a young female (ovulate) cone showing the megasporangium with an archegonium containing an egg cell.
5. Examine a mature female cone and locate the two seeds on the upper surface of each scale. How are the seeds dispersed?
6. Obtain a soaked pine seed. Use a scalpel to cut it open longitudinally, and examine it with a dissecting microscope. Locate the embryo sporophyte embedded in the tissue of the female gametophyte. What tissues contribute to the formation of a pine seed?
7. ***Complete item 5 on the laboratory report.***

Flowering Plants

Flowering plants are the most advanced plants and the most successful land plants.

Their reproductive patterns show some major adaptations over gymnosperms.

1. Reproductive structures are grouped in **flowers** that usually contain both microsporangia and megasporangia.
2. Pollination is usually by wind in grasses, but it is by insects in most monocots and dicots. Insects are attracted to flowers by color and nectar.
3. Portions of the flower develop to form a **fruit** that encloses the seeds and enhances seed dispersal by wind in some plants but by animals in most.

Study the life cycle of a flowering plant illustrated in Figure 29.5 and described here:

A. The sporophyte plant produces flowers with **anthers** (microsporangia) and **ovules** (megasporangia).
B. Many microspores are produced in the anthers by meiosis; only one functional megaspore is formed in each ovule by meiosis.
C. Each microspore divides mitotically to form a **pollen grain** (male gametophyte); each megaspore divides mitotically within its ovule to form a female gametophyte containing one **egg** cell and two **polar nuclei**.
D. Pollen is carried from anther to stigma by wind or insects. The pollen grain forms a **pollen tube** that grows down through the **style** into the **ovary** and then into an **ovule**. Two **sperm nuclei** migrate down the pollen tube and into the ovule.

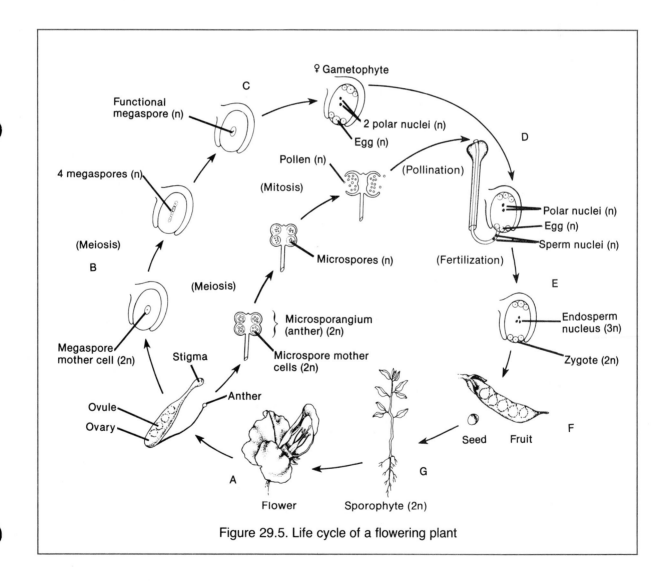

Figure 29.5. Life cycle of a flowering plant

E. One sperm nucleus fertilizes the egg producing a diploid zygote; the other sperm nucleus unites with the two polar nuclei forming a triploid (3n) **endosperm nucleus**.

F. Within the ovule, the zygote becomes a multicellular embryo and the endosperm provides stored food for the embryo. The ovule, embryo, and food constitute a **seed**. The ovary enlarges and matures to form a **fruit** that encloses the seeds.

G. Fruits may be disseminated by animals or wind. When soil temperatures and moisture are correct, the dormant sporophyte embryo in the seed germinates. It uses the stored food in the **cotyledons**, or **endosperm**, to develop roots and a shoot and becomes an independent sporophyte plant.

Assignment 6

1. Study Figure 29.5. Color-code haploid stages red and diploid stages blue.
2. ***Complete item 6 on the laboratory report.***

Flower Structure

The basic structure of a flower is shown in Figure 29.6, but this fundamental organization has many variations.

The **receptacle** supports the flower on the stem, and the reproductive structures are enclosed in two whorls of modified leaves. The inner whorl consists of **petals** that are usually colored to attract pollinating insects. The outer whorl consists of **sepals** that are typically smaller than the petals and are usually green in color.

The **stamens** are the male portions of the flower. Each stamen consists of an **anther** supported by a **filament**. Anthers contain the microsporangia that produce pollen. The **pistil** is the female portion, and it consists of three parts. The basal portion is the **ovary**, which contains **ovules** (megasporangia). The tip of the pistil is the **stigma**, which receives pollen and secretes enzymes promoting pollen germination. The **style** is a slender stalk that joins stigma and ovary. Nectar is secreted near the base of the ovary.

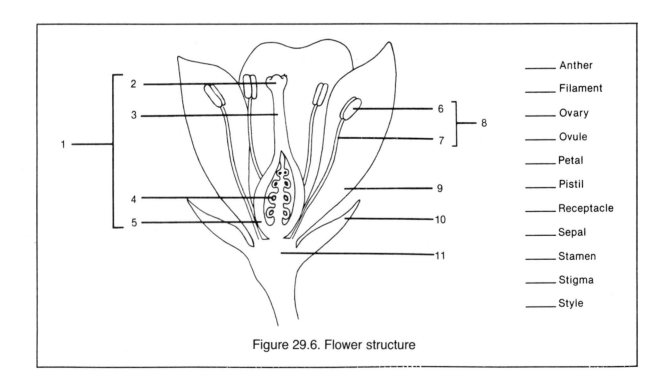

Figure 29.6. Flower structure

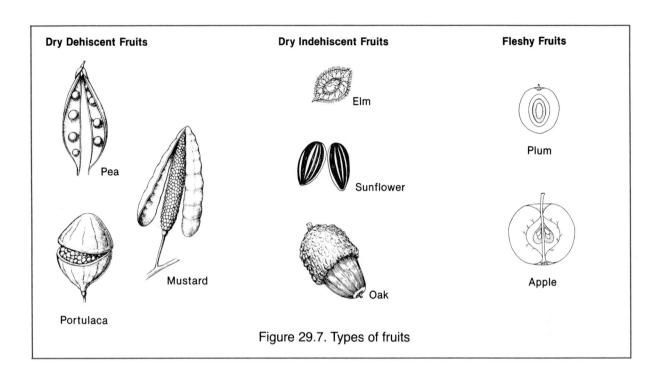

Dry Dehiscent Fruits

Pea

Mustard

Portulaca

Dry Indehiscent Fruits

Elm

Sunflower

Oak

Fleshy Fruits

Plum

Apple

Figure 29.7. Types of fruits

Fruits and Seeds

As the seeds develop in the ovary, the ovary grows and ripens to form a fruit that provides protection for the seeds and facilitates seed dispersal. These are the three basic types of fruits as shown in Figure 29.7.

1. **Dry dehiscent fruits** split open when sufficiently dry to cast out the seeds, sometimes with considerable force. Pea and bean pods are examples.
2. **Dry indehiscent fruits** do not open, and the ovary wall tightly envelops the seed. Acorns and fruits of corn and other cereals are examples.
3. **Fleshy fruits** remain moist for a considerable period of time and are usually edible and colored. Animals scatter the seeds by feeding on the fruits.

A seed consists of a protective **seed coat** that is derived from the wall of the ovule, stored nutrients, and a dormant **embryo sporophyte**. In monocots and some dicots, the stored nutrients compose the endosperm. In most dicots, the embryonic leaves, **cotyledons**, contain many of the stored nutrients, and the endosperm is reduced. Seeds are able to withstand unfavorable conditions and tend to germinate only when favorable conditions exist.

Materials

Colored pencils
Representative flowers, fruits, seeds
Lily flowers
Corn fruits
Bean seeds, soaked
Prepared slides of lily anthers, x.s.
Iodine (I_2 + KI) solution

Assignment 7

1. Label Figure 29.6. Color-code the male parts yellow and the female parts green.
2. Compare a lily flower with Figure 29.6 to locate the parts. Observe the anthers with a dissecting microscope, and note the pollen. Make a wet-mount slide of pollen and observe it with a compound microscope. Is the pollen structure like pine pollen? Make a cross section of the ovary and observe the ovules. How are they attached?
3. Examine a prepared slide of lily anther, x.s. Observe the pollen and microsporangia.
4. Examine the representative flowers and note their variations of the basic flower structure. Make water-mount slides of their pollen and examine them microscopically to compare their structure.

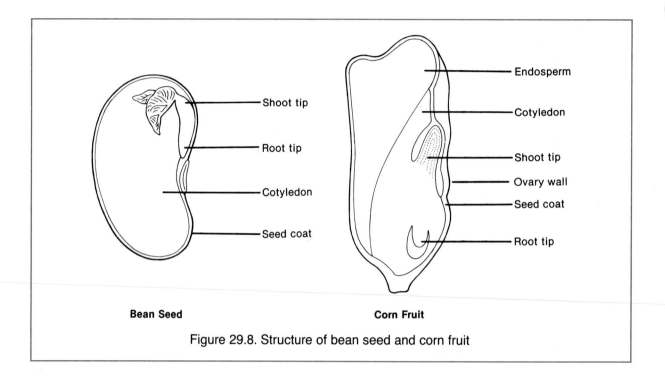

Bean Seed   **Corn Fruit**

Figure 29.8. Structure of bean seed and corn fruit

5. Examine the representative fruits and classify them according to type. How do they aid seed dispersal?

6. Color-code the embryo green and the sites of stored nutrients yellow in Figure 29.8.

7. Split open a soaked bean seed, and locate the parts shown in Figure 29.8. Use a scalpel to cut the soaked corn fruit, and locate the labeled parts. Examine the structures with a dissecting microscope.

8. Use a scalpel to gently scrape the exposed surface of the bean cotyledon and the attached embryo sporophtye.

9. Add a drop of iodine solution to the cotyledon and embryo and also to the cut surface of the corn fruit. After 1 min, rinse with water and blot dry. What nutrient is evident? Where is it located?

10. Examine the representative seeds, and note the variations in size and shape.

11. ***Complete the laboratory report***.

30

REPRODUCTION IN VERTEBRATES

OBJECTIVES

On completion of the laboratory session, you should be able to:
1. Describe the basic reproductive patterns exhibited by vertebrates and indicate the advantages and disadvantages of each.
2. Describe how reptiles solved the problem of reproduction in a terrestrial environment.
3. Identify stages of gametogenesis on charts and prepared microscope slides and describe the stepwise process.
4. Identify on charts or models the components of human male and female reproductive systems and state the function of each.
5. Indicate the mode of action and relative effectiveness of birth control methods.
6. Define all terms in bold print.

REPRODUCTIVE PATTERNS

Vertebrates occur in both aquatic and terrestrial environments, and their reproductive patterns are reflective of the environment that each group has colonized. Thus, some vertebrate groups exhibit internal fertilization, while others use external fertilization.

External fertilization occurs in an aquatic environment. It involves the release of both sperm and eggs into the water where they unite to form the zygote. Most fish and amphibians utilize external fertilization.

Internal fertilization involves the deposition of sperms in the female reproductive tract where fertilization occurs. **Copulation** is usually involved in this process. Reptiles, birds, and mammals utilize internal fertilization.

All animals exhibit **gametic sexual reproduction**, and with few exceptions, it is the only method of reproduction in animals. Separate sexes are the general rule, but a few forms are **hermaphroditic**, possessing both male and female sex organs. **Asexual sporulative reproduction** is restricted to the primitive, nonmotile sponges, and **vegetative reproduction** by budding or fragmentation also is associated with simple animals with limited motility: sponges, coelenterates, and flatworms.

Types of Gametic Reproduction

The three basic patterns of gametic sexual reproduction in vertebrates are based on (1) the site of embryonic development and (2) the source of nutrients for the embryo. Table 30.1 indicates the most common reproductive patterns for the five classes of vertebrates.

Oviparous reproduction is characterized by external embryonic development following external or internal fertilization. Nutrients for the embryo are contained within the fertilized

TABLE 30.1
Common Reproductive Patterns of Vertebrates

	Fertilization		Reproductive Pattern			Reproductive Habitat		Eggs			
	Ext.	Int.	Ovip.	Ovovip.	Vivip.	Aquatic	Terr.	Size	Number	Protection	Stored Food
Fish	√	Very few	√	Very few		√		Sm.	Many	No shell	Small amt.
Amphibians	√		√			√		Sm.	Many	No shell	Small amt.
Reptiles		√	√	Very few			√	Lg.	Moderate number	Shell and int. mem.	Large amt.
Birds		√	√				√	Lg.	Few	Shell and int. mem.	Large amt.
Placental mammals		√			√	Few	√	Very sm.	Very few	Membranes in mother's uterus (no shell)	None

egg. This pattern occurs in fish, amphibians, reptiles, birds, and egg-laying mammals.

Viviparous reproduction involves both fertilization and development within the female reproductive tract, and the developing embryo receives nutrients from the female parent. This pattern is characteristic of mammals.

Ovoviviparous reproduction occurs when fertilization and development of the embryo take place in the female reproductive tract, but the embryo receives nutrients only from the egg and not from the female parent. This pattern is not common in vertebrates but does occur in some sharks and snakes.

Reproduction and Land Colonization

Among terrestrial vertebrates, amphibians have a limited distribution because they must return to water to reproduce. The frog is a suitable example. Female and male simultaneously release eggs and sperm, respectively, into the water where fertilization occurs. After embryonic development is complete, a larva (tadpole) hatches from the egg. Larvae swim by using the tail in a fishlike manner, and they have functional gills. After suitable growth, the larvae metamorphose into the adult by developing

lungs and appendages, and by reabsorbing the gills and tail.

Reptiles were the first truly terrestrial vertebrates because they solved the problem of reproduction without returning to water. Their successful adaptations include internal fertilization and the **amniote egg**. The reptilian egg has a leathery shell that prevents excessive water loss and allows an exchange of oxygen and carbon dioxide between the developing embryo and the atmosphere. An adequate supply of stored nutrients (yolk and albumin) enables the development of the embryo to hatching. In addition, special protective membranes enclose the embryo.

The features of the amniote egg are shown in Figure 30.1. Note the **extra-embryonic membranes**. The **amnion** surrounds the embryo and contains the **amniotic fluid** that provides the embryo with its own "private pond" in which to develop. Instead of returning to water for embryonic development, reptiles "brought the water to the embryo." The **yolk sac** envelops the yolk and absorbs nutrients for the embryo. The **allantois** is an embryonic urinary bladder, but it also spreads out against the outer membranes and serves as a gas-exchange organ. The **cho-**

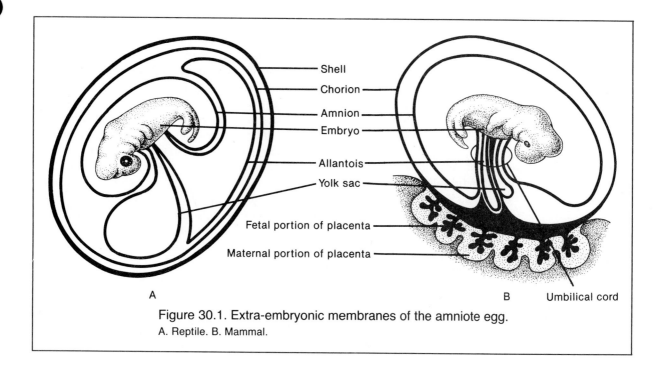

Figure 30.1. Extra-embryonic membranes of the amniote egg.
A. Reptile. B. Mammal.

rion is the outermost membrane and encompasses all of the others.

The bird egg is very similar to a reptilian egg, but it has a hard calcareous shell. The same basic pattern of extra-embryonic membranes develops as the embryo grows. Birds provide more parental care than reptiles and incubate the eggs.

Except for the primitive platypus and spiny echidna, mammals do not lay eggs. Fertilization is internal, and the embryo develops in the uterus of the female parent. The embryo becomes attached to the uterus by a **placenta** that is formed primarily by the chorion and allantois. The stalk of the allantois is the primary constituent of the **umbilical cord**, although the yolk sac and amnion are also components. The umbilical cord carries blood of the embryo to and from the placenta. Maternal and embryo bloods are separated by thin membranes in the placenta where an exchange of materials occurs:

$$\text{Maternal} \xrightarrow{\text{nutrients and } O_2} \text{embryo}$$
$$\text{blood} \xleftarrow{\text{wastes and } CO_2} \text{blood}$$

Thus, mammals have capitalized on the extra-embryonic membranes "invented" by reptiles.

Materials

Fish and amphibian eggs, fresh or preserved
Frog life cycle, preserved or models
Frog larvae, living
Chicken eggs, fresh
Chick embryos showing extra-embryonic membranes
Pregnant cat or pig uterus

Assignment 1

1. ***Complete items 1a to 1d on Laboratory Report 30 that begins on page 459.***
2. Examine the fish, amphibian, and bird eggs. In what ways do the eggs provide protection against environmental hazards and provide for the nutritional needs of the embryo? What is the relationship between parental care and survival of the young?
3. Examine the frog life cycle. Observe living larvae in the aquarium. Note their swimming and feeding behavior.
4. Examine the demonstration of bird and mammalian embryos with extra-embryonic membranes. Note the relationship of the membranes to the embryo. In the mam-

malian embryos, note the placenta and umbilical cord.

5. ***Complete item 1 on the laboratory report***.

HUMAN REPRODUCTIVE SYSTEMS

You will study the human reproductive systems as examples of reproductive systems in mammals. Refer to Figures 30.2 and 30.3.

Materials

Colored pencils
Models of male and female reproductive systems
Model of pregnant female torso
Preserved human embryos and fetuses of various ages
X-ray films of fetuses *in utero*

Male Reproductive System

The male gonads are paired **testes** that are held in the saclike **scrotum**. This arrangement holds the testes outside the body cavity and at a temperature of 94 to 95°F, which is necessary for the production of viable **spermatozoa**. Muscles in the wall of the scrotum relax or contract to position the testes farther from or closer to the body, and in this way regulate the temperature of the testes.

A testis contains numerous **seminiferous tubules** that produce the spermatozoa. **Interstitial cells** are located between the tubules and secrete **testosterone**, the male hormone responsible for the sex drive and the development of the sex organs and secondary sexual characteristics. The secretion of the **interstitial cell stimulating hormone (ICSH)** by the hypophysis (pituitary gland) activates the interstitial cells.

The **penis**, the male copulatory organ, contains three cylinders of spongy **erectile tissue** that fill with blood during sexual excitement to produce an erection. A circular fold of tissue, the **prepuce** (foreskin) covers the **glans penis**. For hygienic reasons, the prepuce of male babies is often removed by a surgical procedure called **circumcision**.

Mature, but inactive, sperm are carried down the seminiferous tubules to the **epididymis**, a long, coiled tube on the surface of the testis.

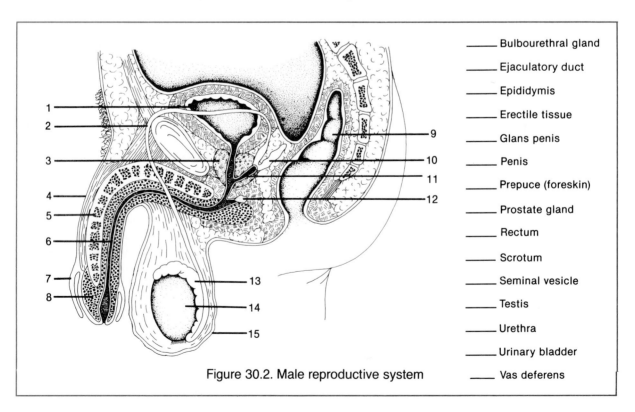

_____ Bulbourethral gland

_____ Ejaculatory duct

_____ Epididymis

_____ Erectile tissue

_____ Glans penis

_____ Penis

_____ Prepuce (foreskin)

_____ Prostate gland

_____ Rectum

_____ Scrotum

_____ Seminal vesicle

_____ Testis

_____ Urethra

_____ Urinary bladder

_____ Vas deferens

Figure 30.2. Male reproductive system

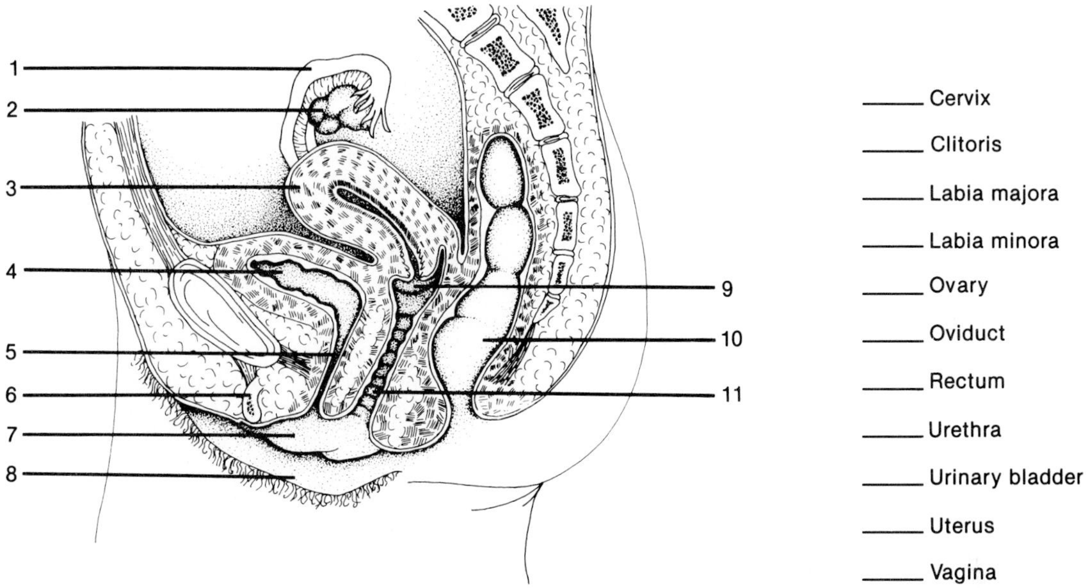

Figure 30.3. Female reproductive system

Labels at right:
_____ Cervix
_____ Clitoris
_____ Labia majora
_____ Labia minora
_____ Ovary
_____ Oviduct
_____ Rectum
_____ Urethra
_____ Urinary bladder
_____ Uterus
_____ Vagina

Sperm are stored here until they are propelled through the reproductive tract by wavelike contractions during **ejaculation**.

The **bulbourethral glands** open into the urethra below the prostate gland and secrete an alkaline liquid that neutralizes the acidity of the urethra prior to ejaculation. At the male climax, sperm pass from the epididymis into the **vas deferens**, a duct that exits the scrotum and enters the body cavity via the inguinal canal. It continues across the surface of the urinary bladder to join with the **ejaculatory duct** within the **prostate gland** that is located around the urethra just below the bladder. Alkaline secretions from the **seminal vesicles** are mixed with the sperm just before the vas deferentia enter the prostate gland, where sperm-activating prostatic secretions are added. Muscular contractions force **semen**, the mixture of sperm and glandular secretions, out through the urethra.

Female Reproductive System

The external female genitalia consist of (1) two folds of skin surrounding the vaginal and urethral openings, the **labia majora** (outer folds) and **labia minora** (inner folds), and (2) the **clitoris**, a nodule of erectile tissue homolo-gous to the penis in the male. Collectively, these structures are called the **vulva**.

The **vagina** is a collapsible tube extending 4 to 6 in. from the external opening to the uterus. It serves as both the female copulatory organ and the birth canal. The **uterus** is a pear-shaped organ located over and posterior to the urinary bladder. The **cervix** of the uterus extends a short distance into the upper end of the vagina.

A pair of **ovaries**, the female gonads, are located laterally to the uterus where they are supported by ligaments. One egg is released from alternate ovaries about every 28 days. The egg is picked up by the expanded end of the **oviduct** and carried toward the uterus by beating cilia of the ciliated epithelium lining the oviduct.

Ovaries secrete two female hormones. **Estrogen** is responsible for the development of the sex organs, the female sex drive, the secondary sex characteristics, and the buildup of the uterine lining. **Progesterone** prepares the uterine lining for the implantation of an early embryo. In turn, ovarian function is controlled by hormones released from the hypophysis (pituitary gland). Consult your text for a discussion of the ovarian and uterine cycles.

Assignment 2

1. Label and color code Figures 30.2 and 30.3.
2. Locate the parts of the male and female reproductive systems on the models provided.
3. Examine the model of a pregnant female torso. Note the positioning of the uterus with the developing fetus. Compare the X-ray films of fetuses with the models.
4. Examine the preserved human embryos and fetuses and compare the degree of development with the age of the embryo or fetus. An embryo is the stage between the second and eighth weeks of development. A fetus is the developmental stage from the eighth week to birth.
5. *Complete item 2 on the laboratory report*.

Gametogenesis

The formation of gametes is called gametogenesis. It includes meiotic cell division, which reduces the chromosome number of gametes to one half that of somatic cells. For example, the diploid (2n) chromosome number in humans is 46, and the haploid (n) gametes contain only 23 chromosomes. If you need to review meiosis, see Exercise 27.

The patterns of gametogenesis described here and illustrated in Figures 30.4 and 30.5 are typical of mammals, although slight variations may occur among individual species.

Spermatogenesis

Sperm formation occurs in the seminiferous tubules of the testes. **Spermatogonia** (2n) are the outermost cells of the tubule. They divide mitotically to form a **primary spermatocyte** and a replacement spermatogonium. The primary spermatocyte divides meiotically to yield two **secondary spermatocytes** (n) after meiotic division I and four **spermatids** (n) after meiotic division II. The spermatids attach to "nurse cells" and mature into spermatozoa. Maturation includes the loss of most of the cytoplasm and the formation of a flagellum from a centriole. The sperm head consists mostly of the cell nucleus. Sperm are carried along the seminiferous tubules and reach the epididymis in about 10 days.

Oogenesis

Prior to the birth of a female child, some **oogonia** of the germinal epithelium surrounding each ovary enlarge, become surrounded by follicular cells, and move into the ovary. These oogonia (2n) become the **primary oocytes** (2n) that enter prophase of meiosis I before oogenesis is arrested. The primary oocytes remain inactive at this stage until puberty.

At puberty, the **follicle stimulating hormone (FSH)** and the **luteinizing hormone (LH)** secreted by the pituitary gland stimulate primary oocytes and follicle cells to further growth. Each month one follicle develops more rapidly than the others to become a mature or **Graafian follicle** filled with fluid containing a large amount of **estrogen** secreted by the follicular cells. Meiotic division I proceeds to produce (1) a **secondary oocyte** (n) that receives most of the cytoplasm and (2) a much smaller **first polar body** that remains attached to the secondary oocyte.

Ovulation occurs when the mature follicle ruptures and ejects the follicular fluid and secondary oocyte through the ovary wall. The secondary oocyte enters the oviduct and is carried toward the uterus by the ciliated epithelium. Note that while it is common to speak of the "egg" as being released by the ovary in ovulation, a secondary oocyte is actually released. After ovulation, the empty follicle becomes the **corpus luteum** that produces progesterone to maintain the uterine lining.

No further division occurs unless the secondary oocyte is penetrated by a sperm. If this occurs, the secondary oocyte completes meiotic division II to form the **egg** (n) and another polar body. The first polar body also may complete meiosis II to form an additional polar body. Subsequently, egg and sperm nuclei fuse to form the diploid zygote. The polar bodies disintegrate.

Materials

Prepared slides of:
  cat testis, sectioned
  cat ovary, sectioned, with Graafian follicle
  cat ovary, sectioned, with corpus luteum
  human sperm

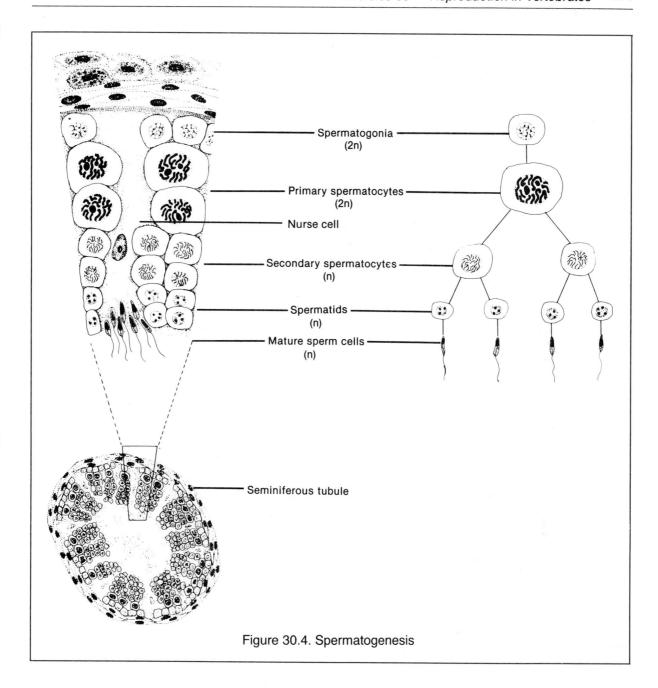

Figure 30.4. Spermatogenesis

Assignment 3

1. Examine a prepared slide of cat testis. Locate the seminiferous tubules, which produce spermatozoa, and the interstitial cells, which produce testosterone, the male hormone. Compare your observations with Figure 30.4, and locate the cells involved in spermatogenesis.

2. Examine the prepared slide of human sperm. Note their small size. About 350 million sperm are released in an ejaculation.

3. Examine a prepared slide of cat ovary. Compare your slide with Figure 30.5. Locate the germinal epithelium, a primary follicle with a primary oocyte, and a Graafian follicle containing a secondary oocyte.

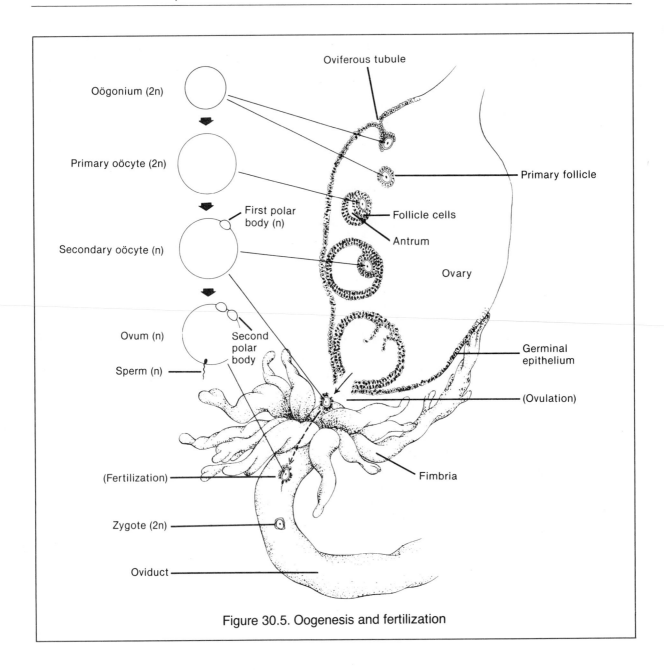

Figure 30.5. Oogenesis and fertilization

4. Examine a prepared slide of cat ovary showing a corpus luteum, which secretes progesterone to maintain the uterine lining.
5. ***Complete item 3 on the laboratory report***.

Birth Control

The control of fertility is one of the major concerns of modern society, not only because of the desire to prevent unwanted pregnancies but also because of the great need to curb a human population growth rate that is out of control.

Table 30.2 indicates the effectiveness of common birth control methods. Descriptions of some of these methods follows.

Tubal ligation is a surgical procedure in which a small section of each oviduct is removed and the cut ends are tied.

Vasectomy is a surgical procedure in which a small section of each vas deferens is removed and the cut ends are tied.

The pill consists of synthetic estrogens and progesterones that inhibit development of ovarian follicles and ovulation.

An **intrauterine device (IUD)** is a metal or plastic device placed in the uterus by a physician, and it remains there for long periods of time. An IUD prevents implantation of an embryo. Studies have shown that the use of an IUD increases the probability of pelvic inflammatory disease and the probability of female sterility.

A **diaphragm** is a dome-shaped device inserted into the vagina and placed over the cervix prior to sexual intercourse. A **cervical cap** is similar to a diaphragm. Both are used with spermicides.

A **sponge** is a spongelike device that is inserted into the vagina and placed against the cervix prior to sexual intercourse. The sponge contains a spermicide.

A **condom** is a thin, tight-fitting sheath of latex or lamb intestine that is worn over the penis during sexual intercourse. It is more effective when used with a spermicide. Only latex condoms provide some protection against sexually transmitted diseases.

Spermicides are chemicals that are lethal to sperm. They are marketed as foams, jellies, creams, and suppositories.

The **rhythm method** involves abstention from sexual intercourse during a woman's fertile period that extends from a few days before to a few days after ovulation. It requires a woman to take her temperature each morning before arising to detect the 0.5 to 1.0°F rise in body temperature that occurs just prior to ovulation.

Withdrawal is the removal of the penis from the vagina just prior to ejaculation. Effectiveness is limited because some sperm are often emitted from the penis prior to ejaculation.

A **douche** is the rinsing out of the vagina after sexual intercourse. It is not very effective

**TABLE 30.2
Effectiveness of Birth Control Methods**

Method	Pregnancies per 100 Sexually Active Women per Year
Abstinence	0
Tubal ligation	0
Vasectomy	0.4
Pill	2
IUD only	5
Condom (high quality)	10
Diaphragm plus spermicide	13
Sponge plus spermicide	17
Rhythm method	24
Spermicide only	25
Withdrawal	26
Condom (poor quality)	30
Douche	60
No birth control method	90

Source: Cecie Starr, *Biology, Concepts and Applications*, (San Francisco, Wadsworth), 1991; p. 484.

since sperm can enter the uterus within 1.5 minutes after being deposited in the vagina.

Materials

Demonstration table of birth control devices and spermicides

Assigment 4

1. Examine the various birth control methods set up as a demonstration. Read the directions for use that accompanies each device or spermicide.
2. *Complete item 4 on the laboratory report*.

31

FERTILIZATION AND EARLY DEVELOPMENT

OBJECTIVES

On completion of the laboratory session, you should be able to:
1. Describe activation and cleavage and identify activated eggs and cleavage stages when observed with the microscope.
2. Describe and identify a blastula and a gastrula.
3. Describe the formation of the germ layers and identify the germ layers in a late gastrula.
4. Indicate the adult tissues and organs formed from the germ layers.
5. Define all terms in bold print.

The union of gametes and early embryological development are difficult to study in many animals, especially chordates. Echinoderms are good subjects for such a study, however, because the gametes are easy to procure and minimal care is needed for the adults and embryos. In addition, embryological development in echinoderms is similar to that in chordates.

The penetration of an egg by a sperm is called **activation**. The subsequent fusion of egg and sperm nuclei is **fertilization**, and this process forms the diploid **zygote**. A series of mitotic divisions called **cleavage** begins and produces progressively smaller cells. As cleavage progresses, a solid ball of cells, the **morula**, is formed, and continued cleavage produces the **blastula**, a hollow ball of cells. This concludes the cleavage process. The blastula is not much larger than the zygote.

Mitotic divisions continue and transform the blastula into a **gastrula** by a process called gastrulation. This stage of early development is completed by the formation of the three **embryonic tissues** or **germ layers**.

In this exercise, you will study activation and early development in the sea urchin. Follow the directions carefully.

PROCUREMENT OF GAMETES

Several sea urchins may be needed to find a male and female since sex cannot be easily determined by external examination.

Materials

Beakers, 50 ml and 100 ml
Dropping bottles
Finger bowls
Hypodermic needles, 22 gauge
Hypodermic syringes, 5 ml
Medicine droppers
Syracuse dishes
Potassium chloride solution, 0.5 M
Seawater at 20°C
Sea urchins

Procedure

1. Inject 1 ml of the 0.5 M potassium chloride solution into each of three to four sea urchins. Insert the hypodermic needle through the membranous region around the mouth as shown in Figure 31.1.

2. Place the urchins on paper towels with the oral (mouth) side down. Watch for the release of the gamete secretions from the aboral surface. The sperm secretion is white, and the egg secretion is pale buff in color.

3. As soon as a female starts shedding, place her on a beaker full of cold seawater, aboral side down so that the aboral surface is in the water. Release of all the eggs will take several minutes. See Figure 31.2.

4. After the eggs have been released, discard the female sea urchin. Swirl the eggs and water to wash the eggs and allow them to settle to the bottom of the beaker. Pour off the water and add fresh seawater. Repeat this washing procedure twice. It will facilitate the activation process.

5. After the final washing, swirl the water to disperse the eggs. Then pour about 25 ml of seawater and eggs into each of five to six finger bowls and keep them at 20°C until used. The eggs will remain viable for two to three days at 20°C when the bowls are stacked to reduce evaporation.

6. Allow several minutes for the sperm secretion to accumulate on the aboral surface of a male. Then remove the secretion with a medicine dropper and place it in a Syracuse dish. See Figure 31.3. Undiluted sperm secretion in a covered dish will be viable for two to three days at 20°C. Prepare a sperm solution, just before use, by placing 2 drops of sperm secretion in 25 ml of seawater and dispense in a dropping bottle.

7. Keep gametes at 20°C until used.

ACTIVATION

Chemicals called **gamones** are released by both the sperm and eggs of sea urchins. They serve to attract sperm to the eggs and also enable the attachment of a sperm to the egg membrane. Similar substances are released by

Figure 31.1. Injection of sea urchin

Figure 31.2. Shedding female on beaker of seawater

Figure 31.3. Removing sperm with a dropper

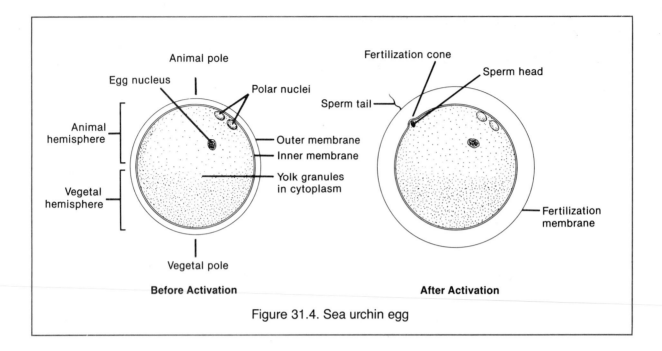

Figure 31.4. Sea urchin egg

human gametes. The first sperm to attach near the **animal pole** of the egg causes activation. The egg extends a **fertilization cone** (really an activation cone) through the egg membranes to engulf the sperm head and draw it into the egg. The sperm tail remains outside the egg. A rapid release of substances from cytoplasmic vesicles into the space between the **inner** and **outer egg membranes** immediately follows and results in the rapid inflow of fluid into this space. The accumulation of fluid pushes the outer membrane farther outward and prevents penetration of another sperm. The outer membrane is now called the **fertilization membrane** (really an activation membrane). Study Figure 31.4. Subsequently, the egg and sperm nuclei fuse to form the diploid nucleus of the zygote in the process of **fertilization**.

EARLY EMBRYOLOGY

After fertilization, the **zygote** begins a series of mitotic divisions that produce successively smaller cells. These divisions are called **cleavage**, and the first division occurs about 45 to 60 min after activation. Succeeding divisions occur at approximately 30-min intervals. The cells formed by the cleavage divisions are called **blastomeres**. See Figure 31.5.

The first cleavage division passes through the **animal** and **vegetal poles** of the zygote to yield cells of equal size. The second division also passes through both poles to form 4 cells of equal size.

The third division is perpendicular to the polar axis and forms 8 cells. The cells of the **animal hemisphere** are slightly smaller than those of the **vegetal hemisphere** due to the greater concentration of **yolk** in the vegetal hemisphere. The fourth division forms 16 cells; however, it forms 4 large cells and 4 tiny cells in the vegetal hemisphere. The tiny cells are called **micromeres**.

The **blastula** is formed about 9 hr after activation and is no longer enveloped by the fertilization membrane. The inner cavity, the **blastocoele**, is filled with fluid. Cilia develop on the outer surfaces of the cells, and their beating provides rotational motion of the blastula.

Continued division of the cells results in the inward growth (invagination) of cells at the vegetal pole led by the micromeres. The **early gastrula** consists of two cell layers. The inner cell layer, the **endoderm**, forms the **embryonic gut** (archenteron), which opens to the exterior via the **blastopore**. The outer cell layer is the

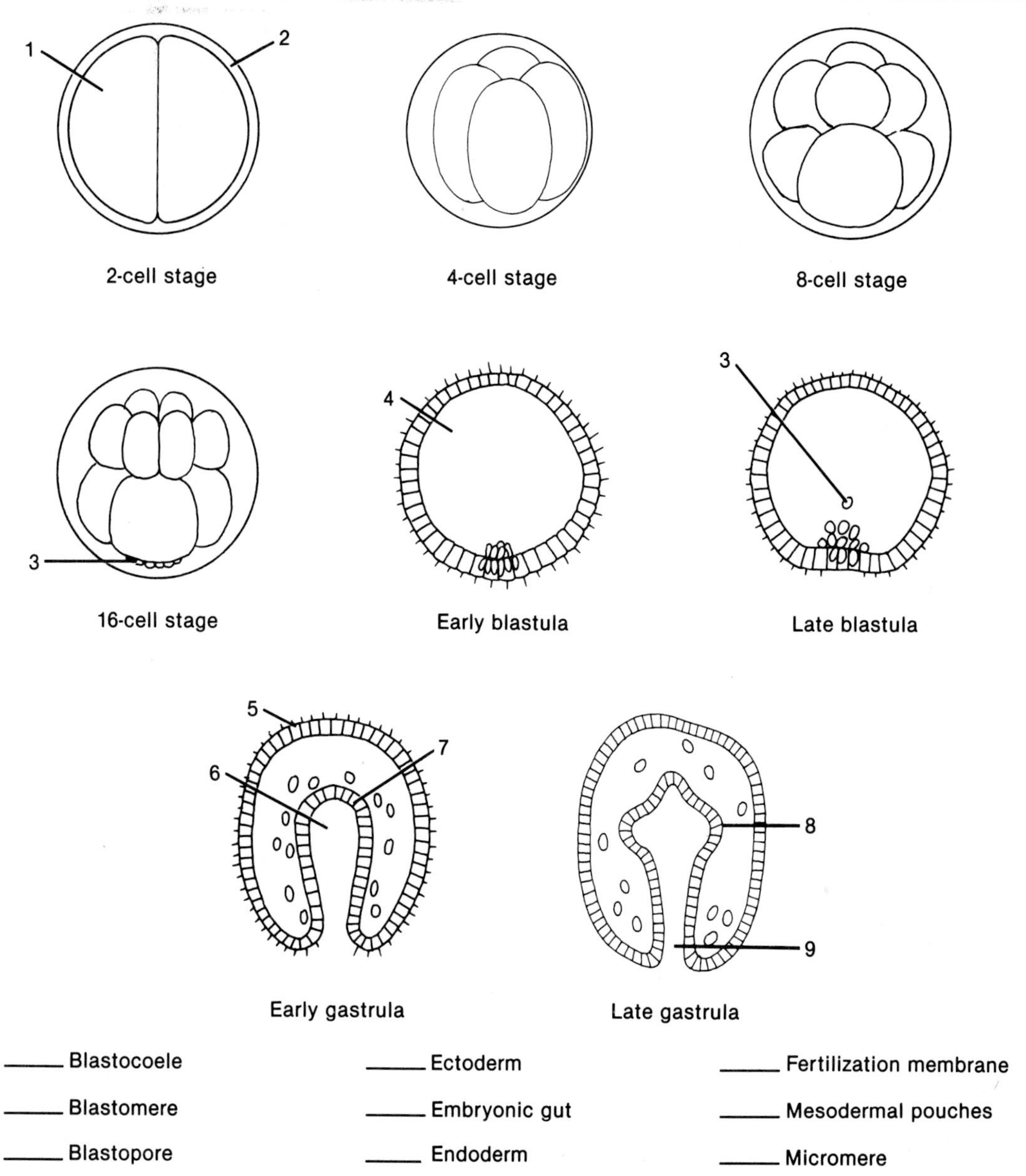

Figure 31.5. Early development in the sea urchin

2-cell stage

4-cell stage

8-cell stage

16-cell stage

Early blastula

Late blastula

Early gastrula

Late gastrula

_____ Blastocoele   _____ Ectoderm   _____ Fertilization membrane

_____ Blastomere   _____ Embryonic gut   _____ Mesodermal pouches

_____ Blastopore   _____ Endoderm   _____ Micromere

ectoderm. In the **late gastrula**, pouches bud off the endoderm to form the **mesoderm**. All three germ layers (embryonic tissues) are now present, and all later-appearing adult tissues and organs are derived from them. See Table 31.1. In both echinoderms and chordates, the blastopore becomes the anus, and a mouth forms later from another opening at the other end of the embryonic gut. In roundworms, mollusks, annelids, and arthropods, the blastopore becomes the mouth and an anal opening forms later.

Materials

Colored pencils
Depression slides and cover glasses
Glass marking pen

TABLE 31.1
Examples of Tissues and Organs Formed from the Germ Layers in Humans

Endoderm	Mesoderm	Ectoderm
Linings of the	Skeleton	Epidermis, including hair and nails
Urinary bladder	Muscles	Inner ear
Digestive tract	Kidneys	Lens, retina, and cornea of the eye
Respiratory tract	Gonads	Brain, spinal cord, nerves, and adrenal
Liver	Blood, heart, and blood vessels	medulla
Pancreas	Reproductive organs	
Thyroid, parathyroids, and	Dermis of the skin	
thymus		

Medicine droppers
Toothpicks
Ward's culture gum
Developing embryos at 3, 6, 12, 24, 48, and
96 hr
Unfertilized eggs in finger bowl of seawater at
20°C
Sperm solution in dropping bottle at 20°C
Prepared slides of sea urchin blastula, gastrula, and larval stages

Assignment 1

Complete item 1 on Laboratory Report 31 that begins on page 463.

Assignment 2

1. *Complete item 2a on the laboratory report.*
2. Place one drop of the egg and seawater mixture (8 to 12 eggs) in a depression slide and observe without a cover glass. Use the $10 \times$ objective. Compare the eggs with Figure 29.4. *Draw two to three eggs in the space for item 2b on the laboratory report.*
3. While the slide is on the microscope stage, add one drop of the sperm mixture at the edge of the depression, record the time, and quickly observe with the $10 \times$ objective. Note how the motile sperm cluster around the eggs. Why? Observe the rapid formation of the fertilization membrane. When most of the eggs have been activated, record the time. *Draw two to three activated eggs in the space for*

item 2b on the laboratory report.

4. Use a toothpick to place a small amount of Ward's culture gum around the depression on the slide and add a cover glass. This prevents evaporation of water but allows passage of O_2 and CO_2. Write your initials on the slide with a glass marking pen.
5. Place your slide, egg mixture, and sperm mixture in the refrigerator at 20°C.
6. *Complete item 2c on the laboratory report.*
7. Label Figure 31.5. Color the cells as follows to distinguish the embryonic tissue:
   ectoderm—blue
   mesoderm—red
   endoderm—yellow
8. Examine your slide of activated eggs at 15- to 20-min intervals, and try to observe the division of the zygote. Keep the slide at 20°C when not observing it.
9. Prepare and observe, in age sequence, slides of different stages of sea urchin development. Make only one slide at a time. Label it by age and with your initials. Keep it at 20°C when not observing it. Examine these slides at 15- to 20-min intervals to observe a cell division. Compare your observations with Figure 31.5.
10. Compare the living blastula, gastrula, and larval stages with the prepared slides.
11. When finished, clean the slides, cover glasses, microscope stage, and objectives to remove all traces of seawater.
12. *Complete item 2 on the laboratory report.*

EXPERIMENTS IN ECHINODERM DEVELOPMENT

Cells of invertebrate embryos other than echinoderms have undergone irreversible chemical and structural changes by the four-cell stage. When the four cells are separated, each is different and incapable of developing into a normal blastula and gastrula and death results. The following experiments were done to investigate (1) the developmental potential of cells in a four-cell sea urchin embryo and (2) any differences between the animal and vegetal hemispheres of the egg.

Experiment 1

The fertilization membranes of embryos at the four-cell stage were removed, and the embryos were placed in calcium-free water to dissolve the "glue" holding the cells together. The cells separated, and each one developed into a normal blastula and gastrula. See Figure 31.6.

Experiment 2

Unfertilized sea urchin eggs were carefully cut in half with a microscopic glass needle along the equator of the cells to form two cells from each embryo, one consisting of the animal hemisphere and the other the vegetal hemisphere.

Since either half would develop with or without the egg nucleus, the presence of the egg nucleus in one of the halves was unimportant. Each cell was activated by a sperm to initiate cleavage.

The animal hemisphere cells formed abnormal blastulas of ciliated ectodermal cells and then died. The vegetal hemisphere cells formed abnormal blastulas composed of cells that were not ectoderm and then formed abnormal gastrulas with a primitive gut of endodermal cells before they died. See Figure 31.7.

Assignment 3

1. Study each experiment and establish a conclusion from the results. Do sea urchin cells at the four-cell stage possess identical developmental potential? What causes identical human twins? Are the animal and vegetal hemispheres of a sea urchin egg structurally and chemically different?
2. **Complete item 3 on the laboratory report**.

CHORDATE DEVELOPMENT

The early development of amphioxus, a primitive chordate, is similar to that observed in the sea urchin. Homologous stages are easily recog-

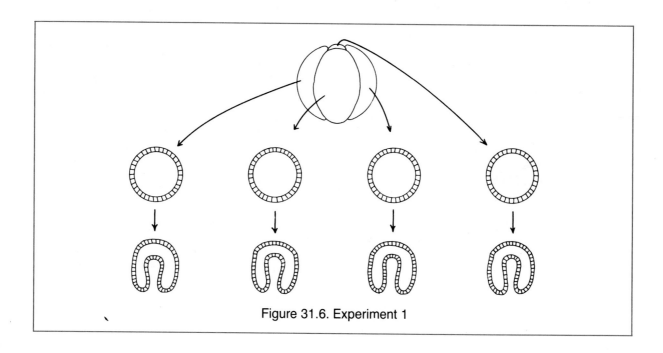

Figure 31.6. Experiment 1

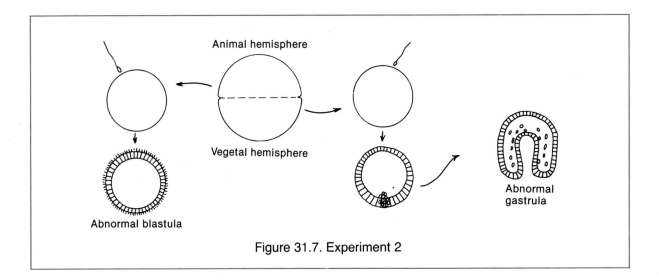

Figure 31.7. Experiment 2

nized since the eggs of both organisms contain relatively little yolk. The homologous stages are more difficult to recognize in eggs with more yolk, especially in bird eggs. This problem exists because cell divisions are slower in regions where yolk is abundant and the cells become much larger.

Materials

Colored pencils
Prepared slides of:
    amphioxus embryonic development
    frog embryo, x.s., neural tube stage

Assignment 4

1. Study Figure 31.8. Note the similarity of development in amphioxus with that of the sea urchin up to the gastrula. Note that (a) the mesoderm is formed from pouches that bud off the endoderm, (b) the mesoderm destined to be the notochord is located be-

tween the mesodermal pouches, and (c) the neural plate is formed by the dorsal part of the ectoderm and folds up to form the neural tube.
2. Color the embryonic tissues in Figure 31.8K-O:
    ectoderm—blue
    neural tube—green
    mesoderm—red
    endoderm—yellow
3. Examine a prepared slide of amphioxus development and locate stages like those in Figure 31.8.
4. Examine a prepared slide of frog embryo, x.s., in the neural tube stage and compare it with a similar stage in amphioxus development shown in Figure 31.8. Locate the ectoderm, neural tube, mesoderm, coelom, endoderm, and embryonic gut. Note the large, yolk-filled endodermal cells forming the ventral part of the gut wall.
5. ***Complete item 4 on the laboratory report.***

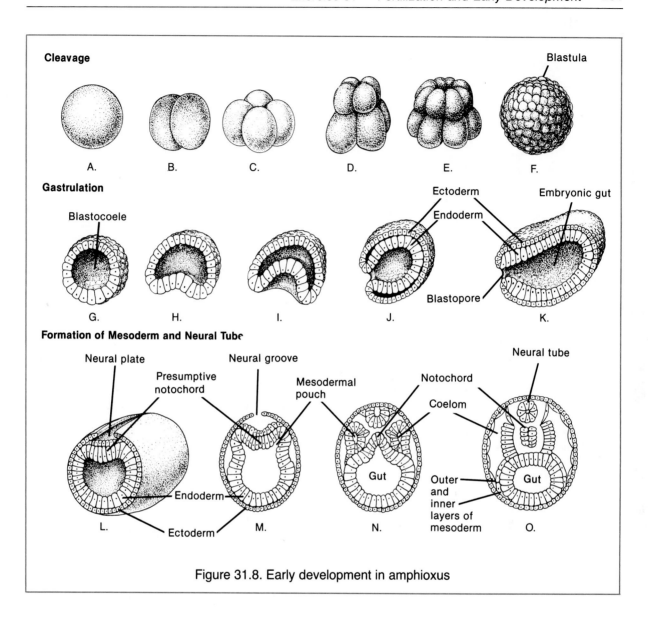

Cleavage

A.   B.   C.   D.   E.   F.

Blastula

Gastrulation

Blastocoele

Ectoderm
Endoderm
Embryonic gut
Blastopore

G.   H.   I.   J.   K.

Formation of Mesoderm and Neural Tube

Neural plate
Presumptive notochord
Neural groove
Mesodermal pouch
Notochord
Neural tube
Coelom
Endoderm
Ectoderm
Outer and inner layers of mesoderm
Gut
Gut

L.   M.   N.   O.

Figure 31.8. Early development in amphioxus

EARLY EMBRYOLOGY
OF THE CHICK

The amniote eggs of reptiles and birds contain a large amount of yolk that serves as a nutrient for the embryo. The yolk never divides but is eventually surrounded by a yolk sac and absorbed. Because of the large amount of yolk, the pattern of development in reptilian and bird eggs is quite different from that of the sea urchin in which the entire fertilized egg cell divides. See Figure 32.1.

STRUCTURE OF A BIRD EGG

The structure of a bird egg is uniquely adapted for the development of the embryo and chick. It consists of the **true egg** and accessory structures. The basic structure is shown in Figure 32.2.

The true egg is surrounded by the **vitelline membrane** and is an extremely large cell containing the fat-rich **yolk** that serves as a nutrient for the embryo. The **blastodisc** is a small, whitish circular area located at the animal pole, and it contains the egg nucleus. All the other components are accessory structures.

The **calcareous shell** protects the embryo from mechanical injury and allows a free exchange of gases with the environment. The inner and outer **shell membranes** allow the passage of gases but prevent evaporative water loss. An air space exists at the rounded end of the egg between the separated inner and outer shell membranes. **Albumin**, the white of the egg, is a clear, viscous protein that serves as a nutrient for the developing embryo. The **chalaza** is a strand of dense protein that suspends the true

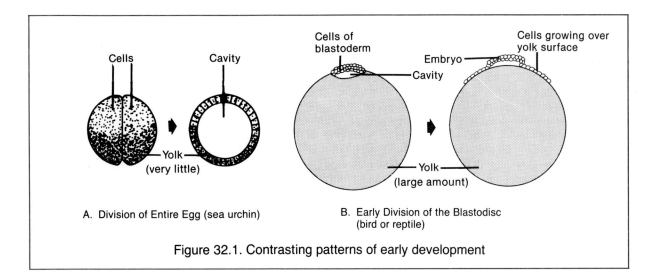

Figure 32.1. Contrasting patterns of early development

A. Division of Entire Egg (sea urchin)

B. Early Division of the Blastodisc (bird or reptile)

egg inside the shell and keeps the blastodisc upward.

Materials

Dissecting instruments
Finger bowls
Paper towels
Scissors, fine-tipped
Unfertilized chicken eggs

Assignment 1

1. Obtain an unfertilized chicken egg from the stock table. *Keeping the same surface of*

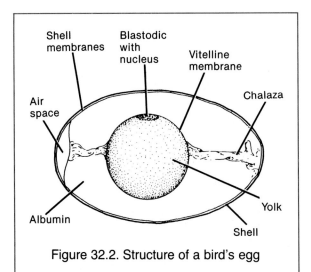

Figure 32.2. Structure of a bird's egg

the shell upward, place it in a finger bowl partially filled with paper toweling to support the egg.

2. Draw a circle with a 2-cm diameter on the upper surface of the shell over the anticipated location of the blastodisc. Use a dissecting needle and forceps to pick away bits of the shell within the circle. Try not to rupture the shell membranes. When the shell has been removed, observe the appearance of the shell membranes. Then use fine-tipped scissors to cut away the shell membranes and remove them.

3. Observe the true egg. The whitish blastodisc should be visible. If necessary, enlarge the opening to see the attachment points of the chalaza.

4. When you have completed your observations, gently break the shell and spill the contents into a finger bowl of water. Observe the attachment of the chalaza to the true egg. Locate the air space at the rounded end of the shell.

5. ***Complete item 1 on Laboratory Report 32 that begins on page 465***.

DEVELOPMENT WHILE IN THE OVIDUCT

After **copulation**, sperm swim to the upper end of the oviduct and contact the ovulated egg. The egg nucleus undergoes its second meiotic di-

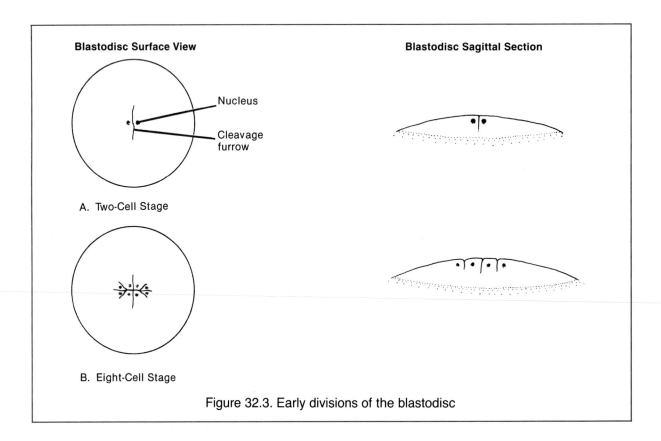

Blastodisc Surface View

Nucleus

Cleavage furrow

A. Two-Cell Stage

B. Eight-Cell Stage

Blastodisc Sagittal Section

Figure 32.3. Early divisions of the blastodisc

vision after a sperm has penetrated the blastodisc. As the fertile egg (zygote) descends the oviduct, the albumin, shell membranes, and a shell are secreted around it. Concurrently, embryonic cell divisions occur in the blastodisc. When an egg is laid (about 20 hr after fertiliza-

tion), the embryo is already at the early gastrula stage.

Figure 32.3A shows the enlarged blastodisc in surface and sagittal views. The zygote nucleus has divided for the first time, and a cleavage furrow separates the two resultant nuclei. This

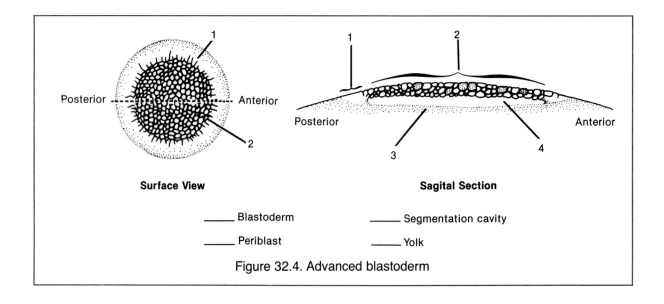

Posterior — Anterior

Posterior

Anterior

Surface View

_____ Blastoderm

_____ Periblast

Sagital Section

_____ Segmentation cavity

_____ Yolk

Figure 32.4. Advanced blastoderm

stage is **homologous** to the two-cell stage of the sea urchin embryo. The eight-cell stage is shown in Figure 32.3B. In the sagittal view, note how the furrows between nuclei extend only partially through the blastodisc.

Continued cell divisions change the central part of the blastodisc into a cellular **blastoderm** several cells in thickness (Figure 32.4). A cavity forms between the blastoderm and the underlying yolk. The peripheral ring of noncellular blastodisc is now called the **periblast**.

The blastoderm and segmentation cavity are considered by some authorities to be homologous to the blastula and blastocoele of sea urchins. The blastoderm cannot form a hollow ball of cells like the sea urchin blastula, however, because the blastoderm develops on the surface of the yolk.

Figure 32.5 shows the stage of embryonic development at the time an egg is laid. The upper layer of small cells (sagittal view) has become **ectoderm** tissue; large granular cells have formed a lower layer of **endoderm** tissue toward the posterior end of the embryo. The endodermal tissue at this stage is crescent shaped in the surface view and represents the partly formed roof of an incomplete **embryonic gut**. The former segmentation cavity is now the cavity of the gut. After the egg is laid, the embryo does not develop any further until it is incubated.

A tubular gut is eventually achieved in the chick embryo, but unlike the rapid gut formation by invagination in the sea urchin, the process is delayed in the chick embryo because of its flattened form.

Assignment 2

1. Study Figure 32.3 and note the incomplete cleavage. Why does this occur?
2. Study and label Figures 32.4 and 32.5.
3. ***Complete item 2 on the laboratory report***.

DEVELOPMENT AFTER INCUBATION BEGINS

Identify in Figure 32.6 the structures described below. A few hours after incubation is started, a **primitive streak** begins to form along the anterior-posterior axis of the embryo by the anterior migration of endoderm cells.

After 16 hr of incubation, the primitive streak is complete and extends entirely across the blastoderm. Note how cells from the blastoderm have spread out over the yolk in an oval pattern (Figure 32.6). The primitive streak consists of an elongated depression, the **primitive groove**, bounded on either side by **primitive ridges**. At the anterior end of the groove is the **primitive**

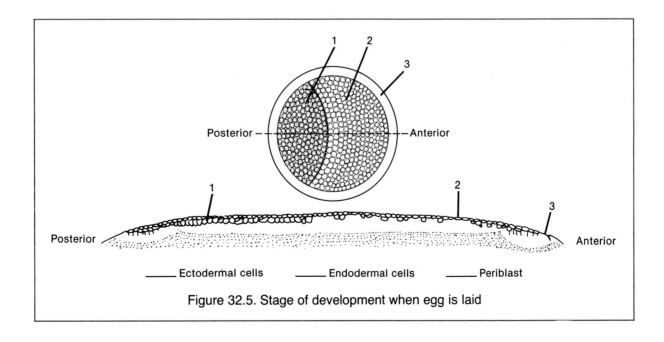

_____ Ectodermal cells   _____ Endodermal cells   _____ Periblast

Figure 32.5. Stage of development when egg is laid

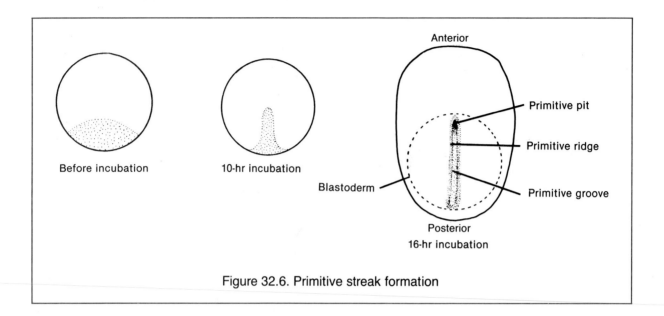

Figure 32.6. Primitive streak formation

pit. The primitive streak and pit together are homologous to the blastopore of the sea urchin embryo.

Cells from the surface of the blastoderm migrate into the primitive groove and pit and spread out in a layer between the ectoderm and endoderm. These migrating cells form **mesoderm** tissue and the **notochord**. Subsequent embryonic development occurs anteriorly from the primitive pit.

The primitive streak and primitive pit will become the anus of the chick embryo; the embryo

therefore develops anterior to the streak. As ectoderm, endoderm, and mesoderm tissues spread anteriorly and laterally over the yolk, they rise up and form a **headfold** in the head region (Figure 32.7) The headfold contains an endoderm-lined cavity that represents the beginning of a **tubular foregut**. Note how relatively small the primitive streak now appears compared to the rest of the embryo.

In Figure 32.8, the neural tube has been modified to form four parts of the developing **brain**, the **optic cups** that will become the eyes, and

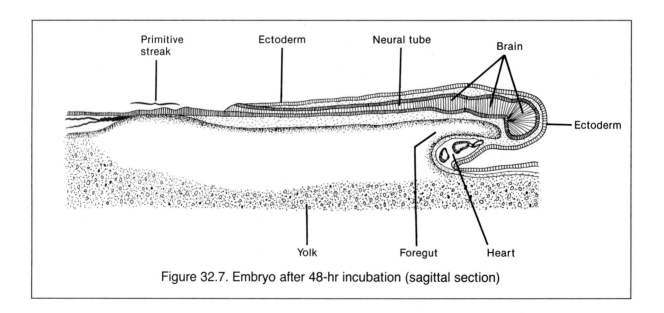

Figure 32.7. Embryo after 48-hr incubation (sagittal section)

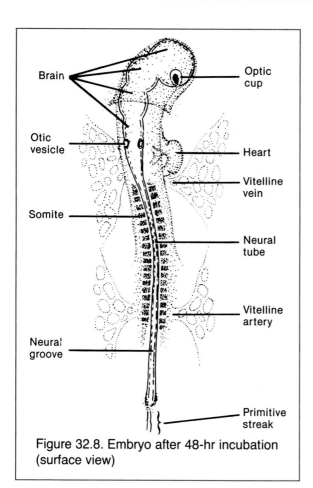

Figure 32.8. Embryo after 48-hr incubation (surface view)

Labels: Brain, Optic cup, Otic vesicle, Heart, Vitelline vein, Somite, Neural tube, Vitelline artery, Neural groove, Primitive streak

Materials

Colored pencils
Models of chick embryonic development
Prepared slide of chick embryo, w.m., 18, 24, 36, 48, and 96 hr
Prepared slides of chick embryo, x.s., 24 hr

Assignment 3

1. Examine a model of the primitive streak stage and compare it to Figure 32.6.
2. Examine a prepared slide of chick embryo, w.m., after 18 hr of incubation. Locate the structures shown in Figure 32.6.
3. Examine the demonstration slide of chick embryo, x.s., 24 hr, which shows mesodermal cells migrating from the primitive groove laterally between ectoderm and mesoderm.
4. Study the models of chick development between 18 and 96 hr of development and examine the demonstration slides.
5. Study Figures 32.7 and 32.8 to understand the structure of a 48-hr chick embryo. Correlate the sagittal section with the whole mount. Color the structures as follows:
     ectoderm—blue
     neural tube—green
     mesoderm—red
     endoderm—yellow
6. Examine prepared slides of 48-hr chick embryo, w.m., and sagittal section. Locate the structures shown in Figures 32.7 and 32.8.
7. Study Figure 32.9 and note the developmental changes between 48 and 96 hr. Color the structures as in step 5.
8. Examine a slide of chick embryo, w.m., 96-hr. Locate the structures shown in Figure 32.9.
9. ***Complete item 3 on the laboratory report***.

STUDY OF A 48-HR EMBRYO

After you understand the structure of a 48-hr chick embryo, observe a living 48-hr embryo. Work in pairs and follow the directions carefully.

Materials

Desk lamp
Incubator, 39°C

the **otic cups** that will become the ears. Posterior to the brain, the neural tube will form the spinal cord. The tubular **heart** is starting to twist, a step toward its development into four chambers. The segmentally arranged **somites** will form the musculature of the body. The **vitelline arteries** carry blood from the embryo to the yolk surface, which facilitates gas exchange and nutrient absorption. Blood is returned to the embryo via the **vitelline veins**.

Refer to Figure 32.9 to see the continuing development in a 96-hr embryo. The more rapid anterior development has caused the brain and head to be much larger than the rest of the body. The **limb buds** and **tail bud** are evident, and the liver and kidneys are starting to form. Note the presence of nonfunctional **pharyngeal slits**.

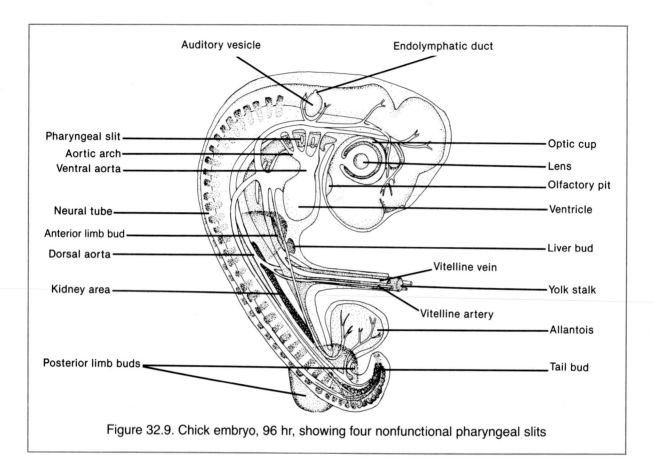

Figure 32.9. Chick embryo, 96 hr, showing four nonfunctional pharyngeal slits

Warming oven, 39°C
Depression slides
Dissecting needles
Filter paper rings with an opening 4 mm ×
  6 mm
Finger bowls
Forceps, label type, square nosed
Forceps, sharp tipped
Scissors, fine tipped
Ringer's solution in dropping bottles
Fertile eggs, 48 hr

Assignment 4

1. Obtain a fertilized egg that has been incubated 48 hr. *Keeping the same surface of the shell upwards,* place it in a finger bowl on paper toweling so that it will not roll. Place a lighted lamp over the egg to keep it warm.

2. Draw a circle with a 2-cm diameter on the shell over the presumed location of the embryo. Using sharp-tipped forceps and a dissecting needle, chip away the shell. Then, using fine-tipped scissors, cut the shell membranes within the circle to expose the embryo. Make the opening larger, if necessary.

3. Locate the embryo, which will appear as a faint white streak in the center of an oval surface. When the large end of the egg shell is to the left, the head of the embryo will be directed away from the observer.

4. Using label forceps, place a filter paper ring over the embryo so that the embryo is visible within the hole in the ring. See Figure 32.10.

5. Allow time for the ring to become wet so that it will adhere to the embryonic membranes. Then use fine-tipped scissors to cut through the embryonic membranes around the outside of the ring. Do not cut deeply into the yolk as this will tend to contaminate your preparation with excessive yolk.

6. Obtain a depression slide from the warming oven and place 2 drops of warm Ringer's solution in the depression. Using the label forceps, lift the paper ring with ad-

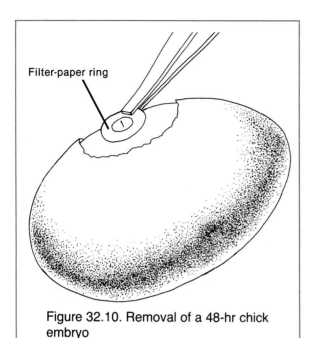

Figure 32.10. Removal of a 48-hr chick embryo

Filter-paper ring

hered embryo and place it in the depression on the slide.

7. Examine the embryo at $40\times$ and $100\times$ with a compound microscope, and identify the structures shown in Figure 32.8.

8. ***Complete item 4 on the laboratory report***.

GENERALIZATIONS

The chemical and structural organization (polarity) of the bird egg influences its pattern of early embryonic development, as was the case in sea urchin development. When incubation begins, certain factors in the external environment (outside the eggshell) exert an influence on the internal environment of the cells of the chick embryo. In turn, the internal environment of these cells selectively activates the genes that control the pattern of development.

33

HEREDITY

OBJECTIVES

On completion of the laboratory session, you should be able to:
1. Explain Mendel's principle of segregation and principle of independent assortment and give examples of each.
2. Solve simple genetic problems involving dominance, recessiveness, codominance, sex linkage, and polygenes.
3. Determine gametes from genotypes where genes are linked or nonlinked.
4. Perform a chi-square analysis.

The **inherited characteristics** of a diploid organism are determined at the moment of sperm and egg fusion. The zygote (2n) receives one member of each chromosome pair from each parent. The genetic information that determines the hereditary traits is found in the structure of the **DNA molecules** in the **chromosomes**. A short segment of DNA that codes for a particular protein constitutes a **gene**, a hereditary unit. Both genes and chromosomes occur in homologous pairs in diploid organisms. See Figure 33.1.

In the simplest situation, an inherited trait, such as flower color, is determined by a single pair of genes. The members of a **gene pair** may be identical (e.g., each codes for purple flowers) or may code for a different variation of the trait (e.g., one codes for purple flowers, and the other codes for white flowers). Again, in the simplest case, only two forms of a gene exist. Alternate forms of a gene are called **alleles**.

Geneticists use symbols (usually letters like "P" or "p") to represent alleles when solving genetic problems. When both members of a gene pair consist of the same allele, such as PP or pp, the individual is **homozygous** for the expressed trait. When the members of the gene pair consist of unlike alleles, such as Pp, the individual is **heterozygous** (hybrid) for the expressed trait.

The genetic composition of the gene pair (e.g., PP, Pp, or pp) is known as the **genotype** of the individual. The observable (expressed) form of a trait (e.g., purple flowers or white flowers) is called the **phenotype**.

An understanding of inheritance patterns enables the prediction of an **expected ratio** for the occurrence of a trait in the progeny (offspring) of parents of known genotypes.

MENDEL'S PRINCIPLES

Gregor Mendel, an Austrian monk, worked out the basic patterns of simple inheritance in 1860, long before chromosomes or genes were associated with inheritance. Mendel's work correctly identified the existence of the units of inheritance now known as genes.

Mendel proposed two principles concerning the activity of genes, and these principles form the basis for the study of inheritance. Look for evidence of these principles as you work through this exercise. In modern terms, these principles may be stated as follows:
1. The **principle of segregation** states that (1) genes occur in pairs and exist un-

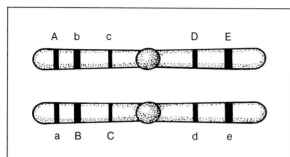

Figure 33.1. Diagrammatic representation of homologous chromosomes and genes

TABLE 33.1
Genotypes and Phenotypes for Flower Color in Garden Peas

Genotypes	Phenotypes
PP	Purple
Pp	Purple
pp	White

changed in the heterozygous state and (2) members of a gene pair are segregated (separated) from each other during gametogenesis, ending up in separate gametes.

2. The **principle of independent assortment** states that genes for one trait are assorted (segregated into the gametes) independently from genes for other traits. This principle applies *only* to traits whose genes are located on different chromosome pairs (i.e., the genes are not linked).

DOMINANT-RECESSIVE TRAITS

When a gene pair consists of two alleles, and one is expressed and the other is not, the expressed allele is **dominant**. The unexpressed allele is **recessive**. Many traits are inherited in this manner. For example, the following traits in garden peas exhibit a dominant/recessive pattern of inheritance. In each case, the dominant trait is in italics, and the dominant allele is capitalized in the genotype.

Flower color: *Purple flowers* (PP, Pp) or white flowers (pp)
Seed color: *Yellow seeds* (YY, Yy) or green seeds (yy)
Seed shape: *Round seeds* (RR, Rr) or wrinkled seeds (rr)
Plant height: *Tall plants* (TT, Tt) or dwarf plants (tt)

Table 33.1 shows the genotypes and phenotypes that are possible for purple or white flow-

ers in peas. Note that the dominant allele is assigned an uppercase P, while the recessive allele is represented by a lowercase p. Only one dominant allele is required for the expression of purple flowers. In contrast, both recessive alleles must be present for white flowers to be expressed in the phenotype. *This relationship is constant for all dominant and recessive alleles.*

Solving Genetic Problems

Consider this genetic problem: What are the expected genotype and phenotype ratios (probabilities) in the progeny of purple-flowering and white-flowering pea plants when each parent is homozygous? Steps used to solve genetic problems such as this are listed in Table 33.2. Note how the steps are used in Figure 33.2 to solve this problem.

The determination of the gametes is a critical step. Recall that meiosis separates homologous chromosomes (and genes) into different gametes. Thus, the members of the gene pair are separated into different gametes. In Figure 33.2, the gametes formed by each parent are identical since each parent is homozygous.

In setting up the Punnett square, the gametes of one parent are placed on the vertical axis, and the gametes of the other parent are placed on the horizontal axis. The number of squares composing a Punnett square depends on the number of different gametes formed by the parents. Four squares are used in Figure 33.2 to enable you to understand the setup, although only one square is actually needed.

All possible combinations of gametes are simulated by recording the gametes on the vertical axis into each square to their right and those on the horizontal axis into each square below them. Note that uppercase letters (dominant alleles) always compose the first letter in each gene pair.

TABLE 33.2
Steps Used to Solve Standard Genetics Problems

1. Be sure that you understand what you are to solve. Write down what is known.
2. Write out the cross using genotypes of the parents.
3. Determine the possible gametes that may be formed.
4. Use a Punnett square to establish the genotypes of all possible progeny.
5. Determine the genotype ratio of the progeny. Count the number of identical genotypes and express them as a ratio of the total genotypes (e.g., 1/4 PP : 2/4 Pp : 1/4 pp).
6. Use the information obtained in step 1 to determine the phenotype ratio from the genotypes (e.g., 3/4 purple flowers to 1/4 white flowers).

When the Punnett square is complete, the individual squares contain the expected genotypes of progeny in the F_1 (first filial) generation. The genotype and phenotype ratios may then be determined.

Test Cross

It is usually not possible to distinguish between homozygous and heterozygous phenotypes exhibiting a dominant trait, but they may be determined by using a **test cross**. In a test cross, the individual exhibiting the dominant phenotype is crossed with an individual exhibiting the recessive phenotype. Recall that an individual exhibiting a recessive phenotype is *always* homozygous for that trait. If all progeny exhibit the dominant trait, the parent with the dominant phenotype is homozygous. If half the progeny exhibit the dominant trait and half exhibit the recessive trait, the parent with the dominant phenotype is heterozygous.

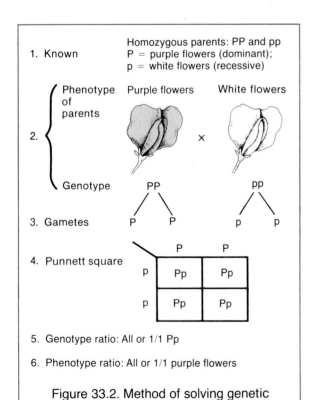

1. Known — Homozygous parents: PP and pp
P = purple flowers (dominant);
p = white flowers (recessive)

2. {
Phenotype of parents — Purple flowers × White flowers

Genotype — PP   pp

3. Gametes   P   P   p   p

4. Punnett square

	P	P
p	Pp	Pp
p	Pp	Pp

5. Genotype ratio: All or 1/1 Pp

6. Phenotype ratio: All or 1/1 purple flowers

Figure 33.2. Method of solving genetic crosses

Assignment 1

Complete item 1 on Laboratory Report 33 that begins on page 467.

Materials

Trays of tall-dwarf and green-albino corn seedlings from monohybrid crosses

Assignment 2

1. Using Figure 33.2 as a guide, determine the expected progeny in the F_2 generation by crossing two members of the F_1 generation, both of which are monohybrids (Pp). A monohybrid is heterozygous for one trait. Complete the Punnett square in Figure 33.3.
2. Once you have completed the Punnett square, determine the genotypes by counting the identical genotypes and recording them as fractions of the total number of genotypes. The different types of genotypes are then expressed as a proportion to establish the expected ratios.

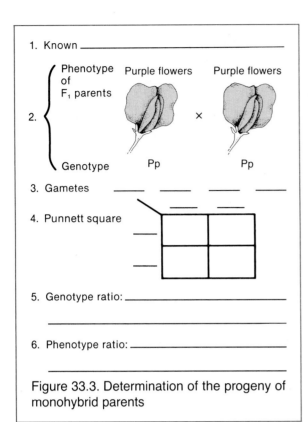

1. Known _____

2. { Phenotype of F₁ parents — Purple flowers × Purple flowers

   Genotype — Pp   Pp }

3. Gametes ____ ____

4. Punnett square

5. Genotype ratio: _____

   _____

6. Phenotype ratio: _____

   _____

Figure 33.3. Determination of the progeny of monohybrid parents

1/4 PP : 2/4 Pp : 1/4 pp
or
1 PP : 2 Pp : 1 pp

The phenotype ratio may then be determined from the genotype ratio. Remember that the presence of a single dominant allele in a genotype produces a dominant phenotype.

3/4 purple-flowering plants : 1/4 white-flowering plants

These ratios are always obtained in a monohybrid cross where the gene consists of only two alleles and one allele is dominant.

3. ***Complete items 2a–2d on the laboratory report.***

4. Examine the corn seedlings. ***Complete items 2e–2j on the laboratory report.***

5. Examine Table 33.3 that illustrates several easily detectable human traits exhibiting a simple dominant/recessive mode of inheritance. ***Complete item 2 on the laboratory report.***

CODOMINANCE

The alleles of some genes are always expressed in the phenotype and are never recessive. Such alleles exhibit **codominance**. The inheritance of color in snapdragons is a suitable example. When a homozygous red-flowering snapdragon is crossed with a homozygous white-flowering snapdragon, the F₁ progeny always have pink flowers since both alleles are expressed.

Sickle-cell anemia in humans is inherited in this manner. The structure of the hemoglobin molecule is controlled by a single gene pair consisting of two alleles: Hb^A for normal hemoglobin and Hb^S for sickle-cell hemoglobin. A person who is homozygous for sickle cell is afflicted with the disease and sometimes dies early in life. The heterozygote has relatively few abnormal hemoglobin molecules, shows no ill effects, and has an increased resistance to malaria.

Assignment 3

1. Determine the expected genotype and phenotype ratios in the F₁ progeny of a cross between homozygous red- and homozygous white-flowering snapdragons. Use R and r for the alleles coding for red and white flowers, respectively.
2. Determine the expected genotype and phenotype ratios of the F₂ generation.
3. ***Complete item 3 on the laboratory report.***

MULTIPLE ALLELES

Some traits are controlled by genes with more than two alleles. The inheritance of ABO blood groups in humans is an example. Three alleles are involved: I^A codes for type A blood, I^B codes for type B blood, and i codes for type O blood. Table 33.4 shows the relationship between genotypes and phenotypes. Note that alleles I^A and I^B are both dominant over i, but that they are codominant to each other.

Assignment 4

Complete item 4 on the laboratory report.

TABLE 33.3
Dominant and Recessive Phenotypes for a Few Human Traits

Trait	Dominant Phenotype	Recessive Phenotype
Ear lobes	Free	Attached
Pigment distribution	Freckles	No Freckles
Hairline	Widow's peak	Straight
Little finger	Bent	Straight
Tongue roller	Yes	No

TABLE 33.4
Phenotypes and Genotypes of the A, B, O Blood Types

Blood Type	Genotype
O	ii
A	$I^A I^A$ or $I^A i$
B	$I^B I^B$ or $I^B i$
AB	$I^A I^B$

DIHYBRID CROSS

In garden peas, yellow seed color is dominant over green seed color, and round seed shape is dominant over wrinkled seed shape. The genes for seed shape and seed color are located on separate chromosomes. Consider a cross of a plant producing yellow-round seeds with a plant producing green-wrinkled seeds. Both plants are homozygous for both traits (seed color and seed

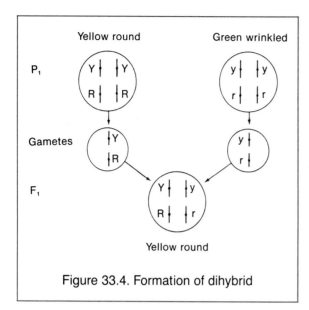

Figure 33.4. Formation of dihybrid

shape). What kinds of seeds will be produced by the progeny?

Based on what is known, you can establish the genotype of the parents and write out the cross:

Yellow-round $\times$ green-wrinkled
YYRR $\qquad$ yyrr

Recalling that the members of a gene pair are separated into different gametes and recogniz-

ing that one member of each gene pair must be in each gamete, you can determine the gametes. Since both parents are homozygous for both traits, each can produce only one type of gamete. Now, you can set up and complete a Punnett square:

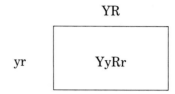

All the progeny are yellow-round and heterozygous for both traits. They are **dihybrids**. Figure 33.4 shows the cross in a way that depicts the chromosomes and genes involved.

What are the expected genotype and phenotype ratios in the F_2 generation? To answer this question, a cross must be made of two F_1 dihybrids with the same phenotype (yellow-round seeds) and genotype (YyRr).

The most difficult part of this problem is determining the gametes. Figure 33.5 and Table 33.5 show two ways to determine the gametes.

Materials

Trays of corn seedlings from a dihybrid cross: tall-green/dwarf-albino

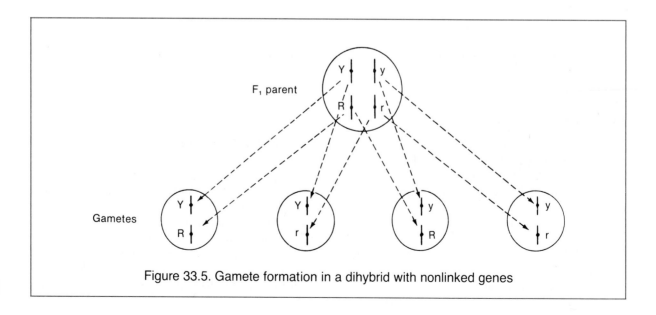

Figure 33.5. Gamete formation in a dihybrid with nonlinked genes

TABLE 33.5
Gamete Determination in a Dihybrid with Nonlinked Genes

Parent Genotype	First Gene Pair	Second Gene Pair		Possible Gametes
YyRr		R	=	YR
	Y			
		r	=	Yr
		R	=	yR
	y			
		r	=	yr

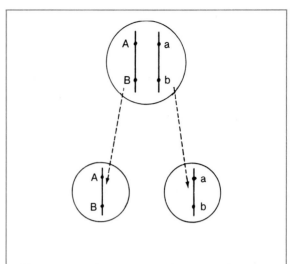

Figure 33.6. Gametogenesis in linked genes

Assignment 5

1. ***Complete item 5a on the laboratory report***. The expected phenotype ratio in dihybrid crosses when each trait is determined by two alleles, one of which is dominant, and when the genes are located on separate chromosomes is always 9 : 3 : 3 : 1.

2. ***Complete item 5 on the laboratory report***.

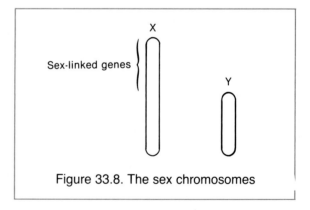

Figure 33.7. Sex inheritance in humans

LINKED GENES

Each chromosome contains many genes that are linked together in a definite sequence. When members of a chromosome pair are separated in gamete formation, the genes of each chromosome tend to remain linked together as a unit. See Figure 33.6.

Sex-Linked Traits

In humans, sex is determined by a single pair of sex chromosomes. Females possess two X chromosomes (XX), and males possess an X and a Y (XY). Sex in humans is inherited as shown in Figure 33.7.

The Y chromosome is shorter than the X and lacks some of the genes present on the X chromosome. Those genes that are present on the X chromosome but absent on the Y chromosome are the **sex-linked genes** that control inheritance of sex-linked traits. See Figure 33.8.

Figure 33.8. The sex chromosomes

For the sex-linked genes, a female is diploid and a male is haploid. Therefore, a recessive allele on the X chromosome of a male will be expressed, whereas the recessive allele must be present on both X chromosomes of a female to be expressed. Common sex-linked traits in humans are red-green color blindness and hemophilia.

Assignment 6

1. Contrast the number of classes of gametes formed by dihybrids when the genes are linked and nonlinked.
2. ***Complete item 6 on the laboratory report***.

Pedigree Analysis

Now that you understand the fundamentals of simple inheritance patterns, it is possible to trace a trait in a pedigree (family tree) to determine if it is inherited in a simple dominant/recessive or a sex-linked pattern of inheritance.

Assignment 7

1. ***Complete item 7 on the laboratory report***.
2. Examine your family photos to see if you can trace in your family the traits noted in Table 33.3.

POLYGENIC INHERITANCE

Traits inherited as dominants or recessives are qualitative traits. For example, seeds of pea plants are either round or wrinkled, yellow or green, and the flowers are either red or white. General observation of plants and animals suggests that some traits are not inherited in this manner, that is, some are quantitative in nature. For example, people are not either short or tall but show a gradation of heights typical of a normal (bell-shaped) curve. Such traits exhibit **polygenic inheritance** in which (1) several genes control the same trait and (2) codominance is evident among the alleles. What other human traits seem to be determined by polygenes?

In wheat, red seeds are codominant with white seeds, and the seeds of an F_1 hybrid are all medium red. Seed color is controlled by two pairs of genes, and the F_2 progeny show the usual genotype ratio expected in a dihybrid cross; however, the phenotype ratio is 15 red to 1 white. Four shades of red are present in the progeny. See Figure 33.9 in which alleles R_1 and

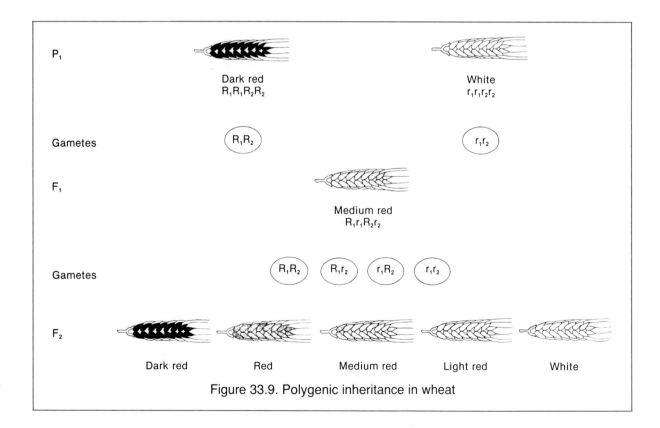

P_1 Dark red $R_1R_1R_2R_2$ White $r_1r_1r_2r_2$

Gametes R_1R_2 r_1r_2

F_1 Medium red $R_1r_1R_2r_2$

Gametes R_1R_2 R_1r_2 r_1R_2 r_1r_2

F_2 Dark red   Red   Medium red   Light red   White

Figure 33.9. Polygenic inheritance in wheat

R_2 code for red seeds and alleles r_1 and r_2 code for white seeds.

Assignment 8

1. Work out the cross of F_1 wheat plants in Figure 33.9 to determine the genotype ratio and to understand the basis of the phenotype ratio.
2. **Complete item 8 on the laboratory report.**

CHI-SQUARE ANALYSIS

To this point in the exercise, you have learned how to predict the expected genotype and phenotype ratios of progeny. However, biologists must verify the expected ratio of a cross to establish the pattern of inheritance. This is done by using the **chi-square** (χ^2) test. This statistical test indicates the probability (ρ) that differences between the expected ratio and the actual ratio are due to chance alone or whether use of a different hypothesis (expected ratio) would be more appropriate to explain the results (observed ratio). The formula for the chi-square test is $\chi^2 = \Sigma(d^2/e)$ where

χ^2 = chi square
Σ = sum of
d = deviation (difference) between expected and observed results
e = expected results

Consider a monohybrid cross involving flower color in garden peas. The predicted phenotype ratio is 3 purple-flowering plants to 1 white-flowering plant. Thus, if 100 plants were produced from the cross, 75 should have purple flowers and 25 should have white flowers. Table 33.6 shows the results of such a cross and the calculation of chi square.

Comparing the calculated value of chi-square (χ^2) with the values in Table 33.7 is necessary to determine the probability (ρ) that the deviation from the expected ratio is either (1) by chance and verifies the predicted ratio or (2) greater than chance and does not support the predicted ratio. Note that the chi-square values are arranged in columns headed by probability values and in horizontal rows by phenotype classes minus one (C − 1).

The two classes of progeny in the example are purple flowers and white flowers. Since 2 classes − 1 = 1, you must look for the calculated chi-square value in the first horizontal row of values. A χ^2 value of 0.48 falls between the columns of 0.50 and 0.20 probability. This means that by random chance the deviation between the expected and actual results will occur between 20% and 50% of the time. Thus, the predicted ratio for the progeny is supported. Probabilities greater than 5% ($\rho \geq 0.05$) are generally accepted as supporting the hypothesis (expected ratio), while those of 5% or less indicate that the results could not be due to chance.

Materials

Corn ears with kernels showing a ratio of:
  3 purple to 1 white
Corn ears with kernels showing a ratio of:
  9 smooth-yellow to 3 smooth-white to 3 shrunken-yellow to 1 shrunken-white
Corn ears formed from experimental crosses

Assignment 9

1. Examine a corn ear with both purple and white kernels that have resulted from

TABLE 33.6
Chi-Square Determination

Phenotype	Actual Results	Expected Results	Deviation (d)	(d²)	(d²/e)
Purple flowers	78	75	3	9	9/75 = 0.12
White flowers	22	25	3	9	9/25 = 0.36
					$\Sigma(d^2/e)$ = 0.48
					χ^2 = 0.48

TABLE 33.7
Chi-Square Values

C − 1	Deviation Insignificant Hypothesis Supported					Deviation Significant Hypothesis Not Supported	
	Probability (*p*)						
	.99	.80	.50	.20	.10	.05	.01
1	.00016	0.64	.455	1.642	2.706	3.841	6.635
2	.0201	.446	1.386	3.219	4.605	5.991	9.210
3	.115	1.005	2.366	4.642	6.251	7.815	11.341
4	.297	1.649	3.357	5.989	7.779	9.488	13.277

monohybrid cross. The predicted ratio is 3 purple to 1 white. Count the purple and white kernels to determine the actual ratio. Mark the row of kernels where you start counting with a pin stuck into the cob under the first kernel. Then do a chi-square analysis of the results in Table 33.8.

2. Examine a corn ear formed by an F_2 dihybrid so that a 9 : 3: 3 : 1 phenotype ratio of the kernels is expected. The two independent phenotypes involved are smooth-shrunken and yellow-white. Smooth and yellow are dominant. Determine the actual

results, and do a chi-square analysis using Table 33.9.

3. ***Complete items 9a to 9d on the laboratory report***.

4. Examine the corn ears from experimental crosses to determine the phenotype ratios of the kernels (progeny) and to determine whether the cross was a monohybrid cross, monohybrid test cross, dihybrid cross, or dihybrid test cross. For each ear, use the following steps:

a. Determine the number of phenotypic classes in the progeny. Construct a chi-square analysis table.

TABLE 33.8
Chi-Square Analysis of Progeny from a Monohybrid Cross

Phenotype	Actual Results	Expected Results	Deviation (d)	(d^2)	(d^2/e)	
Purple kernels	_____	_____	_____	_____	_____	= _____
White kernels	_____	_____	_____	_____	_____	= _____
					$\Sigma(d^2/e)$ =	_____
					χ^2 =	_____

p falls between _____ and _____.   Hypothesis is _____.

TABLE 33.9
Chi-Square Analysis of Progeny from a Dihybrid Cross

Kernel Phenotype	Actual Results	Expected Results	Deviation (d)	(d^2)	(d^2/e)
Smooth-yellow	_____	_____	_____	_____	_____ = _____
Smooth-white	_____	_____	_____	_____	_____ = _____
Shrunken-yellow	_____	_____	_____	_____	_____ = _____
Shrunken-white	_____	_____	_____	_____	_____ = _____

$$\Sigma(d^2/e) = \underline{\hspace{2cm}}$$

$$\chi^2 = \underline{\hspace{2cm}}$$

p falls between _____ and _____.   Hypothesis is _____.

b. Count and record the number of progeny in each class.
c. Inspect the number of progeny in each class and form a hypothesis as to which type of class is probably involved. Record the expected numbers of progeny in each class, as predicted by the hypothesis, in the chi-square table.
d. Perform a chi-square analysis to determine if the data support your hypothesis.
5. *Complete the laboratory report.*

MOLECULAR AND CHROMOSOMAL GENETICS

Chromosomes are responsible for transmitting the hereditary material from cell to cell in cell division and from organism to progeny in reproduction. This is why the distribution of replicated chromosomes in mitotic and meiotic cell divisions is so important in eukaryotic cells. The genetic information is contained in the structure of **deoxyribonucleic acid (DNA)**, which forms the hereditary portion of the chromosomes.

DNA AND THE GENETIC CODE

DNA is a long, thin molecule consisting of two strands twisted in a spiral arrangement to form a double helix somewhat like a twisted ladder. See Figure 34.1. The sides of the ladder are formed of sugar and phosphate molecules, and the rungs are formed by nitrogenous bases joined together by hydrogen bonds.

Each strand of DNA consists of a series of **nucleotides** joined together to form a polymer of nucleotides. Each nucleotide of DNA is formed of three parts: (1) a deoxyribose (C_5) sugar, (2) a phosphate group, and (3) a nitrogenous base. Four kinds of nitrogenous bases are in DNA. The purine bases (double-ring structure) are **adenine** (A) and **guanine** (G). The pyrimidine bases (single-ring structure) are **thymine** (T) and **cytosine** (C). Note the **complementary pairing** of the bases in Figure 34.1. Can you discover a pattern to their pairing? It is the sequence of nucleotides with their respective purine or pyrimidine bases that contains the genetic information of the DNA molecule.

DNA Replication

Each DNA molecule is able to replicate itself during interphase of the cell cycle and thereby maintain the constancy of the genetic information in new cells that are formed. **Replication** begins with the breaking of the weak hydrogen bonds that join the nitrogen bases of the nucleotides. This results in the separation of the DNA molecule into two strands of nucleotides. See Figure 34.2. Each strand then serves as a **template** for the synthesis of a complimentary strand of nucleotides that is formed from nucleo-

301

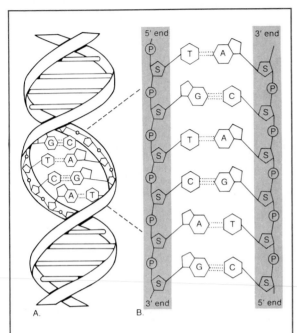

Figure 34.1. DNA structure.

A. The double helix of a DNA molecule. Complementary pairing of the nitrogenous bases joins the sides like rungs of a twisted ladder. A = adenine, G = guanine, C = cytosine, and T = thymine. B. If a DNA molecule is untwisted, it would resemble a ladder in shape. The sides of the ladder are formed of deoxyribose sugar and phosphate, and the rungs consist of nitrogenous bases.

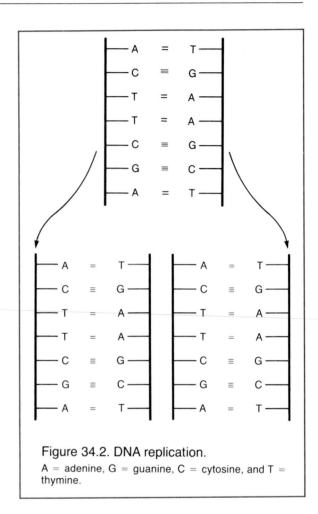

Figure 34.2. DNA replication.

A = adenine, G = guanine, C = cytosine, and T = thymine.

tides available in the nucleus. The complementary pairing of nitrogen bases determines the sequence of nucleotides in the new strands and results in the formation of two DNA molecules that are identical. Since each new DNA molecule contains one "old" strand and one "new" strand, replication is said to be **semiconservative**. Occasionally, errors are made during replication, and such errors are a type of **mutation**. Of course, replication is controlled by a series of enzymes that catalyze the process.

Materials

Colored pencils
DNA, RNA, and protein synthesis kits

Assignment 1

1. Color-code the nitrogenous bases in Figure 34.1 and circle one nucleotide.

2. *Complete items 1a to 1c on Laboratory Report 34 that begins on page 473.*
3. Use a DNA kit to construct a segment of a DNA molecule that matches the base sequence of the DNA segment in item 1c on the laboratory report.
4. *Complete item 1d on the laboratory report.*
5. Use a DNA kit to construct a segment of a DNA molecule that matches the "old" nonreplicated DNA segment in item 1d on the laboratory report. Then separate the strands and construct the replicated strands as shown in item 1d.

RNA Synthesis

DNA serves as the template for the synthesis of **ribonucleic acid (RNA)**. RNA differs from DNA in three important ways: (1) it consists of a single strand of nucleotides, (2) its nucleotides contain ribose sugar instead of deoxyribose sugar, and (3) **uracil** (U) is substituted for thymine as one of the four nitrogenous bases.

To synthesize RNA, a segment of a DNA molecule untwists and the hydrogen bonds between the nucleotides are broken. The nucleotides of one strand pair with complementary RNA nucleotides in the nucleus. When the RNA nucleotides are joined by sugar-phosphate bonds, the RNA strand is complete, and it separates from the DNA strand. Few or many RNA molecules may be formed before the DNA strands reunite.

Assignment 2

1. ***Complete item 2 on the laboratory report.***
2. Use a DNA-RNA kit to synthesize an RNA molecule with a base sequence identical to the hypothetical RNA molecule in item 2 on the laboratory report.

Protein Synthesis

The genetic information of DNA functions by determining the kinds of protein molecules that are synthesized in the cell. A sequence of three bases—a base triplet—in a DNA molecule has been shown to code indirectly for an amino acid. By controlling the sequence of amino acids, DNA determines the kind of protein produced. Recall that enzymes are proteins and that the chemical reactions in a cell are controlled by enzymes. Thus, DNA indirectly controls cellular functions by controlling enzyme production.

Each of the three types of RNA molecules plays an important role in protein synthesis. **Messenger RNA (mRNA)** is a complement of the genetic information of DNA. A **transcription** of the genetic information in DNA is made when mRNA is synthesized. A base triplet of mRNA is called a **codon**, and the codons for the 20 amino acids have been determined. The genetic information is carried by mRNA as it passes from the nucleus to the **ribosomes**, sites of protein synthesis in the cytoplasm.

Transfer RNA (tRNA) carries amino acids to the ribosomes. At one end of a tRNA molecule are three nitrogenous bases, an **anticodon**, which is complementary to a codon of the mRNA. At the other end is an attachment site for 1 of the 20 types of amino acids.

The codon of mRNA and anticodon of tRNA briefly join to place a specific amino acid in its position in a polypeptide chain. This interaction takes place on the surface of a ribosome containing the necessary enzymes for the reaction. **Ribosomal RNA (rRNA)**, the third type of RNA, is an integral component of ribosomes, and it plays an important role in decoding the mRNA message to enable protein synthesis. The formation of an amino acid chain is the **translation** of the genetic information.

Figure 34.3 depicts the interaction of mRNA, tRNA, and rRNA in the formation of a polypeptide. Note how the sequence of the amino acids is controlled by the pairing of the codons and anticodons. It may be simplified as follows:

Transcription   **Translation**

Base triplets → codons → polypeptide
of DNA   of RNA

The Genetic Code

In protein synthesis, the genetic information inherent in the sequence of base triplets in DNA is transcribed into the sequence of codons in mRNA which, in turn, are translated into the sequence of amino acids in a polypeptide chain. In this way, DNA determines both the kinds of amino acids and their sequence in proteins that are synthesized.

There are 64 possible combinations of nucleotide bases in mRNA codons. Their translation is shown in Table 34.1. Note that AUG specifies the amino acid methionine and also is the start signal for protein synthesis. Three codons, UUA, UAG, and UGA, do not specify an amino acid, but they signal the ribosomes to stop assembling the polypeptide chain. Most amino acids are specified by more than one codon, but no codon specifies more than one amino acid. Thus, the code is **redundant** but not **ambiguous**.

Mutations

Base-pair substitution, deletion, or addition constitutes a mutation in a gene. The effect of such mutations is variable, depending on how the mutation is translated via the genetic code.

For example, if a base substitution mutation resulted in a codon change from GCU to GCC,

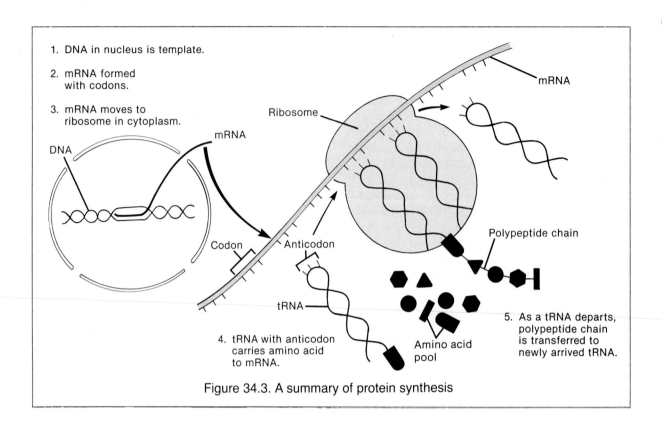

1. DNA in nucleus is template.

2. mRNA formed with codons.

3. mRNA moves to ribosome in cytoplasm.

4. tRNA with anticodon carries amino acid to mRNA.

5. As a tRNA departs, polypeptide chain is transferred to newly arrived tRNA.

Figure 34.3. A summary of protein synthesis

TABLE 34.1
The Codons of mRNA and the Amino Acids That They Specify

AAU AAC	Asparagine	CAU CAC	Histidine	GAU GAC	Aspartic acid	UAU UAC	Tyrosine
AAA AAG	Lysine	CAA CAG	Glutamine	GAA GAG	Glutamic acid	UAA UAG	(Stop)*
ACU ACC ACA ACG	Threonine	CCU CCC CCA CCG	Proline	GCU GCC GCA GCG	Alanine	UCU UCC UCA UCG	Serine
AGU AGC	Serine	CGU CGC		GGU GGC		UGU UGC	Cysteine
AGA AGG	Arginine	CGA CGG	Arginine	GGA GGG	Glycine	UGA UGG	(Stop)* Tryptophan
AAU AUC AUA	Isoleucine	CUU CUC CUA	Leucine	GUU GUC GUA	Valine	UUU UUC	Phenylalanine
AUG	Methionine and start	CUG		GUG		UUA UUG	Leucine

*Signals the termination of the polypeptide chain.

there would be no effect since both specify the amino acid alanine. But if the change was from GCU to GUU, valine would be substituted for alanine in the polypeptide chain and may have a marked effect on the protein. Similarly, if UAU mutated to UAA, it would terminate the polypeptide chain at that point instead of adding tyrosine to the chain. This likely would form a nonfunctional protein.

The mutation involving the addition or deletion of one or two base pairs will cause a **frameshift** in the reading of the codons that may either terminate the polypeptide chain or insert different amino acids into the chain. Usually, addition or deletion mutations have a more disastrous effect than substitution mutations.

Assignment 3

1. Add the bases of the DNA template and the anticodons of tRNA in Figure 34.4.
2. **Complete items 3a to 3c on the laboratory report**.
3. Use a DNA-RNA-protein synthesis kit to synthesize an amino acid sequence as shown in item 3c on the laboratory report starting with the DNA template.
4. **Complete item 3 on the laboratory report**.
5. Use a DNA-RNA-protein synthesis kit to synthesize the amino acid sequences determined in items 3e and 3f on the laboratory report.

HUMAN CHROMOSOMAL DISORDERS

In the study of cell division, you learned that (1) chromosomes occur in pairs in diploid cells, (2) chromosomes are faithfully replicated and equally distributed in mitotic cell division, and (3) cells formed by meiotic cell division receive only one member of each chromosome pair. In both types of cell division, the distribution of the chromosomes is systematically controlled, but errors sometimes occur. In this section of the exercise, you will consider human genetic defects, **chromosomal aberrations**, in which whole chromosomes or large parts thereof are missing or added. Since you now understand the role of DNA, you can appreciate the effect of the deletion or addition of large amounts of DNA on normal cellular function.

The 46 chromosomes of human body cells are classified as 22 pairs of **autosomes**, nonsex chromosomes, and 1 pair of **sex chromosomes**, XX in females and XY in males. Normally, the separation of chromosomes in meiosis of gametogenesis places 22 autosomes and 1 sex chromosome in each gamete, but occasionally errors

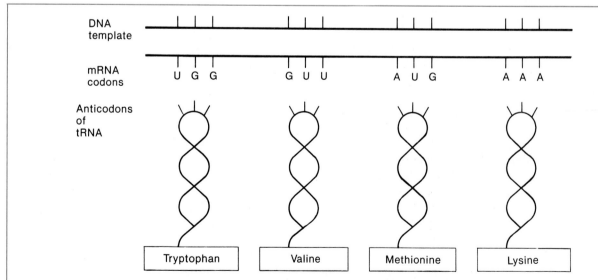

Figure 34.4. Interaction of DNA, mRNA, and tRNA in protein synthesis. Complete the figure by adding the bases in the anticodons of tRNA and the base triplets of DNA.

TABLE 34.2
Examples of Chromosomal Abnormalities

Chromosomal Abnormality	Effect
Trisomy 18	E syndrome; usually fatal within 3 mo due to multiple congenital defects
Trisomy 21	Down's syndrome: mental retardation, short and incurved fifth finger, marked creases in palm, characteristic facial appearance
Deletion from short arm of chromosome 5	Cri-du-chat syndrome: mental and physical retardation, round face, plaintive catlike cry, death by early childhood
XXY	Klinefelter's syndrome (male): underdeveloped testes, breasts enlarged, usually sterile, and mentally retarded
XO	Turner's syndrome (female): underdeveloped ovaries, no ovulation or menstruation

place both members of a chromosome pair in the same gamete. As a result, another gamete lacks this chromosome entirely. If either of these gametes participates in fertilization, the resulting zygote will have an abnormal chromosome number and will be minimally or severely affected, depending on the chromosome involved. Severe defects or death usually occur. For example, if the zygote contains an extra copy of chromosome 21, the presence of three 21 chromosomes (trisomy 21) results in Down's syndrome. See Table 34.2. In contrast, the loss of a chromosome 21, like the loss of any autosome, is lethal.

Chromosomal abnormalities also may stem from the **translocation** of a portion of one chromosome to a member of a different chromosome pair, resulting in reduced or extra chromosomal material in a gamete. Table 34.2 indicates a few disorders caused by chromosomal abnormalities.

Cytogeneticists are able to determine some of the abnormalities among chromosomes by examining them at the metaphase stage of mitosis. A photograph of the spread chromosomes is taken and enlarged. Then, the chromosomes are cut out one at a time from the photo and sorted on an analysis sheet to form a **karyotype**, an arrangement of chromosome pairs by size that allows determination of the chromosome number and any abnormalities that can be visually identified.

Examine the normal male karyotype in Figure 34.5. Note that the chromosomes are sorted into seven groups, A through G, on the basis of size and the location of the centromere. This process

separates the sex chromosomes, X and Y, from each other in the karyotype.

Materials

Scissors
Forceps
Prepared slide of human chromosomes
Rubber cement or glue stick

Assignment 4

1. Examine a prepared slide of human chromosomes. Note their small size and replicated state.
2. Study Figure 34.5 to understand how chromosomes are sorted in preparing a karyotype.
3. Prepare a karyotype from the chromosome spread for each patient in Figure 34.6, one patient at a time. Cut out the chromosomes one by one and place them on the karyotype analysis forms on the laboratory report with the centromere on the dashed line. Do not glue them in place until you are sure of their correct positions. Use the size of the chromosome and the position of the centromere to determine the correct position of each chromosome. Refer to Figure 34.5 as needed.
4. If an abnormality exists, refer to Table 34.2 to determine the specific defect.
5. ***Complete item 4 on the laboratory report.***

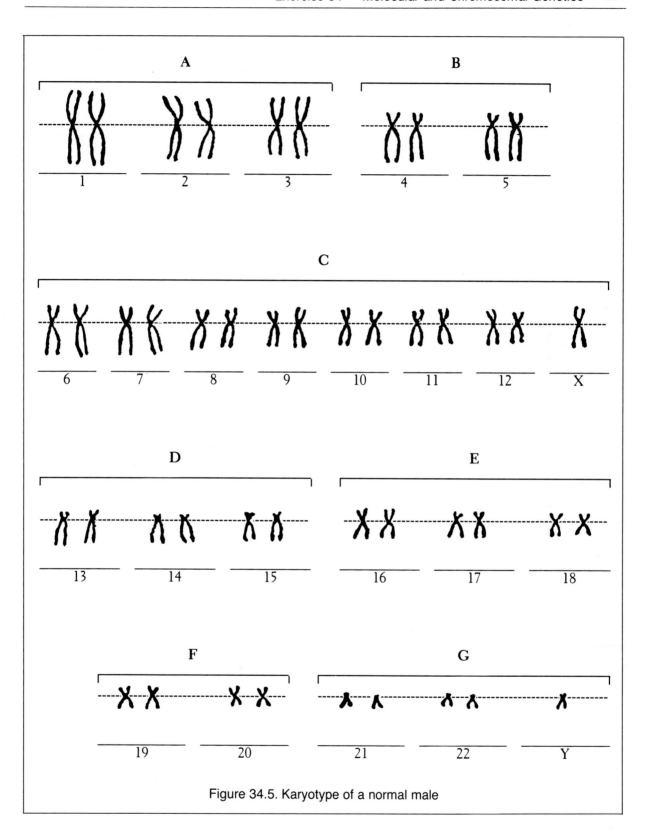

Figure 34.5. Karyotype of a normal male

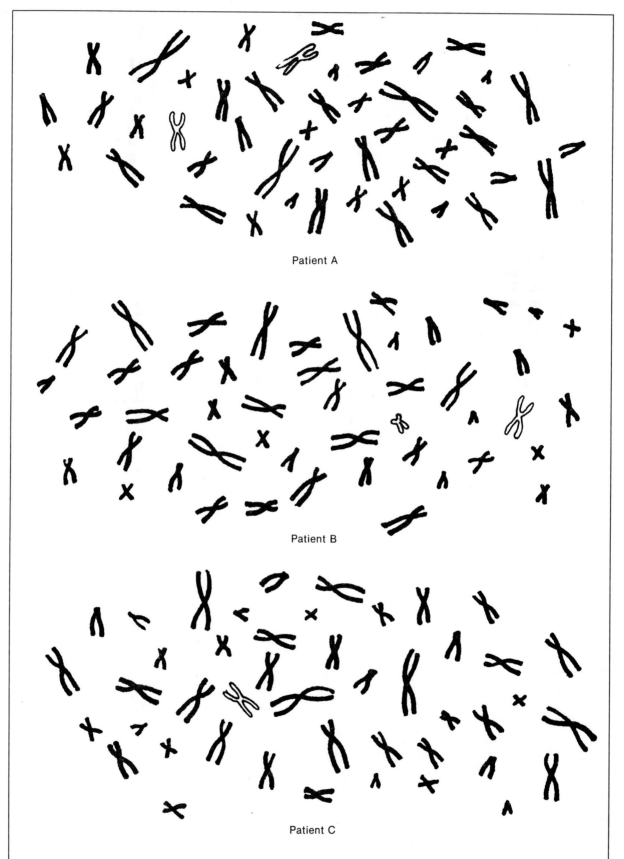

Figure 34.6. Metaphase chromosomes of three patients. Sex chromosomes are shown in outline.

35

EVOLUTION

In 1858, Darwin presented his theory of evolution by natural selection, supported by a mass of evidence from various sources. Much of his evidence is still considered valid today. It was drawn from the fields of **paleontology**, **comparative morphology**, **comparative embryology**, and **biogeography**. Important new evidence has since been obtained from the fields of **biochemistry**, **physiology**, and **genetics**.

The fact that evolution has occurred since the first appearance of living organisms is well established, although biologists may quibble over the exact mechanisms producing it. Two current hypotheses are gradualism (microevolution) and punctuationalism (macroevolution). The **grad-**ualistic hypothesis** proposes that new species result from an accumulation of continual, successive, small genetic changes. The **punctuational hypothesis** proposes that species are rather static except during relatively brief periods when substantive genetic changes occur to form new species and higher taxonomic categories. These two processes may both play a role in evolution.

THE FOSSIL RECORD

The fossil record shows changes that have occurred over millions of years and establishes the **phylogeny** (evolutionary relationships) in the descent from simpler to more complex organisms. Of course, the fossil record is incomplete. Considering that extremely few organisms become fossilized, that relatively few fossils survive normal geologic processes, and that the discovery of such fossils is limited, the fossil record is surprisingly good.

A fossil may be prints or casts left by organisms or an organism itself whose tissues have been replaced by minerals. The age of fossils is determined by establishing the age of the sedimentary rock in which they are found. Age-dating methods using the normal decay of radioactive elements have been especially useful.

A summary of the fossil record for a few major organismic groups is shown in Figure 35.1. The length of the bars indicates the appearance and persistence of the groups, and the width of the bars shows the number of species present.

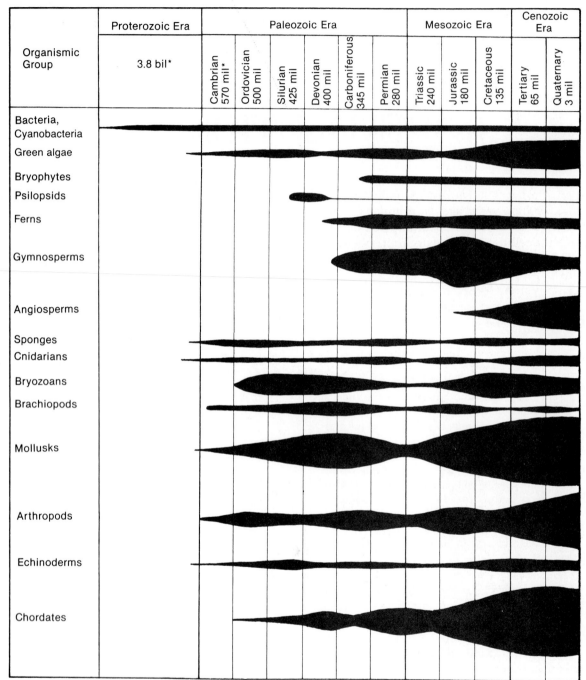

Organismic Group	Proterozoic Era	Paleozoic Era						Mesozoic Era			Cenozoic Era	
	3.8 bil*	Cambrian 570 mil*	Ordovician 500 mil	Silurian 425 mil	Devonian 400 mil	Carboniferous 345 mil	Permian 280 mil	Triassic 240 mil	Jurassic 180 mil	Cretaceous 135 mil	Tertiary 65 mil	Quaternary 3 mil
Bacteria, Cyanobacteria												
Green algae												
Bryophytes												
Psilopsids												
Ferns												
Gymnosperms												
Angiosperms												
Sponges												
Cnidarians												
Bryozoans												
Brachiopods												
Mollusks												
Arthropods												
Echinoderms												
Chordates												

*Approximate beginning date.

Figure 35.1. Fossil record of major organismic groups

The geographic distribution of fossil and modern organisms is largely explained by the movement of the continents. The earth's crust is divided into seven major plates and a number of smaller plates. The relative movement of these plates accounts for movement of the continents and variations in climate. Figure 35.2 shows the relative positions of the land masses at selected times in geologic history. Refer to this figure as you consider the key events of the geologic era and periods.

The **Proterozoic era** was dominated by prokaryotic forms. The earliest fossils are bacteria from 3.5 billion years ago. Unicellular green algae and fungi appeared about 900 million years ago. Multicellular algae date from about 650 million years ago, and multicellular soft-bodied animals appeared about 600 million years ago. Note the number of organismic groups that appeared during the Proterozoic.

During the **Cambrian period**, the land masses were located in tropical latitudes. *Gondwana* was a very large land mass that wrapped around about half of the earth. *Laurentia* was the smaller land mass that later became part of North America. The warm, shallow seas flourished with multicellular algae and marine arthropods (crustaceans), especially trilobites. Other marine animalas included sponges, coelenterates, brachiopods, mollusks, and echinoderms.

In the **Ordovician**, marine algae and invertebrates were dominant. Trilobites, primitive marine arthropods, were abundant, and the radiation of brachiopods and bryozoans was especially rapid. Cephalopods and jawless fish, the first vertebrates, appeared. Later in this period, Gondwana began moving toward the south pole.

In the **Silurian**, Gondwana was concentrated near the south pole, and later began its movement northward. The colonization of land by plants began. The psilopsids, which include club mosses and horsetails, were probably the first vascular plants. Jawed fish appeared, and arthropods invaded the land. Many marine crustaceans and cephalopods became extinct.

The **Devonian** was dominated by fish. The various fish groups appeared, including lungfish and lobe-finned fish. By the end of the period, the first amphibians had evolved from lobe-finned fish, and the vertebrate invasion of the land began. Among land plants, club mosses and

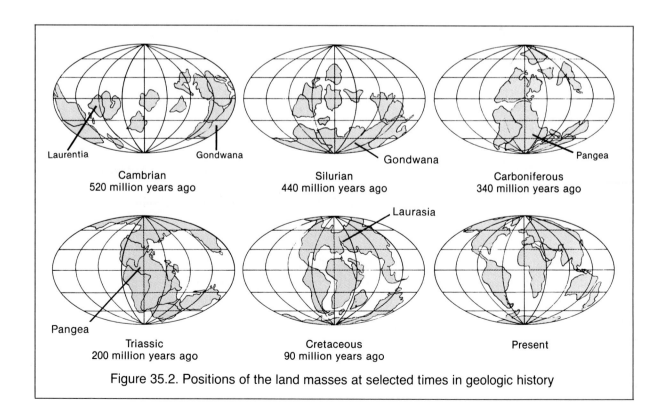

Figure 35.2. Positions of the land masses at selected times in geologic history

horsetails were common, and ferns made their appearance, apparently evolving from psilopsids Later in the period, seed ferns and gymnosperms appeared, apparently evolving from ferns.

The **Carboniferous** was the age of amphibians when immense swamp forests of club mosses, horsetails, and ferns dominated the low-lying land masses. The covering of these coastal forests by sand and silt have formed the coal deposits of today. Insects radiated rapidly, and the reptiles arose from amphibians. The appearance of bryophytes in this period, long after the evolution of vascular plants, is an exception to the general pattern of evolutionary events.

About 300 million years ago, the land masses joined together to form a giant, unstable land mass called *Pangea* that remained intact until late in the Mesozoic era. At the beginning of the **Triassic**, a period of mountain building and reduction of the land-sea interface (due to the land masses joining together) produced cooler climates and resulted in the extinction of many species. Over 75% of the marine invertebrate species disappeared with the reduction of the shallow seas, and many land forms became extinct as well. Species of reptiles increased, but species of amphibians decreased. This large-scale extinction is known as the Permo-Triassic Crisis.

The **Mesozoic era** was the age of dinosaurs. They appeared in the **Triassic**, reached their peak in the early **Cretaceous**, and then rather quickly became extinct. The radiation of reptiles produced many diverse groups. Therapsids gave rise to the first mammals in the late **Triassic**. Birds evolved from thecodont reptiles in the **Jurassic**. Gymnosperms and ferns were the dominant land plants. Angiosperms appeared in the **Triassic**, possibly evolving from seed ferns, and became common in the Cretaceous as gymnosperms and ferns declined.

In the late Mesozoic, Pangea broke up, and the land masses began to disperse toward their present positions. The **Laramide Revolution**, a time of volcanic activity and mountain building, contributed to a cooler climate. The oceans receded from the land masses. Fragmentation of the extensive tropical forests led to the emergence of grasslands in the mid-Cenozoic.

The **Cretaceous extinction** occurred at this time (65 million years ago). Over half of the ma-rine species and many land forms, including dinosaurs, became extinct. Evidence suggests that an asteroid collided with the earth at about this time, but just how this contributed to the mass extinction is not clear.

The climatic change and the extinctions opened up new vistas for surviving forms. Angiosperms became the dominant land plants, and mammals and birds became the dominant vertebrates in the **Tertiary**.

Materials

Representative fossils

Assignment 1

1. Study Figures 35.1 and 35.2. Note the initial appearance of each group and the variation in the number of fossil species.
2. Examine the fossils available. Mark the position of each fossil on Figure 35.1 and **complete items 1a to 1e on Laboratory Report 35 that begins on page 481**.
3. Study Figure 35.3, and compare it with the preceding discussion. Use Figure 35.3 to label the vertebrate phylogenetic tree, and add the geologic periods in Figure 35.4.
4. Complete Figure 35.5 by adding the geologic periods in the left column and labeling each of the lines in the plant phylogenetic tree.
5. **Complete item 1 on the laboratory report**.

EVIDENCE FROM VERTEBRATE EMBRYOLOGY

Figure 35.6 depicts vertebrate embryos in comparable stages of development. The general similarity impressed early biologists who considered it as prime evidence of the relationship of vertebrate groups. For example, human embryos exhibit pharyngeal pouches at one stage of development, although they never have functional gills.

Materials

Models of vertebrate embryos at comparable stages

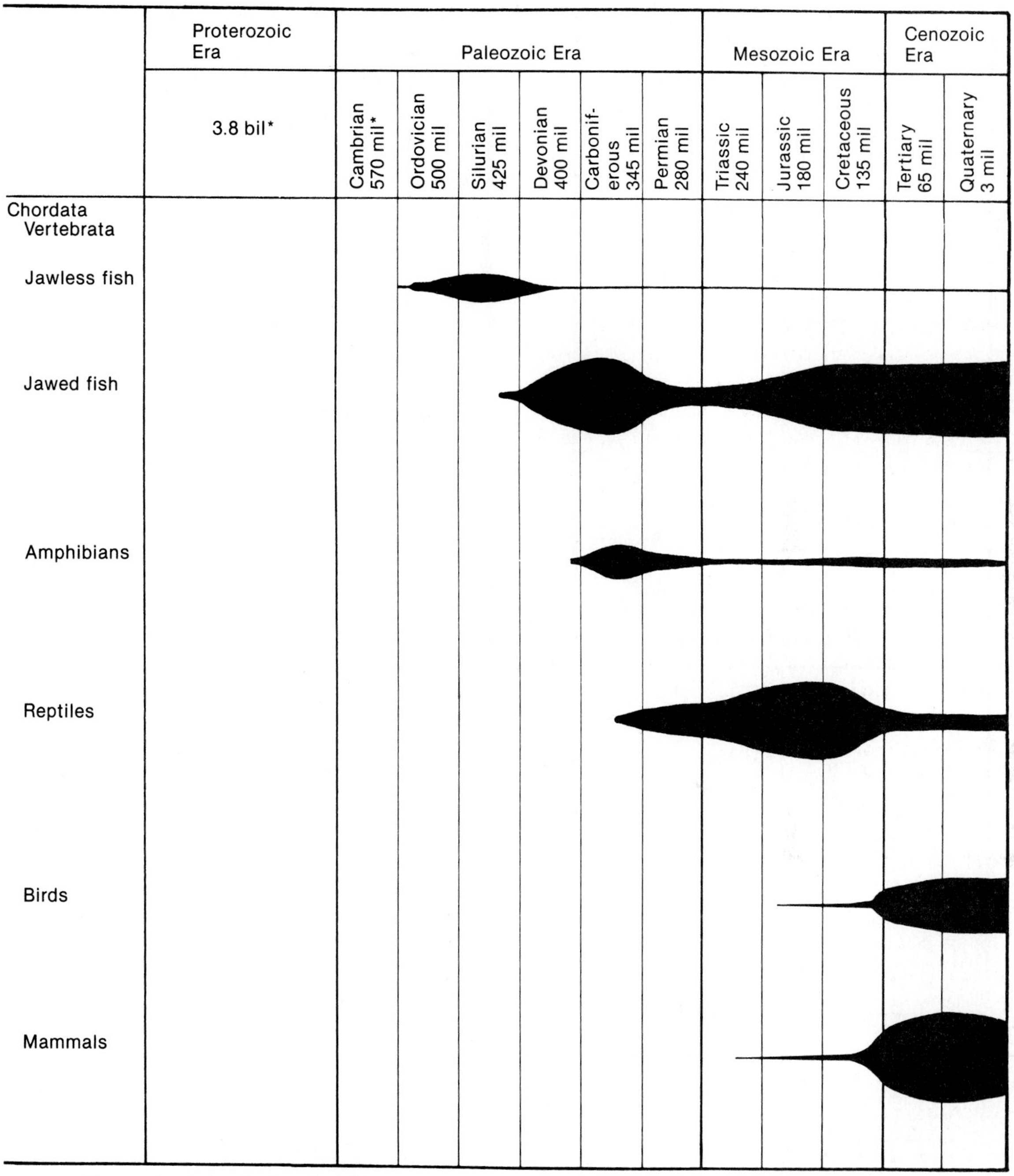

*Approximate beginning date for periods.

Figure 35.3. Fossil record of vertebrates

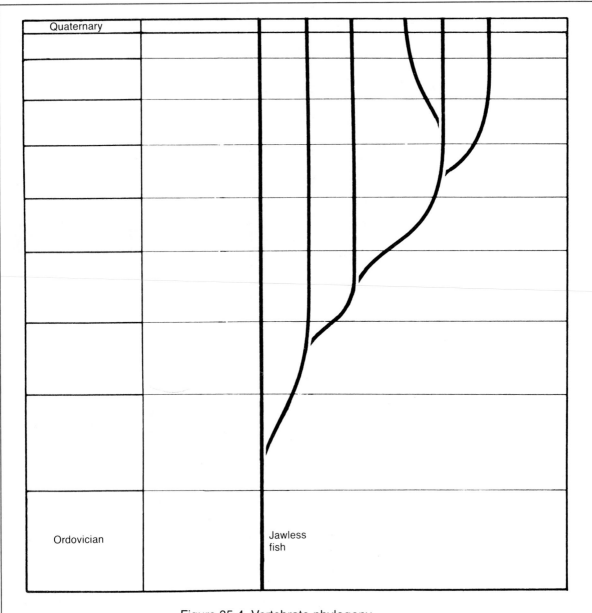

Quaternary

Ordovician

Jawless
fish

Figure 35.4. Vertebrate phylogeny

Prepared slides of:
chick embryo, w.m., 96 hr
pig embryo, w.m., 10 mm

Assignment 2

1. Compare the models of vertebrate embryos with each other.
2. Examine the prepared slides of chick and pig embryos. Compare them with Figure 35.6. Locate the pharyngeal pouches.

3. ***Complete item 2 on the laboratory report***.

EVIDENCE FROM VERTEBRATE MORPHOLOGY

Figure 35.7 illustrates the basic structure of the pectoral appendages of several vertebrate groups. Even after animals have adapted to

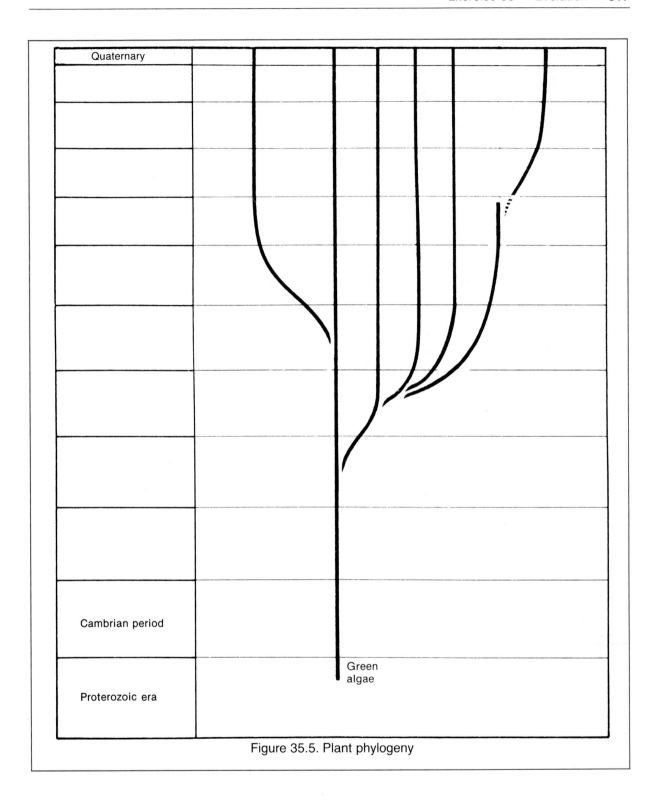

Quaternary

Cambrian period

Green
algae

Proterozoic era

Figure 35.5. Plant phylogeny

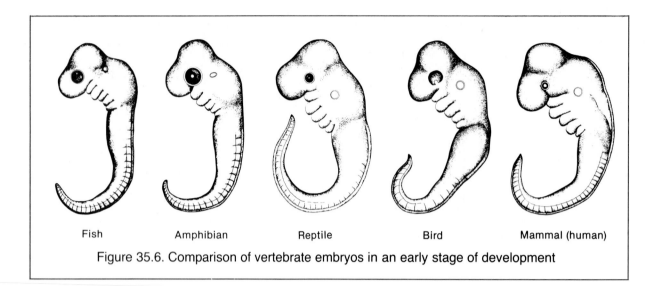

Fish   Amphibian   Reptile   Bird   Mammal (human)

Figure 35.6. Comparison of vertebrate embryos in an early stage of development

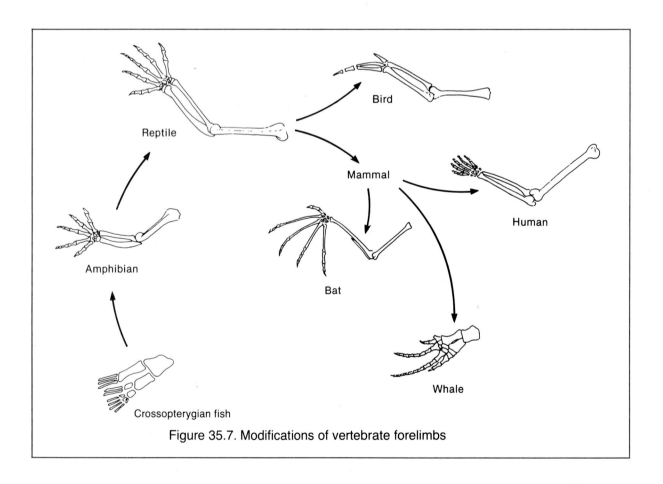

Figure 35.7. Modifications of vertebrate forelimbs

specialized ways of life, the basic similarity of their limb bones is evident and is testimony to their relationship and common ancestry. Structures that have arisen by descent from a common ancestor are called **homologous structures**.

Materials

Colored pencils
Articulated skeletons of representative vertebrates
Model showing homology in anterior appendages
Model of foot evolution in the horse

Assignment 3

1. Study Figure 35.7 and note the homology of the limb bones. For each limb, color the humerus blue, the ulna red, and the radius green.
2. Examine the anterior appendages of the skeletons available and locate the homologous bones. Observe the structural changes to the basic vertebrate plan exhibited by each specimen. Note how the bat and bird have evolved different wing structures.
3. Study Figure 35.8. The fossil record of the horse is one of the more complete records for any animal. Note its phylogeny and the

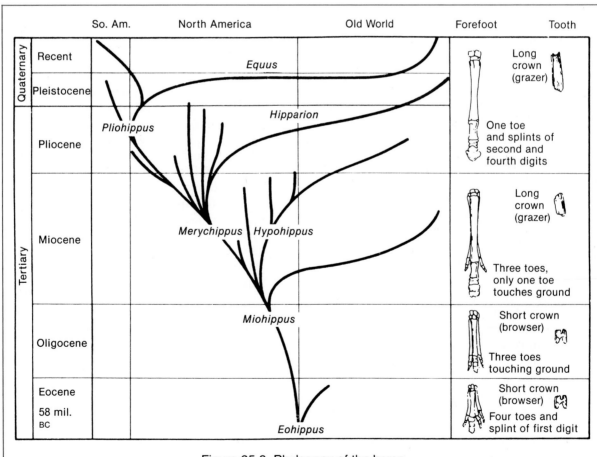

Figure 35.8. Phylogeny of the horse

progressive changes in foot and tooth structure. Examine the models of evolutionary changes in the feet of horses.

4. ***Complete item 3 on the laboratory report***.

EVIDENCE FROM BIOCHEMISTRY

Basic chemical similarities, such as deoxyribonucleic acid (DNA), the genetic code, and adenosine triphosphate (ATP), are common to all living things and leave little doubt as to the unity of nature.

Both structural and chemical similarities decrease with increasing remoteness of relationships. The closer the relationship is between two species, the greater is the number of identical proteins and nucleotides in DNA that they have in common. Comparative biochemical studies of cytochrome c, nucleotide sequences in DNA, amino acid sequences in hemoglobin, and serum proteins all support the fact of evolution.

In this section, you will perform an experiment that simulates the antigen-antibody reactions that are used to indicate the degree of similarity among blood serum proteins of vertebrates. These reactions are based on the fact that the immune system produces specific **antibodies** (proteins) against specific **antigens**, foreign proteins, that enter the body. An antibody produced against a specific antigen will combine with that antigen and will react less well with a closely related antigen.

In the research laboratory, the procedure is performed as follows.

1. Human serum proteins (antigens) are injected into a rabbit. In a few weeks, the rabbit's immune system produces large numbers of antibodies against the injected human serum proteins; that is, it forms antihuman antibodies. The rabbit's immune system is now sensitized to human serum proteins.

2. A sample of rabbit blood is drawn, and the serum is separated out. The serum contains the rabbit-produced antihuman antibodies. If a sample of human serum (which contains serum proteins) is added to it, an antigen-antibody reaction occurs. In this reaction, the rabbit-produced antihuman antibodies attach to the human serum proteins, forming an antigen-antibody complex that precipitates out and that can be measured by visual inspection or special instruments.

3. If a sample of nonhuman vertebrate serum is added to the serum from the sensitized rabbit, the amount of precipitation of the antigen-antibody complex indicates the degree of similarity between human and the nonhuman serum proteins, which, in turn, indicates how closely that vertebrate is related to humans.

Materials

Ward's immunology and evolution experiment kit
Toothpicks

Assignment 4

1. Obtain a small plastic tray with several wells (depressions) and a dropping bottle each of synthetic: rabbit serum, human sensitizing serum, human serum proteins, and "unknown" test sera numbered I, II, III, IV, and V.

2. Place 2 drops of synthetic rabbit serum in each well numbered 1 through 6.

3. Add 2 drops of synthetic human serum to each well and mix with a toothpick to "sensitize" the rabbit serum to human serum proteins. *Discard the toothpick.*

4. Add 4 drops of human serum to well 6. Mix with a toothpick and note the reaction. *Discard the toothpick.* The amount of precipitation (white particle formation) of human serum proteins in well 6 serves as a standard against which the precipitations of the "unknown" test sera are to be compared.

5. Add 4 drops of test serum I to well 1 and mix with a toothpick. *Discard the toothpick* and note the reaction.

6. Add 4 drops of test serum II to well 2 and mix with a toothpick. *Discard the toothpick* and note the reaction.

7. In a similar manner, add 4 drops of test sera III, IV, and V to wells 3, 4, and 5, respectively. Mix each with a *separate* toothpick and note the reactions. *Discard each toothpick.* Compare the amount of precipitation in wells 1 to 5 with the amount of

TABLE 35.1
Positive Reactions to Antihuman Antibodies

Serum Source	% Positive Reactions
Human	100
Chimpanzee	96
Gorilla	96
Orangutan	92
Old World monkeys	80
New World monkeys	77
Marmosets	50
Lemurs	20

precipitation in well 6. The test sera represent human, chimpanzee, orangutan, monkey, and cow. You are to determine on the basis of the degree of precipitation the animal from which each serum sample was obtained.

8. ***Complete item 4 on the laboratory report.***

HUMAN EVOLUTION

The pattern of human evolution is not clear, and the fossil record is sparse. Considerable disagreement exists among scientists regarding the position of some of the fossil forms, but they do agree on the following points:

1. Modern human's closest living relatives are the chimpanzee and gorilla.
2. The great apes and hominids (humans and humanlike forms) evolved from a common ancestor.
3. Hominid evolution is characterized by adaptive radiation and a branching evolutionary tree.

Figure 35.9 shows one possible pattern of human evolution, and a discussion of the major fossils in this pattern follows.

Dryopithecus

This primitive ape lived 25 million to 10 million years ago in Africa, Asia, and Europe. The fossils suggest that *Dryopithecus* was primarily a tree dweller and that it also used grassland plants as food. The evolutionary line leading to

modern apes is believed to originate with this form. *Dryopithecus* may be the ancestor of *Ramapithecus*.

Ramapithecus

Meager fossils of this form have been dated at 14 million to 10 million years. The jaws and teeth exhibit both ape and human characteristics. The **canines** are smaller than those of apes, but they are larger than those of humans. The **incisors** are relatively small. Wear on the low-crowned **molars** suggests sequential eruption, a human characteristic. The dental arch is unlike the narrow U-shaped dental arch of apes and could be a primitive form leading toward the parabolic dental arch of hominids.

Some authorities believe *Ramapithecus* is ancestral to the hominid line and place it near the point of ape and hominid separation. A few believe that it is just a progressive Miocene ape. Most agree, however, that it possesses hominid-like characteristics and apparently lived in a savannah (grassland with scattered trees) habitat.

Australopithecus

The earliest definitive hominid fossil is *Australopithecus afarensis* (named Lucy by her discoverers), which is 3 million to 3.7 million years old. *A. afarensis* has been suggested as the ancestor of the later-appearing *Australopithecus* species and the earliest form of the genus *Homo*. In this view, neither *A. africanus* (2 million years ago) nor *A. robustus* (1.8 million years ago) are ancestral to the genus *Homo*. Some authorities believe *A. afarensis* is only an early form of *A. africanus* and that both *A. robustus* and *Homo* evolved from *A. africanus*. Another view is that *Homo* evolved much earlier so that neither *A. africanus* nor *A. afarensis* could be ancestral to *Homo*.

Australopithecus was 4 to 4.5 ft tall, erect, and bipedal and had a brain a bit larger than a chimpanzee. Some evidence suggests that *Australopithecus* used tools and hunted other animals for food. Plants are thought to have formed a major part of the diet, however.

Homo habilis

This is the first human. It lived 2 million to 2.5 million years ago, and it coexisted with *A.*

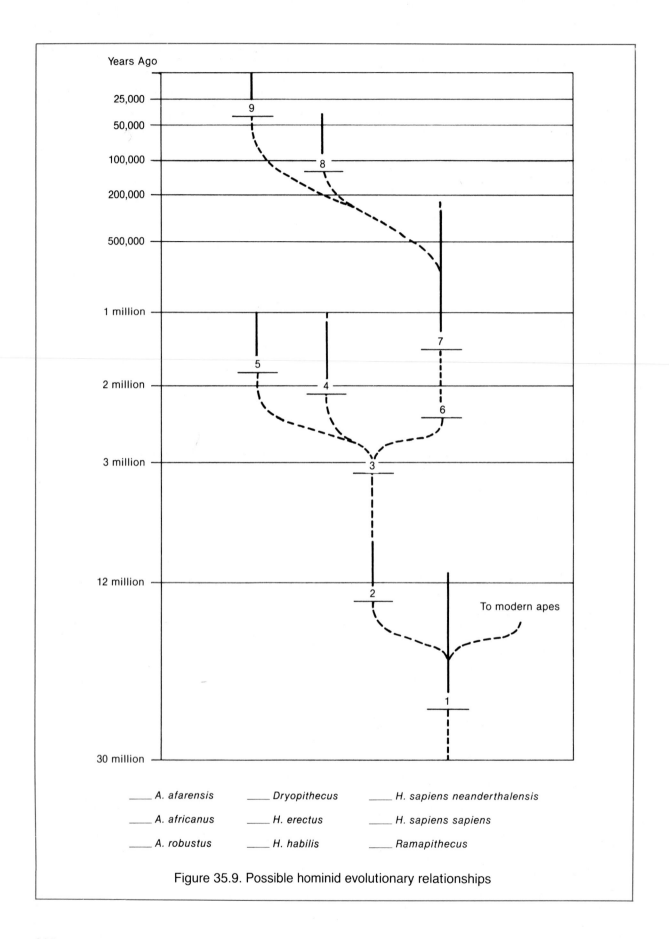

Figure 35.9. Possible hominid evolutionary relationships

africanus and *A. robustus. A. africanus* may be ancestral to *H. habilis,* which in turn is thought to be ancestral to *Homo erectus.* Some authorities believe that *H. habilis* is only an advanced form of *A. africanus.* The skeletal features of *H. habilis,* especially the teeth and cranial size, are more human-like than are those of *Australopithecus.*

Homo erectus

Homo erectus lived from 1.5 million to 300,000 years ago. Numerous fossils of this species have been found in Africa, Asia, and Europe. *H. erectus* was taller (5 ft), larger brained, and more efficiently bipedal than was *Australopithecus.* Evidence indicates construction and use of stone tools, use of fire, and a hunter and gatherer lifestyle. The skull is characterized by (1) heavy browridges, (2) sloping forehead, (3) long, low-crowned cranium, (4) no chin, and (5) protruding face.

Homo sapiens neanderthalensis

This human lived from 125,000 to 40,000 years ago. Neanderthals were stocky with heavy muscles and about 5 ft tall. Skulls exhibit (1) heavy browridges, (2) large protruding faces, (3) no chin, and (4) a sloping, low-crowned, but large, cranium. They constructed stone, bone, and stick tools; used fire, wore clothing; were skilled hunters; and were able to cope with the frigid climate of the glacial periods. They also buried flowers with their dead. They coexisted briefly with modern humans before their extinction. The cause for their extinction is not understood.

Homo sapiens sapiens

Modern humans appeared about 40,000 years ago. The fossil record shows that they ranged throughout western and central Europe, although they may have originated elsewhere. The early modern humans represented by **Cro-Magnon man** were (1) excellent tool and weapon makers, using stone, sticks, bone, and antlers as raw materials; (2) skilled hunters; and (3) fine artists, as evidenced by cave paintings and sculptures. The radiation of these early modern humans ultimately produced the various races observed today.

TABLE 35.2
Cranial Capacity of Hominids

Australopithecus afarensis	500
Australopithecus africanus	450–600
Australopithecus robustus	450–600
Homo habilis	650
Homo erectus	750–1200
Homo sapiens neanderthalensis	1100–1300
Homo sapiens sapiens	1400–1700

Assignment 5

1. Examine Table 35.2.
2. Label Figure 35.9.
3. ***Complete item 5 on the laboratory report***.

SKULL ANALYSIS

Figure 35.10 indicates some of the major differences between gorilla and human skulls. Although this comparison is between two modern species, it does suggest the type of changes that should be present in intermediate forms when a common ancestry is assumed.

An understanding of hominid evolution is based on the study of fossil bones, teeth, and artifacts, such as tools, associated with the specimens. Analysis of the fossils involves detailed measurements and careful comparisons of the fossils. In this section, you will analyze certain skeletal materials and fossil replicas by making measurements and calculating indexes as well as conduct qualitative comparisons of the available specimens.

Indexes are useful for comparative purposes because they overcome the problems caused by differences in the size of the specimens. An index is a ratio calculated by dividing one measurement by another, and it indicates the proportional relationship of the two measurements. Calipers and metersticks are to be used in making the measurements, and all measurements are to be in millimeters. Figures 35.11 and 35.12 show how to make the measurements. The indexes are calculated by dividing one measurement by another (carried to two decimal places) and multiplying by 100.

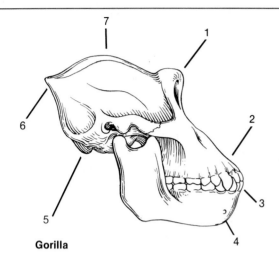 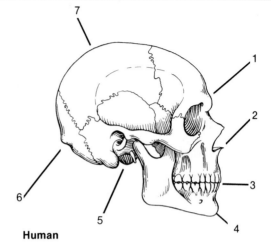

Gorilla

1. Heavy brow ridges; low-crowned, sloping cranium

2. Muzzlelike face

3. Large canines

4. No chin

5. Posterior skull attachment

6. Occipital crest for attachment of heavy neck muscles.

7. Sagittal crest for attachment of heavy jaw muscles.

Human

1. Almost no brow ridges; high forehead and high-crowned cranium

2. Flattened face

3. Small canines

4. Well-developed chin

5. Central skull attachment

6. No occipital crest, neck muscles not as heavy and attached lower

7. No sagittal crest, lighter jaw muscles attached lower

Figure 35.10. Comparison of gorilla and human skulls

Handle the bones and fossil replicas carefully since they are easily damaged. Avoid making marks or scratches on the specimens. The measurements are best made by working in groups of two to four students.

Materials

Calipers
Metersticks
Skull replicas of:
   Australopithecus africanus
   Homo erectus
   Homo sapiens neanderthalensis
   Homo sapiens sapiens (Cro-Magnon)
Skull of modern human

Cranial (Cephalic) Index

When this index is determined for living forms, it is called a **cephalic index**. The term **cranial index** is used in reference to nonliving specimens. An index of less than 75 indicates a longheaded condition. An index of 80 or more identifies a roundheaded individual.

$$\text{Cranial index} = \frac{\text{cranial breadth}}{\text{cranial length}} \times 100$$

Cranial breadth: maximum width of the cranium

Cranial length: maximum distance between the posterior surface and the small prominence (glabella) between the browridges

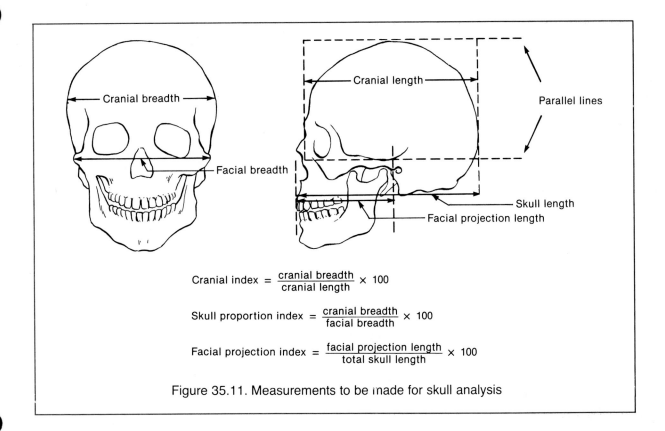

$$\text{Cranial index} = \frac{\text{cranial breadth}}{\text{cranial length}} \times 100$$

$$\text{Skull proportion index} = \frac{\text{cranial breadth}}{\text{facial breadth}} \times 100$$

$$\text{Facial projection index} = \frac{\text{facial projection length}}{\text{total skull length}} \times 100$$

Figure 35.11. Measurements to be made for skull analysis

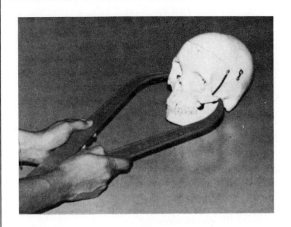

Calipers are used to establish the distance between two points.

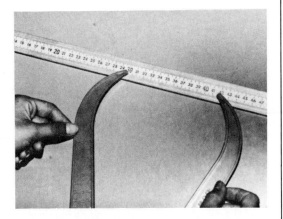

The distance between the two points is determined by placing the caliper tips on the meterstick and measuring the distance between their inside edges.

Figure 35.12. Procedures for making measurements for skeletal analysis

TABLE 35.3
Cephalic Indices of Class Members

1. _____	6. _____	11. _____	16. _____	21. _____	26. _____
2. _____	7. _____	12. _____	17. _____	22. _____	27. _____
3. _____	8. _____	13. _____	18. _____	23. _____	28. _____
4. _____	9. _____	14. _____	19. _____	24. _____	29. _____
5. _____	10. _____	15. _____	20. _____	25. _____	30. _____

Range _____   Average _____

Skull Proportion Index

The skull is composed of the face and the cranium, and the skull proportion index identifies the proportional relationship between these two components. The greater the value of the index, the larger the cranium is in relation to the face.

$$\text{Skull proportion index} = \frac{\text{cranial breadth}}{\text{facial breadth}} \times 100$$

Cranial breadth: maximum breadth of the cranium

Facial breadth: maximum distance between the lateral surfaces of the cheekbones (zygomatic arches)

Facial Projection Index

A projecting, muzzlelike face is a primitive condition among primates. This index identifies the degree of facial projection in a specimen. The greater the value of the index, the greater the degree of facial projection.

$$\text{Facial projection index} = \frac{\text{facial projection length}}{\text{total skull length}} \times 100$$

Facial projection length: distance between the anterior margins of the auditory canal and upper jaw (maxilla)

Total skull length: maximum distance between the posterior surface of the cranium and the anterior margin of the upper jaw (maxilla)

Assignment 6

1. Determine the cephalic index for each member of your laboratory group. Exchange data with all groups in the class and list the indices for each person in Table 35.3. Determine the range and average index value for the class.
2. *Complete items 6a to 6d on the laboratory report*.
3. Determine the (1) cranial index, (2) skull proportion index, and (3) facial projection index for each of the skulls available. *Record your data in item 6e on the laboratory report*.
4. Compare the available skulls as to the following characteristics and rate each on a 1–5 scale. *Record your responses in item 6e on the laboratory report*.

Skull and vertebra attachment
   1 = most posterior
   5 = most central

Browridges
   1 = most pronounced
   5 = least pronounced

Length of canines
   1 = longest
   5 = no longer than incisors

Forehead
   1 = most sloping
   5 = best developed

Chin
   1 = no chin
   5 = best developed

5. *Complete the laboratory report*.

36

EVOLUTIONARY MECHANISMS

OBJECTIVES

On completion of the laboratory session, you should be able to:
1. Identify the factors that produce evolutionary change.
2. Use the Castle-Hardy-Weinberg law for determining genetic stability or change in hypothetical populations.
3. Describe the effect of mutation, gene flow, natural selection, and genetic drift on the gene pool of populations.
4. Define all terms in bold print.

Simply defined, **evolution** is a change in the genetic composition of a population. The genetic composition of a population is usually called the **gene pool**, and it refers to all the alleles in the interbreeding population. Evolution results from a change in the frequency of the alleles in the gene pool produced by one or more of the following factors:

Mutation. The naturally occurring mutations of genes and chromosomes produce variation in the gene pool. These heritable changes serve as the raw material of evolution. Evolution can occur only in populations where preexisting genetic variation exists. Genetic recombination tends to distribute the mutations throughout the gene pool.

Gene Flow. The frequency of alleles in the gene pool may change due to the immigration or emigration of organisms.

Natural Selection. The impact of the environment on the survival and reproduction of genotypes within the population is a major force in changing the frequency of alleles. Selection is always on the phenotypes, but the genetic effect is on the genotypes. The interaction of mutations and natural selection is the primary mechanism causing evolution.

Genetic Drift. The frequency of alleles in the gene pool may change by pure chance. It is a significant evolutionary mechanism in small populations only.

CASTLE-HARDY-WEINBERG LAW

Castle, Hardy, and Weinberg independently determined a simple mathematical expression that establishes a point of reference in evaluating genetic changes in a population. This is known as the **Castle-Hardy-Weinberg (CHW) law** or **equilibrium**, and it is operative whenever these conditions are met:
1. Mutations do not occur.
2. The population size is large.
3. The population is isolated from other populations of the same species (i.e., no immigration or emigration occurs).

4. No selection takes place (i.e., the genotypes are equally viable and fertile).

Contrast this list of conditions with the factors involved in the process of evolution noted previously.

The CHW equilibrium states that *in the absence of forces that change gene frequencies, the frequencies of the alleles in a population will remain constant from generation to generation.*

The frequency (percentage) of alleles in a population may be mathematically expressed as

$$p + q = 1$$

where

p = the frequency of one allele of a gene

q = the frequency of the alternate allele of the gene

For example, consider the gene controlling the ability to taste PTC (phenylthiocarbamide) in a human population meeting the requirements noted previously. The ability to taste PTC is dominant, and the inability to taste PTC is recessive. Therefore, the alleles may be expressed as T (taster) and t (nontaster).

Considering $p + q = 1$, if the frequency (percentage) of the q (t) allele is 50% (0.5), the frequency of the p (T) allele is expressed as

$$p = 1 - q$$

by substitution

$$p = 1 - .5$$
$$p = .5$$

If the allele frequencies are known, a Punnett square may be used to predict the genotype and phenotype frequencies in the next generation when the CHW equilibrium is in effect. Figure 36.1 shows how this is done and also the general equation expressing the genotype frequencies. The **CHW equation** is expressed as

$$p^2 + 2pq + q^2 = 1$$

where

p^2 = the frequency of TT individuals (homozygous dominants)

$2pq$ = the frequency of Tt individuals (heterozygotes)

q^2 = the frequency of tt individuals (homozygous recessives)

Allele frequencies (known) p = frequency of T = .5

q = frequency of t = .5

Set up and complete a Punnett square

Sperm

	.5 T (p)	.5 t (q)
.5 T (p)	.25 TT (p^2)	.25 Tt (pq)
.5 t (q)	.25 Tt (pq)	.25 tt (q^2)

Eggs

The sum of the predicted genotype frequencies equals 1 or 100%.

.25 TT   +   .50 Tt   +   .25 tt   =   1

25% TT   +   50% Tt   +   25% tt   =   100%

When expressed in terms of p and q, we get the equation:

$$p^2 \quad + \quad 2pq \quad + \quad q^2 \quad = \quad 1$$

Figure 36.1. Using a Punnett square to predict the genotype and phenotype frequencies in the next generation of a population where the frequencies of the alleles are known. The genotype frequency expressed in terms of p and q gives us the CHW equation.

For a gene with only two alleles, the CHW equation may be used to calculate the genotype frequencies in a population if the frequency of one allele is known, or to calculate the frequencies of the alleles if the genotypes are known.

Once the allele frequencies have been established, you can predict the genotype and phenotype frequencies in the next generation, if the conditions of the CHW equilibrium are met, by substituting the allele frequencies into the CHW equation instead of using a Punnett square. See Figure 36.2.

Consider a hypothetical population of 75% tasters and 25% nontasters. If you want to determine the percentage of homozygous tasters (TT) and heterozygous tasters (Tt) in the population, you can use the equation described above.

Allele frequencies (known) p = frequency of T = .5

q = frequency of t = .5

CHW equation p^2 + $2pq$ + q^2 = 1

By substitution $(.5)^2$ + $2(.5)(.5)$ + $(.5)^2$ = 1

.25 + .50 + .25 = 1

Genotype frequency
in next generation   25% TT : 50% Tt : 25% tt

Phenotype frequency
in next generation   75% tasters : 25% nontasters

Figure 36.2. Using the CHW equation to predict the genotype and phenotype frequencies in the next generation.

Phenotypes (known)	75% tasters	25% nontasters
Genotypes	Tt and Tt	tt
Percentage of nontasters in population:	tt = 25% = .25	
Frequency of t	q = frequency of t = $\sqrt{q^2}$ by substitution, $q = \sqrt{.25}$ = .5	
Frequency of T	p = frequency of T = $1 - q$ by substitution, $p = 1 - .5$ = .5	
CHW equation	$p^2 + 2pq + q^2 = 1$ substitution allele frequencies,	
	$(.5)^2$ + $2(.5)(.5)$ + $(.5)^2$ = 1 .25 + .50 + .25 = 1	
Genotype frequencies	25% TT : 50% Tt : 25% tt	

Figure 36.3. Using the CHW equation to determine the allele frequencies in a population and then determining the genotype frequencies in that population.

Since nontasters are homozygous recessives, their genotype is known to be tt (q^2). It is used to determine the frequencies of both alleles as shown in Figure 36.3. Once the allele frequencies are known, you can determine the genotype and phenotype frequencies in the population by substituting the allele frequencies into the CHW equation.

Assignment 1

1. Assuming that the conditions of the CHW equilibrium are met, determine the genotype and phenotype frequencies in the two successive generations of a population containing 64% tasters and 36% nontasters. How long will these frequencies persist?
2. Try a piece of PTC paper to see if you are a taster or nontaster.
3. Tabulate the number of tasters and nontasters in your class and calculate the percentage of each. Assuming that the conditions of the CHW law are met, calculate the frequencies of the T and t alleles in your class.
4. ***Complete item 1 on Laboratory Report 36 that begins on page 487.***

NATURAL SELECTION

The theory of natural selection was independently developed by Darwin and Wallace. The basic tenets of natural selection are

1. All organisms are capable of producing more offspring than the environment can support. Therefore, many die before they reach reproductive age.
2. Within each population, considerable variability exists among phenotypes, and most of this variation is genetically based. Because of this variability, some phenotypes are better adapted to the environment than others, resulting in differences in the viability and survival of the phenotypes.
3. Individuals whose traits are more adaptive have a greater chance of reproducing, and their offspring tend to form the largest portion of the population.

Natural selection is the selection *for* those phenotypes that are better adapted, with the result that such phenotypes contribute more surviving offspring to the population than other phenotypes. This means that some phenotypes are selected *against* as well. Note that the environment does the selecting and that natural selection really hinges on **differences in reproductive success**.

Selection Against Dominant Alleles

Suppose that the allele which enables the ability to taste PTC suddenly mutated so that it also caused sterility. Selection would be *against* tasters and *for* nontasters. How would this affect succeeding generations? Using our previous population of 64% tasters and 36% nontasters, the allele frequencies are $p = .4$ and $q = .6$. Substituting into the CHW equation we get

$$(.4)^2 + 2(.4)(.6) + (.6)^2 = 1$$
$$.16 + .48 + .36 = 1$$

Since all tasters are now sterile, the 36% nontasters of the original population are now the total, 100%, of the reproductive population. Will there be any tasters in the next generation?

Selection is rarely 100%. Usually it provides a slight advantage or disadvantage. However, selection is rather rapid on dominant alleles. For example, assume that the 64% tasters in the population have a lowered reproductive success so that they produce only 80% of their expected progeny. What will be the effect on succeeding generations? The contribution (percentage) of tasters and nontasters to the next generation would be

Tasters = $.64 \times .80 = .51$

Nontasters = $1 - .51 = .49$

The genotype frequencies may be determined by calculating the frequency of the alleles and substituting into the CHW equation. Note that as the frequency of the alleles and substituting into the CHW equation. Note that as the frequency of one allele decreases, the frequency of the alternate allele increases.

Frequency of t = $\sqrt{.49} = .7$
Frequency of T = $1 - .7 = .3$
$(.3)^2 + 2(.3)(.7) + (.7)^2 = 1$
$.09 + .42 + .49 = 1$

Selection Against Recessive Alleles

Now let's reverse the previous hypothetical selective pressure. In the population of 64% tas-

ters and 36% nontasters, assume that the nontaster allele has mutated so that it produces sterility in the homozygous state. The effect of this 100% selection against nontasters may be determined for the first generation as follows:

$$(.4)^2 + 2(.4)(.6) + (.6)^2 = 1$$
$$.16 + .48 + .36 = 1$$

Since nontasters will not contribute to the next generation, new allele frequencies must be calculated to determine the genotype and phenotype percentages in the next generation. The 64% tasters constitute the total, 100%, of the reproductive population. Thus,

$$p^2 = .16 \div .64 = .25$$
$$p = \sqrt{.25} = .5, \text{ and}$$
$$q = 1 - .5 = .5$$

Substituting into the CHW equation;

$$(.5)^2 + 2(.5)(.5) + (.5)^2 = 1$$
$$.25 + .50 + .25 = 1$$

Phenotype percentages in the next generation would be

75% tasters : 25% nontasters

Assignment 2

1. Determine the effect of 20% selection against tasters for five generations. If continued, would tasters ever become extinct? Use the data in Table 36.1 in your calculations.
2. If a computer and an appropriate program are available, determine the effect of this selective pressure to see if the dominant allele can be eliminated from the population.
3. ***Complete items 2a–2c on the laboratory report***.
4. Determine the effect of 100% selection against nontasters for the next five generations. Will nontasters ever become extinct?
5. If a computer and an appropriate program are available, determine if nontasters will persist in the population after 10, 100, and 1000 generations.
6. ***Complete item 2 on the laboratory report***.

TABLE 36.1
Table of Squares and Square Roots

n	n^2	$\sqrt{n}$	n	n^2	$\sqrt{n}$	n	n^2	$\sqrt{n}$
1	1	1.000	34	1156	5.831	68	4624	8.246
2	4	1.414	35	1225	5.916	69	4761	8.307
3	9	1.732	36	1296	6.000	70	4900	8.367
4	16	2.000	37	1369	6.083	71	5041	8.426
5	25	2.236	38	1444	6.164	72	5184	8.485
6	36	2.450	39	1521	6.245	73	5329	8.544
7	49	2.646	40	1600	6.325	74	5476	8.602
8	64	2.828	41	1681	6.403	75	5625	8.660
9	81	3.000	42	1764	6.481	76	5776	8.718
10	100	3.162	43	1849	6.557	77	5929	8.775
11	121	3.317	44	1936	6.633	78	6084	8.832
12	144	3.464	45	2025	6.708	79	6241	8.888
13	169	3.605	46	2116	6.782	80	6400	8.944
14	196	3.742	47	2209	6.856	81	6561	9.000
15	225	3.873	48	2304	6.928	82	6724	9.055
16	256	4.000	49	2401	7.000	83	6889	9.110
17	289	4.123	50	2500	7.071	84	7056	9.165
18	324	4.243	51	2601	7.141	85	7225	9.220
19	361	4.359	52	2704	7.211	86	7396	9.274
20	400	4.472	53	2809	7.280	87	7569	9.327
21	441	4.583	54	2916	7.348	88	7704	9.381
22	484	4.690	55	3025	7.416	89	7921	9.434
23	529	4.796	56	3136	7.483	90	8100	9.487
24	576	4.899	57	3249	7.550	91	8281	9.539
25	625	5.000	58	3364	7.616	92	8464	9.592
26	676	5.099	59	3481	7.681	93	8649	9.644
27	729	5.196	60	3600	7.746	94	8836	9.695
28	784	5.292	61	3721	7.810	95	9025	9.747
29	841	5.385	62	3844	7.874	96	9216	9.798
30	900	5.477	63	3969	7.937	97	9409	9.849
31	961	5.568	64	4096	8.000	98	9604	9.899
32	1024	5.657	65	4225	8.062	99	9801	9.950
33	1089	5.745	66	4356	8.124	100	10000	10.000
			67	4489	8.185			

GENETIC DRIFT

Sometimes evolution occurs by genetic drift in small populations due to the change in allele frequencies by pure chance. In this section, you will simulate the random-mating process by drawing the combinations of alleles from the gene pool. Work in pairs.

Materials

Vials of beads, 100 red and 100 white
Beaker, 250 ml

Assignment 3

1. Obtain two vials of red beads and white beads, respectively. These beads represent the alleles: red = T, white = t.
2. Place 50 beads of each color into a beaker and mix them to form the gene pool. The frequency of each allele is .5 (50%). This is the "0" generation. ***Record the allele frequencies in item 3a on the laboratory report***. What frequency of genotypes and phenotypes are expected in the next generation according to the CHW equilibrium?

TABLE 36.2
Genetic Drift F₁ Generation

	TT	*Tt*	*tt*
Tabulation			
Genotype Frequency			
Phenotype Frequency			

New allele frequencies:

T_____ t_____

3. Simulate random mating by drawing two beads from the beaker with your eyes closed. Record the genotype of this "individual" by placing a mark in the correct space in Table 36.2. Repeat this process until you have drawn 20 pairs of beads representing 20 individuals in the small population. From these results, determine the frequency of the genotypes and phenotypes and the frequencies of the T and t alleles in the first generation.

4. Using the calculated frequencies of alleles T (red beads) and t (white beads) that you have determined for the first generation, create a new gene pool of 100 beads that reflects these percentages. For example, if T = .6 and t = .4, place 60 red beads and 40 white beads in the beaker to form the new gene pool.

5. Using the previously described procedures for (a) drawing 20 pairs of beads from the gene pool to simulate random mating in each generation, (b) calculating the genotype and phenotype ratios and the frequencies of the alleles, and (c) reforming the gene pool in accordance with the frequencies of the alleles in the preceding generation, determine the allele frequencies for five generations. **Record your tabulations and calculated frequencies in a table similar to Table 36.2 and the allele frequencies in item 3a on the laboratory report**.

6. If a computer and an appropriate program are available, determine the allele frequencies after genetic drift for 10, 50, and 100 generations.

7. **Complete the laboratory report**.

PART V

ECOLOGY AND BEHAVIOR

ECOLOGICAL RELATIONSHIPS

Life on earth is maintained in a delicate balance by the interaction of ecological interrelationships. All organisms, including humans, are subject to the ecological laws that govern the relationships of organisms with their environment. Humans, more than other organisms, have the ability to modify the environment to their own choosing. Such power must be used with care because its misuse can yield disastrous results.

Ecology is the study of the interrelationships between organisms and their environment, including both biotic and abiotic components. It is primarily concerned with the study of populations, communities, ecosystems, and the biosphere.

- A **population** is a group of interbreeding members of the same species living in a defined area.
- A **community** is composed of all populations within a defined area.
- An **ecosystem** consists of both the community and the abiotic components of a defined area. Soil, air, water, temperature, and organic debris are examples of abiotic components.
- The **biosphere** consists of all ecosystems of the world, that is, all regions where organisms exist.

ENERGY FLOW

Members of a community may be classified according to their functional roles. **Producers** consist of photosynthetic and chemosynthetic **autotrophs** that convert inorganic nutrients into the organic chemicals of protoplasm. In most ecosystems, photosynthesis is the most important process in this conversion. Producers provide all of the organic molecules and chemical energy for the entire community of organisms.

Consumers are **heterotrophs** that depend on the organic compounds formed by producers for their nutrient needs. The primary (first level) consumers are **herbivores** that feed directly on the producers: plants. Secondary and tertiary consumers are **carnivores** that feed on other animals. Some animals are **omnivores** that feed on both plants and animals. The **top carnivore** of a community is not preyed on by any other organism.

Waste products of organisms and dead organisms are decayed (decomposed) by the **decomposers**, primarily fungi and bacteria. They extract the last bit of energy from the organic molecules and convert them into inorganic molecules that can be used once again by the producers.

Chemical energy of organic nutrients flows through the community from producers to consumers to decomposers. The transfer of energy follows the laws of thermodynamics. The **first law of thermodynamics** states that energy is neither created nor destroyed but only changed in form. The **second law of thermodynamics** states that usable energy is reduced with each energy transfer.

The quantitative relationships of producers and consumers based on energy flow may be depicted in an **ecological pyramid**. The three types of ecological pyramids are (1) a pyramid of numbers, (2) a pyramid of biomass (dry weight), and (3) a pyramid of Calories (energy). Generally, each of these pyramids shows a decrease in quantity from the base (producers) to the apex (top carnivore). In unique situations, pyramids of numbers and biomass may be inverted, but the pyramid of Calories never is inverted. See Figure 37.1.

In general, only about 10% of the energy in one trophic level (layer in the pyramid) is transferred to the next trophic level. The remaining 90% is respired or unassimilated in the process. Thus, herbivores contain about 10% of the Calories in the plants consumed. Primary carnivores are reduced to 1% of the energy stored in producers.

Assignment 1

1. Study Figure 37.2 and note the pathway of energy flow. Determine the producers, the herbivores, and the primary, secondary, and tertiary carnivores. Label the figure.
2. ***Complete item 1 on the Laboratory Report 37 that begins on page 489.***

BIOGEOCHEMICAL CYCLES

Although energy flows through the community only once, the chemicals composing organisms cycle repeatedly between the abiotic components and the community. For example, the molecules composing your body were previously composing bodies of plants and other animals. The molecules pass along the food web and ultimately reach the decomposers that break them down into inorganic molecules that, in turn, can be used again by the producers.

The **carbon cycle** shown in Figure 37.3 illustrates the cycling of carbon from the inorganic carbon in the atmosphere to the organic carbon of organisms and its return to the atmosphere. The cycle is continuous.

Nitrogen is an indispensable part of amino acids and proteins. The **nitrogen cycle** is shown in Figure 37.4. Plants are able to use inorganic nitrogen in the form of nitrates, but they depend on certain bacteria to convert organic debris into this usable form. The most common pathway is the conversion of organic nitrogen first into ammonia, then into nitrites (NO_2^-), and finally into nitrates (NO_3^-) that can be used by plants.

Certain bacteria are able to fix (convert) atmospheric nitrogen into organic nitrogen. *Rhizobium* is a nitrogen-fixing bacterium found in nodules on the roots of legumes (e.g., beans, peas, alfalfa). The bacteria and plants live in a **mutualistic symbiosis** since both organisms

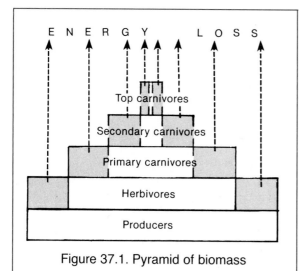

Figure 37.1. Pyramid of biomass

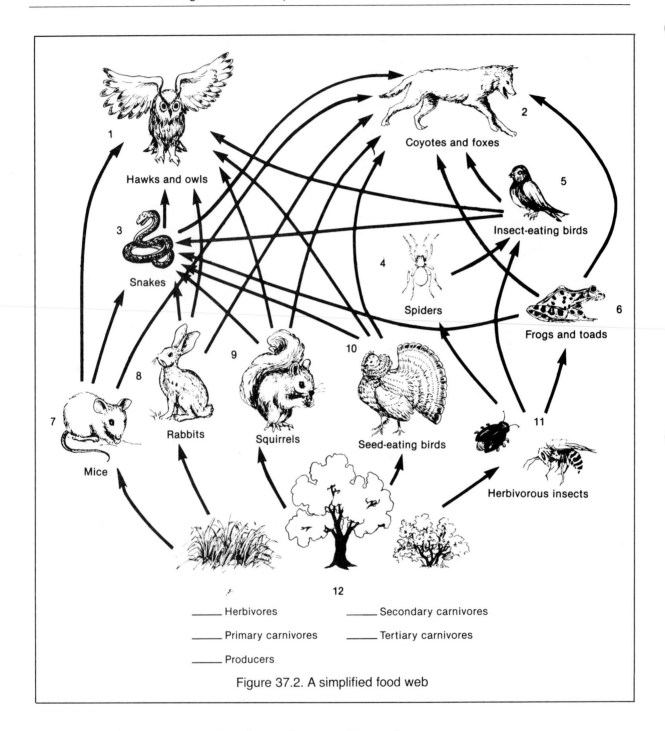

Herbivores _____   Secondary carnivores _____

Primary carnivores _____   Tertiary carnivores _____

Producers _____

Figure 37.2. A simplified food web

benefit from the association. *Azotobacter* is a bacterium that fixes nitrogen nonsymbiotically.

Materials

Beaker, 400 ml
Microscope slides
Slide holder

Bunsen burner
Methylene blue in dropping bottle
Clover and grass roots
Prepared slide of *Rhizobium* in a clover nodule

Assignment 2

1. Study Figures 37.3 and 37.4 and identify the processes involved. Label Figure 37.3.

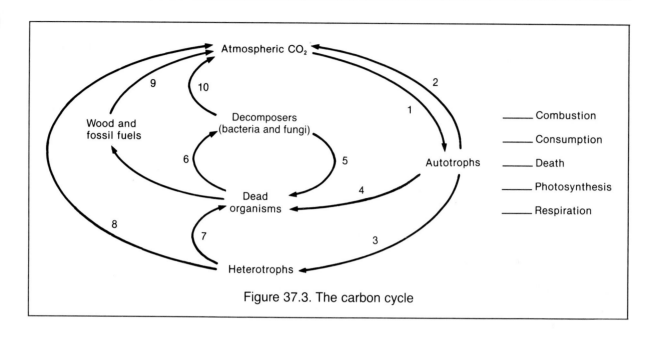

Figure 37.3. The carbon cycle

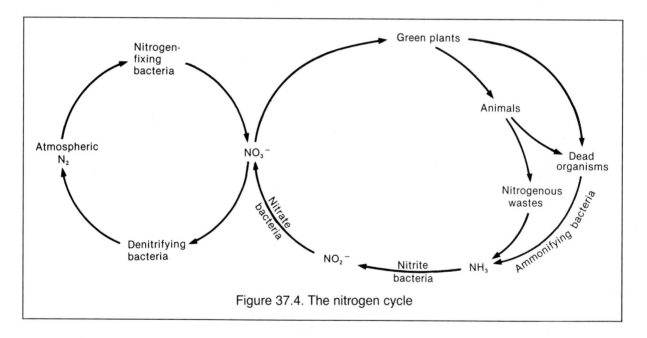

Figure 37.4. The nitrogen cycle

2. Examine the grass and clover roots. Locate the nodules on the clover roots.
3. Prepare a slide of *Rhizobium,* following these steps:
   a. Cut a nodule in half and smear the cut surface on a clean microscope slide.
   b. After the smear has dried, pass the slide quickly through a flame several times.
   c. Add 3 to 5 drops of methylene blue to the slide and let it stand for 3 min. Rinse the slide gently in a beaker of water and observe it with your microscope at 400×.
4. Examine a prepared slide of *Rhizobium* in a clover nodule and compare it with your slide.
5. ***Complete item 2 on the laboratory report.***

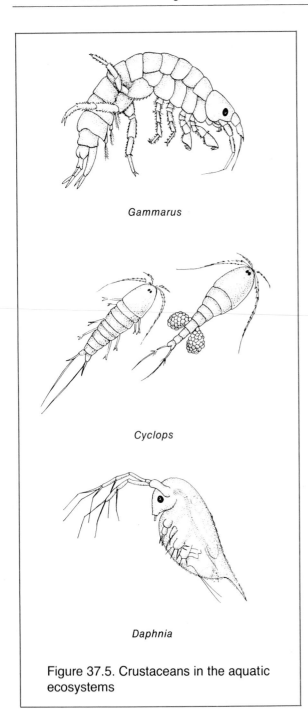

Gammarus

Cyclops

Daphnia

Figure 37.5. Crustaceans in the aquatic ecosystems

ECOSYSTEM ANALYSIS

In this section, you will analyze aquatic microecosystems set up in the laboratory to simulate three lake environments: (1) normal, (2) polluted with acid rain, and (3) polluted with organic materials. Your task is to compare these ecosystems to ascertain the effect of acid and organic pollution on the diversity and density of organisms.

Acid rain results as a side effect of air pollution. Sulfur oxides and nitrogen oxides react with water in the atmosphere to form sulfuric acid and nitric acid, respectively, that ultimately are returned to earth in rain. The burning of high-sulfur coal is a major source of sulfur oxides that lead to acid rain. The burning of oil and gasoline produces nitrogen oxides.

Pure rainwater has a pH of 5.6. Rainwater with a pH less than 5.6 is classified as acid rain. Rainwater with a pH of 1.5 has been recorded. Lakes subjected to acid rain have a lowered pH that severely interrupts the normal biogeochemical cycles so that available nutrients are decreased. Few organisms can tolerate such conditions.

Organic pollution largely results from human activity that increases the amount of fertilizers, sewage, nitrates, and phosphates in lakes and streams. Organic pollution provides an excess of nutrients for the ecosystem and leads to changes in the community of organisms. Polluted waters tend to have a reduced oxygen content which kills off desirable heterotrophic species, especially gamefish. Aquatic organisms vary in their ability to tolerate organic pollution.

In the simulated ecosystems, you will encounter several genera of autotrophs (cyanobacteria, photosynthetic protists, green algae) and three genera of crustaceans (*Cyclops, Daphnia,* and *Gammarus*). Figures 37.5 and 37.6 illustrate organisms likely to be encountered. The crustaceans are omnivores that feed on bacteria, cyanobacteria, algae, and protozoa, and, in a natural ecosystem, they serve as an important food source for small fish.

Materials

Aquatic ecosystems in small aquaria or 1000-ml beakers simulating (1) normal, (2) acid pollution, and (3) organic pollution
Beaker, small
Keys to algae (Ward's)
Medicine droppers
Microscope slides and cover glasses
Pipettes, 5 ml, with rubber bulbs

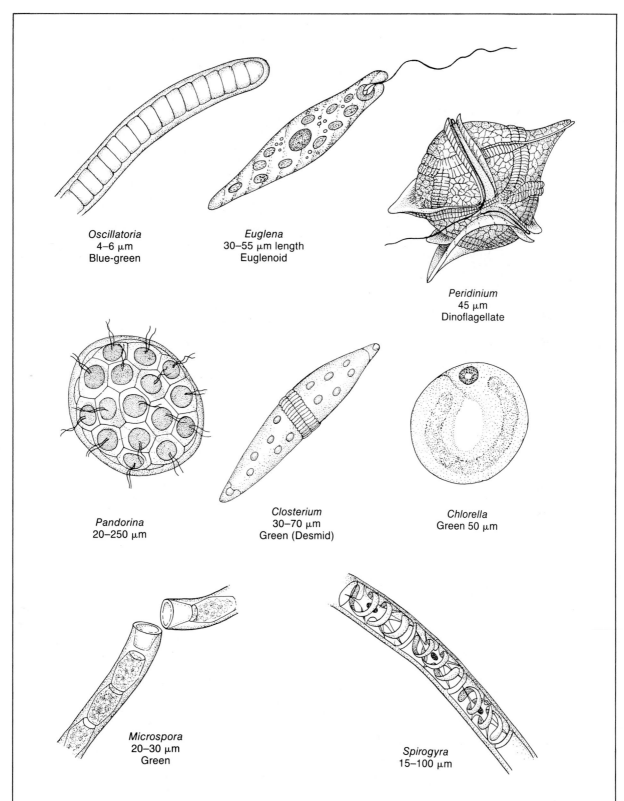

Figure 37.6. Photosynthetic autotrophs that may be in the simulated ecosystems. Organisms are not drawn to scale.

Assignment 3

1. Observe the three microecosystems macroscopically to determine the relative density of autotrophs. This is easily done by noting the degree of clarity or greenish color of the water. ***Record your observations in item 3a on the laboratory report***.
2. Learn the recognition characteristics of the three types of crustaceans in Figure 37.5. *Gammarus* is the largest form and tends to stay near the bottom of the container. *Daphnia* is intermediate in size and swims in the water by pulsating beats of the antennae, as it engulfs microscopic autotrophs. *Cyclops* is the smallest and will be either swimming in the water or attached to filamentous algae or the sides of the container. All these crustaceans can be located with the unaided eye or a hand lens.
3. Locate the crustaceans in the microecosystems. Note the type and relative density of the crustaceans in each one. ***Record your***

observations in items 3a and 3b on the laboratory report.
4. Learn the recognition characteristics of the autotrophs in the microecosystems by studying Figure 37.6 and the specimens under demonstration microscopes set up by your instructor.
5. Determine the autotrophs in each microecosystem by making slides of water from each microecosystem and observing them with your microscope. If filamentous algae are present, remove a small amount with forceps and mount it on a slide for observation. Most of the autotrophs are *very small*. Use the 10× objective to locate a specimen; then switch to the 40× objective to identify it. Use reduced illumination for best results. It may be necessary to make many slides to locate specimens in the microecosystem with the lowest density of autotrophs. ***Record your results in item 3b on the laboratory report***.
6. ***Complete the laboratory report***.

POPULATION GROWTH

38

OBJECTIVES

On completion of the laboratory session, you should be able to:
1. Distinguish between population growth and population growth rate.
2. Compare theoretical and realized population growth curves.
3. Describe the role of the following in population growth:
   a. Biotic potential
   b. Environmental resistance
   c. Density-dependent factors
   d. Density-independent factors
   e. Environmental carrying capacity
4. Compare the growth of human and natural animal populations
5. Compare the growth of human populations in developed and developing countries.
6. Define all terms in bold print.

A **population** is a group of interbreeding members of the same species within a defined area. Populations possess several unique characteristics that are not found in individuals: **density**, **birthrate**, **death rate**, **age distribution**, **biotic potential**, **dispersion**, and **growth form**. One of the central concerns in ecology is the study of population growth and factors that control it.

Natural populations are maintained in a state of dynamic equilibrium with the environment by two opposing factors: biotic potential and environmental resistance. **Biotic potential** is the maximum reproductive capacity of a population

that is theoretically possible in an unlimiting environment. It is never realized except for brief periods of time. **Environmental resistance** includes all **limiting factors** that prevent the biotic potential from being attained.

The limiting factors may be categorized as **density-dependent factors**, the effects of which increase as the population increases. Space, food, water, waste accumulation, and disease are examples of such factors. In contrast, **density-independent factors** exert the same effect regardless of the population size. Climatic factors and the kill of individual predators generally function in this manner.

Assignment 1

Complete item 1 on Laboratory Report 38 that begins on page 493.

Growth Curves

There are two basic types of population growth patterns. **Theoretical population growth** is the growth that would occur in a population if the biotic potential of the species were realized, that is, if all limiting factors were eliminated. This type of growth does not occur in nature except for very short periods of time. **Realized population growth** is the growth of a population that actually occurs in nature. Let's consider both types of population growth in bacteria.

341

Theoretical Growth Curves

In the absence of limiting factors, bacteria exhibit tremendous theoretical population growth potential. Bacterial population size is measured as the number of bacteria in a milliliter (ml) of nutrient broth, a culture medium. Assume that bacterial cells divide at half-hour intervals. If the initial population density were 10,000 bacteria per milliliter of nutrient broth (10×10^3 bacteria/ml), a half hour later the population would be 20,000 bacteria per milliliter (20×10^3 bacteria/ml), and the population would have doubled. See Table 38.1. In another half hour, the population would double again to 40×10^3 bacteria/ml. In an unlimiting environment, this population would double every half hour. If the population size is calculated for each half-hour interval and plotted on a graph, a line joining the points on the graph yields a **theoretical growth curve**.

A **theoretical growth rate curve** may be produced by determining the growth rate for each half-hour interval, plotting these values on a graph, and drawing a line to join the points. The **growth rate** is the change in population size (ΔN) per change in time (Δt), or $\Delta N/\Delta t$. In our example, the growth rate for the first half-hour is

$$\frac{\Delta N}{\Delta t} = (20 \times 10^3) - (10 \times 10^3)$$
$$= (10 \times 10^3) \text{ bacteria/ml/0.5 hr}$$

TABLE 38.1
Theoretical Growth of a Bacterial Population

Time (hr)	Population Density (10^3 bacterial/ml)	Growth Rate (10^3 bacterial/ml)
0.0	10	
0.5	20	_____
1.0	_____	_____
1.5	_____	_____
2.0	_____	_____
2.5	_____	_____
3.0	_____	_____
3.5	_____	_____
4.0	_____	_____

TABLE 38.2
Realized Growth of a Bacterial Population

Time (hr)	Population Density (10^3 bacteria/ml)	Growth Rate (10^3 bacteria/ml)
0.0	10	
0.5	20	_____
1.0	40	_____
1.5	80	_____
2.0	150	_____
2.5	290	_____
3.0	450	_____
3.5	520	_____
4.0	520	_____
4.5	515	_____
5.0	260	_____
5.5	80	_____
6.0	0	_____

And for the second half hour, it is

$$\frac{\Delta N}{\Delta t} = (40 \times 10^3) - (20 \times 10^3)$$
$$= (20 \times 10^3) \text{ bacteria/ml/0.5 hr}$$

Realized Growth Curves

Natural populations exhibit realized growth that results from the opposing effects of biotic potential and environmental resistance. Table 38.2 indicates the growth of a bacterial population in a tube of nutrient broth—a limited environment which results in the ultimate death of the entire population. The population was sampled at half-hour intervals, and no additional nutrients were added.

Assignment 2

1. Determine the theoretical growth of the bacterial population at half-hour intervals and record the data in Table 38.1.
2. Determine the theoretical growth rate of the bacterial population at half-hour intervals and record the data in Table 38.1.
3. ***Plot the theoretical growth and growth rate curves on the appropriate graphs in item 2a on the laboratory report.***

4. Determine the realized growth rate of the bacterial population at half-hour intervals and record the data in Table 38.2.

5. ***Plot the realized growth and growth rate curves on the appropriate graphs in item 2a on the laboratory report.*** Use a different symbol or color for these curves to distinguish them from the theoretical curves.

6. Compare the theoretical and realized population growth curves. The theoretical curve is a J-shaped curve. The realized growth curve starts off in the same manner, but it is soon changed by environmental resistance. As the limiting factors take effect, the curve becomes S shaped. In the artificial environment of a test tube, the bacterial population dies out. In natural populations where materials can cycle and energy is continuously available, the population size levels off where it is at equilibrium with the environment.

7. Study Figure 38.1, which compares generalized theoretical and realized population growth curves. The theoretical growth curve exhibits exponential growth with the slope of the curve becoming ever steeper. In contrast, the effect of environmental resis-

tance causes the realized growth curve to become S shaped.

Several parts of the realized growth curve are recognized. Growth is slow, initially, in the **lag phase** due to the small number of organisms present. This is followed by the **exponential growth phase** when the periodic doubling of the population yields explosive growth. As environmental resistance takes effect, the growth gradually slows, leading to the **inflection point** that indicates the start of the **decreasing growth phase** as the curve turns to the right. Continued environmental resistance causes the curve to level off at the **carrying capacity** of the environment where the population size is in balance with the environment.

The carrying capacity is the maximum population that can be supported by the environment—a balance between biotic potential and environmental resistance. The population remains stable because births equal deaths.

8. ***Complete item 2 on the laboratory report.***

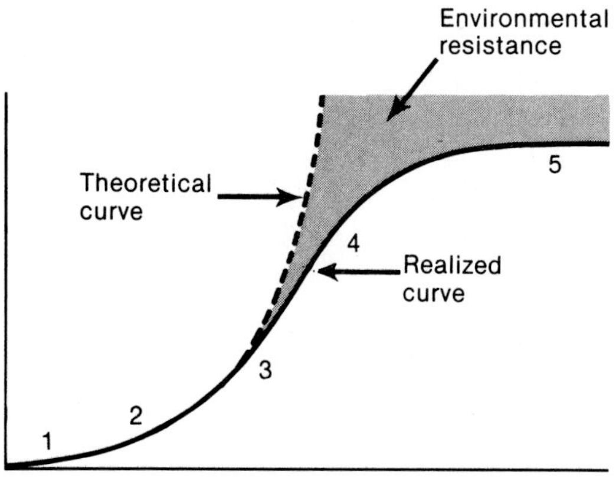

Figure 38.1. Theoretical and realized growth curves. 1 = lag phase, 2 = exponential growth phase, 3 = inflection point, 4 = decreasing growth phase, 5 = carrying capacity.

HUMAN POPULATION GROWTH

One of the major concerns of modern society is the rapid growth of the human population in recent decades. Table 38.3 shows the estimated size of the world population at various times in history and projects a rather conservative growth in the near future.

Growth Rate

Ultimately, a population is limited by a decrease in **birthrate**, an increase in **death rate**, or both. In natural nonhuman populations, an increase in death rate is the usual method. This allows the environment to select for those better adapted individuals who will contribute the greater share of the genes to the next generation. While this mode of selection occurred in preindustrialized human societies, it has been highly modified, but not stopped, by technology in modern societies.

TABLE 38.3
Estimated World Human Population

Year	Population Size (millions)	Growth Rate (millions/year)
8000 B.C.	5	_____
4000	86	_____
1 A.D.	133	_____
1650	545	_____
1750	728	_____
1800	906	_____
1850	1130	_____
1900	1610	_____
1950	2515	_____
1960	3019	_____
1970	3698	_____
1980	4450	_____
1990	5336	_____
2000	6500*	_____

*Conservative projection.

Concern over the growth rate of the human population gained worldwide attention in the 1960s and persists today. The growth rate is the difference between the number of persons born (birthrate) and the number who die (death rate) per year. For humans, the growth rate usually is expressed per 1000 persons. For example, if a human population shows 25 births and 10 deaths per 1000 persons in a year, the growth rate may be determined as follows:

$$\frac{25 - 10}{1000} = \frac{15}{1000} = \frac{1.5}{100}$$
$$= 1.5\% \text{ growth rate}$$

The growth rate reached a peak of 2% in 1965 and declined to 1.7% in 1984, but this slight decline does not mean that the population growth is no longer a concern.

The growth rate is often expressed as the **doubling time**, that is, the time required for the population to double in size. In 1850, the population required 135 yr to double, but this was reduced to only 35 yr in 1965. All of the efforts to reduce population growth only extended the doubling time to 41 yr by 1984.

The doubling time of a population is determined by dividing 70 yr (a demographic constant) by the growth rate. For example, if the world population growth rate is 1.7%, the doubling time would be determined as follows:

$$d = \frac{70 \text{ yr}}{\text{growth rate}} = \frac{70 \text{ yr}}{1.7} = 41 \text{ yr}$$

This means that the entire world population will double in only 41 years. If the present standard of living is to be maintained, the available resources must also double in 41 years. Is this likely? Are the earth's resources infinite?

TABLE 38.4
1989 Human Population Data

Country or Region	Population Size (10^6)	Birth Rate (per 10^3)	Death Rate (per 10^3)	Growth Rate (% inc./yr)	Doubling Time (yr)
World	5201	27	10	1.7	41.2
Developing countries	3990	30	11	_____	_____
Developed countries	1211	15	9	_____	_____
Africa	628	45	15	_____	_____
Asia	3052	28	9	_____	_____
Europe*	497	13	11	_____	_____
Latin America	434	29	8	_____	_____
North America	274	15	9	_____	_____
Oceania	26	20	8	_____	_____
C.I.S. (formerly the USSR)	286	18	11	_____	_____

*Excludes the Commonwealth of Independent States (formerly the USSR).
Source: 1991 Demographic Yearbook. Published by the United Nations.

Obviously, the standard of living varies throughout the world. Developed (industrialized) countries have a relatively high standard of living, and developing countries have a relatively low standard of living. The standard of living is usually expressed as the resources per person and is determined by dividing the gross national product (GNP) by the population size. The contrast in the standard of living in developed and developing countries is evident in the GNP per capita for the United States and India in 1988.

United States = $19,840

India = $329

Assignment 3

1. Use the data in Table 38.3 to ***plot the human population growth curve in item 3a on the laboratory report***.
2. Calculate and record in Table 38.3 the growth rate per year for the time intervals shown.
3. ***Complete items 3b–3i on the laboratory report***.
4. Calculate the growth rate and doubling time for the countries or regions shown in Table 38.4.
5. ***Complete the laboratory report***.

39

ANIMAL BEHAVIOR

Animal behaviors may be divided into two fundamentally distinct types: innate behavior and learned behavior. **Innate behavior** is inherited behavior. It is an automatic and consistent response to a specific stimulus. Innate behavior is performed correctly the first time a specific stimulus is perceived. **Learned behavior** is not inherited. It tends to change or improve with experience, that is, with repeated exposure to the specific stimulus.

Simple animals exhibit only innate behavior. As more complex nervous systems evolved, learned behavior became possible and increasingly important. The more complex the nervous system, the more learned behavior is exhibited by the animal. Learned behavior reaches its peak in humans who exhibit learning and reasoning in complex behavioral responses to stimuli. However, humans also exhibit innate behavior such as reflexes.

There are four types of innate behavior: (1) kineses, (2) taxes, (3) reflexes, and (4) fixed action patterns. A **kinesis** is an orientation behavior in which the animal changes its speed of movement, but not direction, in response to a stimulus. If the animal slows down in response to a stimulus, the response is *positive kinesis*. If it speeds up, the response is *negative kinesis*. Positive kinesis enables animals to stay close to the stimulus, for example, favorable environmental conditions, although they are not actually attracted by it.

Taxes are orientation behaviors in which an animal is either attracted to (positive) or repelled (negative) by the stimulus. In each case, the animal changes direction while the rate of movement may or may not change. Light, heat, cold, gravity, chemicals, and sound are known to elicit such responses in animals. For example, an animal attracted to light exhibits *positive phototaxis,* and an animal repelled by a chemical exhibits *negative chemotaxis.*

Reflexes involve a rapid, stereotyped movement of the body or a body part rather than the orientation of the entire body as in kineses and taxes. Common reflexes in humans include, the knee-jerk reflex, blinking an eye, and coughing.

Fixed action patterns are predictable, stereotyped movements that may be rather complex. The stimulus that causes a fixed action pattern is called a *releaser.* For example, a newly hatched sea gull chick will peck at the red spot on the parent bird's bill causing the parent to regurgitate food for the young chick. The red spot is the releaser stimulating the pecking by the chick, and the pecking of the chick is the releaser stimulating the parent bird to regurgitate food. Fixed action patterns may be so complex and so appropriate that they may seem to be learned upon casual observation.

In this exercise, you will investigate innate behavior in simple animals. Since a variety of stimuli affects animal behavior, conditions except for the stimulus being tested should remain constant in your experiments.

Assignment 1

Complete item 1 on Laboratory Report 39 that begins on page 497.

Taxes in Planaria

Recall that planaria are bilaterally symmetrical animals with special sense organs located at the anterior end. Eyespots are sensitive to light, and the auricles contain receptors for tactile (touch) and chemical stimuli. In this section you will investigate null hypotheses that planaria do not respond to light, water current, chemicals, or gravity.

1. **Phototaxis:** movement toward or away from light.
2. **Rheotaxis:** movement into or away from a current.
3. **Chemotaxis**: movement toward or away from a chemical source.
4. **Geotaxis**: movement toward or away from the force of gravity.

Materials

Brown planaria
Black construction paper
Camel's-hair brushes
Finger bowls
Glass marking pens
Laboratory lamp
Magnetic stirrer
Masking tape
Medicine droppers
Petri dishes
Pond water
Test tubes
Test-tube racks
Cork stoppers for test tubes
Chopped liver

Assignment 2

1. Obtain a petri dish. Fill it half full with pond water. Place five planaria in the dish using small camel's-hair brushes or the modified medicine droppers with enlarged openings. Keep the dish motionless at your lab station, and wait 5 min for the planarians to adapt. Observe the movement of the planaria in the dish under normal classroom lighting. Do the planaria disperse randomly over the bottom of the dish? Does the direction of their movement seem to be random? Do they change direction often? ***Complete item 2a on the laboratory report***.

2. To test the hypothesis that planaria do not exhibit phototaxis, make a light shield of black construction paper and masking tape and place it over one half of the dish. Expose half the dish to normal classroom light while the other half is shaded by the light shield. Observe the movements of the planaria. Do they move directly into the shaded half or into the light? Do they tend to stay in the light or shade? Do they remain motionless in light or shade? After 10 min, record the number of planaria in the lighted half and in the shaded half of the dish. Record your results and those of the entire class. ***Complete item 2b on the laboratory report***.

3. Transfer your planaria to a finger bowl half filled with pond water, and allow 5 min for the planaria to adapt. To test the hypothesis that planaria do not exhibit rheotaxis, place your petri dish of planaria on a magnetic stirrer, and place the stirring piece in the center of the dish. Turn on the stirrer at the *slowest possible speed* to create a gentle, circular flow of water in the dish. If a magnetic stirrer is not available, slowly stir the water with a pencil. Observe the flatworms to see of they exhibit rheotaxis. Record your results and those of the entire class. ***Complete item 2c on the laboratory report***.

4. Remove the light shield and allow 5 minutes for the planaria to adapt. To test the hypothesis that planaria do not exhibit chemotaxis, gently place a small piece of liver in the center of the dish without dis-

turbing the water. Observe the planaria to see if they exhibit chemotaxis. Record your results and those of the entire class. ***Complete item 2d on the laboratory report***.

5. To test the hypothesis that planaria do not exhibit geotropism, fill a vial with pond water and transfer one planarian into the vial. Insert a cork stopper. Use a marking pen to mark the midpoint of the vial.

    Make a cylindrical light shield from black construction paper and masking tape that will loosely slip over the vial. One end of the light shield must be closed to exclude light.

    Lay the vial horizontally on the table. When the planarian moves to near the midpoint of the vial, gently stand the vial upright and slip the light shield over the vial. At 2-min intervals for 6 min, lift the light shield briefly to determine the location of the planarian. Record whether the planarian is located in the lower half or upper half of the vial. If the planarian tends to stay at one end of the vial after 6 min, gently invert the vial and repeat the experiment. Record your results and those of the entire class. ***Complete item 2 on the laboratory report***.

Behavior in Sowbugs

Sowbugs (isopods) or wood lice are typically found under rotting logs and other organic debris in a habitat that is dark and moist. In this section, you will investigate the orientation behavior that causes sowbugs to reside in such habitats. Two major stimuli for this orientation behavior are moisture and light. You will investigate the reactions of sowbugs to these stimuli by testing two null hypotheses.

1. Sowbug locomotion is not affected by light.
2. Sowbug locomotion is not affected by moisture.

Materials

Sowbugs
Black construction paper
Dropping bottles of water
Masking tape

Plastic petri dishes, 14-cm diameter
Filter paper, 14.5-cm diameter

Assignment 3

1. ***Complete items 3a and 3b on the laboratory report***.
2. Obtain a large petri dish and fit a circle of filter paper into the bottom of it.
3. Place five sowbugs into the petri dish and replace the cover. How did they respond to being touched? Place the dish on the table at your workstation under normal classroom lighting. Observe their behavior for several minutes. What is their response to vibrations? Tap the dish and see. Do they seem to move rapidly or slowly? Are they equally distributed over the dish? ***Complete items 3c to 3e on the laboratory report***.
4. To test the hypothesis that sowbug behavior is unaffected by light, make a low, roof-like light shield of black construction paper to shade one half of the petri dish. The light shield should be low but allow you to look under it to make your observations. Shake the dish to activate the sowbugs, and position them in one half of the dish. Place the light shield over the other half of the dish to shade it, and observe their responses. Do their orientation movements seem random? Do they make quick directional movements toward or away from the light? Once in the light or shade do they stay there? Observe the movement of the sowbugs and record their locations at 3-min intervals for 15 min. Record your results and those of the total class. ***Complete items 3f and 3g on the laboratory report***.
5. To test the hypothesis that sowbug behavior is not affected by moisture, add 5 drops of water at a spot near the edge of the filter paper. This will provide a choice for the sowbugs: dry and moist. Shake the dish to get the sowbugs moving, and expose the dish to normal classroom lighting. Observe their movements and locations at 3-min intervals for 15 min. Watch to see if their movements are faster or slower in the moist area as compared to the dry area. Record your results and those of the total

class. ***Complete items 3h and 3i on the laboratory report***.

6. To determine which stimulus (moisture or light) is stronger, you will provide a choice of either moisture or dim light and enable the sowbugs to choose their preferred location. Shake the dish to activate the sowbugs, and place the light shield over the dish to shade the *dry half* of the dish opposite the moist area. Observe the movements and locations of the sowbugs at 3-minute intervals for 15 min. Record your results and those of the total class. ***Complete item 3 on the laboratory report***.

Taxes in Fruit Flies

Drosophila (fruit flies) are much more advanced than flatworms and sowbugs, so their nervous system is more complex. They perceive images with their compound eyes, and tactile and chemical receptors are especially abundant on their antennae and feet.

Materials

Vial (3 in. or longer) of fruit flies
Black construction paper
Glass marking pen
Masking tape
Vials (3 in. or longer) with cork stoppers

Assignment 4

1. Obtain a vial of five fruit flies. Mark the midpoint of the vial with a glass marking pen. Stand the vial up vertically, stopper up, on the table at your lab station. After 3 min, record the location of the flies. Invert the vial and after 3 min record the location of the flies. ***Complete item 4a on the laboratory report***.

2. Using the materials available, design and conduct experiments to determine if the distribution of fruit flies in the vial is due to phototaxis and/or geotaxis. ***Complete item 4 on the laboratory report***.

PART VI

LABORATORY REPORTS

LABORATORY REPORT 1

Student _____

Lab Instructor _____

Orientation

1. Laboratory Procedures and Biological Terms

   a. Describe how you are to prepare for each laboratory session.

     _____

     _____

     _____

   b. Using Appendix A, indicate the literal meaning of the following terms.

     Biology _____

     Morphology _____

     Unicellular _____

     Leukocyte _____

     Organelle _____

     Gastric _____

     Pathology _____

     Dermatitis _____

     Osteocyte _____

     Hypodermic _____

     Chromosome _____

     Pseudoscience _____

     Intercellular _____

     Brachial _____

     Branchial _____

     Extracellular _____

   c. Using Appendix A, construct terms with the following literal meanings.

Study of tissues	_____
Within a cell	_____
Rapid heart rate	_____
Cancer-causing substance	_____
Large molecule	_____

2. Metric System

   a. Indicate the name and value of these metric symbols.

Symbol	Name of Unit	Value of Unit
km	_____	_____
ml	_____	_____
mg	_____	_____
cm	_____	_____
mm	_____	_____
μg	_____	_____

b. Indicate the diameter of a penny in the following units.

_____mm _____ cm _____ m _____ in.

c. Indicate the weight of a 250-ml beaker in the following units.

_____g _____ mg _____ oz

d. List the steps that you used to determine the weight of 20 ml of water.

1. _____
2. _____
3. _____
4. _____
5. _____

e. Indicate the weight of 20 ml of water. _____g

f. A physician directs a patient to take 1 g of vitamin C each day. How many 250-mg tablets must be taken each day? _____

g. Normal body temperature is 37°C (98.6°F). If a patient's temperature is 39.9°C, what is the temperature in degrees Fahrenheit? _____F

h. The U.S. National Research Council recommends that adults eat 0.8 g of protein daily per kilogram of body weight. What is the minimum number of grams of protein that should be eaten by a man weighing 185 pounds?   _____g

What is your weight?   _____ lb; _____ kg

What should be your minimum protein intake?   _____g

i. If the ground beef in a ¼-lb hamburger contains 20% protein and 25% fat, indicate the grams of   fat _____g

                                                                        protein _____g

3. **Observations**

a. Examine the leaves provided in the laboratory and construct a dichotomous key based on their characteristics. To help you get started, first separate the leaves into two groups: (1) needle or scalelike leaves and (2) flat, thin leaves as shown on the next page. Use the next page for the development of your key.

b. After you have completed your key, describe how you approached the problem and the steps or process that you used to construct your key.

_____
_____
_____
_____
_____
_____
_____
_____

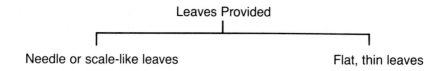

LABORATORY REPORT 2

Student _____

Lab Instructor _____

The Microscope

1. The Compound Microscope

a. List the labels for Figure 2.1

1. _____   9. _____
2. _____   10. _____
3. _____   11. _____
4. _____   12. _____
5. _____   13. _____
6. _____   14. _____
7. _____   15. _____
8. _____

b. Write the term that matches each meaning.

1. Used as handle to carry microscope   _____
2. Lenses attached to the nosepiece   _____
3. Concentrates light on the object   _____
4. Lens you look through   _____
5. Platform on which slides are placed   _____
6. Rotates to change objectives   _____
7. The shortest objective   _____
8. The longest objective   _____
9. Control knob used for fine focusing   _____
10. Control knob used for rough focusing   _____
11. Controls amount of light entering condenser   _____

c. List the power of the ocular and objectives on *your* microscope, and calculate the total magnification for the combinations noted.

Magnification		Total
Ocular	Objective	Magnification
_____	_____	_____
	_____	_____
	_____	_____
	_____	_____

d. Write the term that matches each meaning.

1. Tissue used to clean lenses   _____
2. Objective with the least working distance   _____
3. Slide with an attached cover glass   _____
4. Objective with the largest field   _____

2. Initial Observations

a. Indicate the direction the image moves when the slide is moved:

To the left _____   Away from you _____

b. Indicate the steps to be used in focusing on an object with the high-power objective. _____

_____

_____

c. Describe the best way to relocate an object that is "lost" while viewing with the high-power objective. _____

_____

d. Draw the letter *i* as it appears when observed with each of the following:

Unaided Eye	**4× Objective**	**10× Objective**	**40× Objective**

3. Depth of Field

Based on your observations, draw a side view of a large spine from a fly wing.

40×   **100×**

4. Diameter of Field

a. Indicate the estimated diameter of field at each *total magnification* for your microscope.

Magnification	**Diameter of Field**
_____×	_____mm
_____×	_____mm
_____×	_____mm

b. Indicate the estimated lengths of objects that extend across:

1. Two thirds of the field at 40× _____mm _____ μm
2. 25% of the field at 100× _____mm _____ μm
3. Half of the field at 400× _____mm _____ μm
4. 80% of the field at 40× _____mm _____ μm

c. Determine and record these measurements:

The length of the letter *i* including the dot _____mm _____ μm

The diameter of the dot _____mm _____ μm

5. Applications

a. When viewing the crossed hairs and the top hair is in sharp focus, is the other hair visible or in sharp focus at the following total magnifications?

Magnification	Visible	Sharp Focus
40×	_____	_____
100×	_____	_____
400×	_____	_____

b. At 400×, is the depth of field greater or less than the diameter of the blond hair? _____

c. Estimate the diameter of the blond hair. _____ μm

d. Describe the shape of the blond hair. _____

e. On the basis of your observations, are the following characteristics increased, decreased, or unchanged when magnification is increased?

Illumination _____   Depth of field _____

Working distance _____   Diameter of field _____

f. Draw five to six organisms observed in the pond water slides. Indicate the approximate size, color, and speed of motion of each. Make your drawings large enough to show details.

6. The Dissecting Microscope

a. When using the dissecting microscope, indicate the direction the image moves when the coin is moved

To the left _____   Toward you _____

b. Measure and record the diameter of field at each magnification of the dissecting microscope.

Magnification	Diameter of Field
20×	_____mm
40×	_____mm

7. Review

a. Contrast a prepared slide and a wet-mount slide.

Prepared slide _____

_____

Wet mount slide _____

_____

b. How should prepared slides be handled? _____

_____

c. Describe how to determine the total magnification when using any objective of your microscope.

_____

_____

d. Describe how to determine the length of an object observed with your microscope. _____

_____

e. How should light intensity be adjusted when viewing nearly transparent specimens? _____

_____

LABORATORY REPORT 3

Student _____

Lab Instructor _____

The Cell

1. Prokaryotic Cells

Sketch the appearance of the prokaryotic cells of bacteria and cyanobacteria at 400× as observed from your slides.

Bacteria   **Cyanobacteria**

2. Eukaryotic Cell Structure

a. Write the term that matches each statement.

1. Sites of photosynthesis _____
2. Filaments of DNA and protein _____
3. Control center of the cell _____
4. Sites of aerobic cellular respiration _____
5. Supports plant cells _____
6. Separates the nucleus and cytoplasm _____
7. Sites of protein synthesis _____
8. Sacs of strong digestive enzymes _____
9. Packages materials in vesicles for export from the cell _____
10. Controls passage of materials into and out of the cell _____
11. Semiliquid substance in which cellular organelles are embedded _____
12. Large fluid-filled organelle in mature plant cells _____
13. Channels for the movement of materials within the cell _____
14. Assembles precursors of ribosomes within the nucleus _____
15. Nuclear components containing the genetic code controlling cellular processes _____

b. List the labels for Figures 3.2 and 3.3

Figure 3.2

1. _____   7. _____   13. _____
2. _____   8. _____   14. _____
3. _____   9. _____   15. _____
4. _____   10. _____   16. _____
5. _____   11. _____   17. _____
6. _____   12. _____

Figure 3.3

1. _____
2. _____
3. _____
4. _____
5. _____
6. _____

7. _____
8. _____
9. _____
10. _____
11. _____
12. _____

13. _____
14. _____
15. _____
16. _____
17. _____

3. Onion Epidermal Cells

a. List the labels for Figure 3.4.

1. _____
2. _____

3. _____
4. _____

5. _____
6. _____

b. Diagram three adjacent onion epidermal cells as observed on your slide and label the parts that you see.

c. Determine and record the average length and width of three adjacent cells.
   Width _____ mm   Length _____ mm

d. Examination of your slide will show that a few nuclei do not appear next to a cell wall. Explain that observation. _____
_____
_____

4. *Elodea* Leaf Cells

a. What structure gives a cell its shape? _____
b. Describe the shape of a cell. _____
c. What organelles are green? _____
d. What structure fills the greatest volume in a cell? _____
e. Draw a spine cell at the edge of the leaf at 100×. Label the observed parts.

f. Are the cells in each layer of the leaf about the same size? _____
g. Determine and record the average width and length of three cells in the upper cell layer.
   Width _____ mm   Length _____ mm
h. List the labels for Figure 3.6.

1. _____
2. _____

3. _____
4. _____

5. _____
6. _____
7. _____

5. **Human Epithelial Cells**
   a. What structures observed in onion epidermal cells are *not* present in human epithelial cells.

      _____

   b. Indicate the description that best describes the human epithelial cells.
      Thin and platelike _____   Thick and boxlike _____
   c. Are human epithelial cells thinner than *Elodea* cells? _____
   d. Diagram a few cells from your slides and label the parts observed.

   e. List the labels for Figure 3.7.
      1. _____   2. _____   3. _____

6. **Amoeba**
   a. Diagram the *Amoeba* on your slide and label the parts observed.

   b. Is the cell membrane rigid or flexible? _____
   c. Is the *Amoeba* alive? _____   Why do you think so? _____

      _____

   d. Would you use this same evidence to determine if cells in your body are alive? _____
      Explain. _____

      _____

7. **Levels of Organization**
   a. Define these terms.
      Unicellular organism _____

      _____

      Colonial organism _____

      _____

      Multicellular organism _____

      _____

      Tissue _____

      _____

Organ _____

_____

Organ system _____

_____

b. Diagram a few cells from three different tissues observed from your slides of the buttercup and earthworm.

Buttercup Tissues   **Earthworm Tissues**

8. **Review**

a. Name the two characteristics of living organisms that distinguish them from nonliving things.

_____   _____

b. Indicate the structural and functional unit of life. _____

c. Based on your study and observations, indicate the presence of the following structures in animal and plant cells by placing an X in the appropriate spaces.

Structure	Animal	Green Plant	Nongreen Plant
Cell membrane			
Cell wall			
Central vacuole			
Centrioles			
Chloroplasts			
Chromatin granules			
Cytoplasm			
Endoplasmic reticulum			
Lysosomes			
Mitochondria			
Microtubules			
Nucleolus			
Nucleus			
Ribosomes			

LABORATORY REPORT 4

Student _____

Lab Instructor _____

Chemical Aspects

1. Fundamentals

a. Define these terms.

Element _____

_____

Compound _____

_____

Atom _____

_____

Molecule _____

_____

b. Consult Table 4.2 and indicate the (a) symbol and (b) number of protons, electrons, and neutrons in atoms of these elements.

Element	Symbol	Protons	Number of Electrons	Neutrons
Hydrogen	_____	_____	_____	_____
Oxygen	_____	_____	_____	_____
Nitrogen	_____	_____	_____	_____
Carbon	_____	_____	_____	_____
Chlorine	_____	_____	_____	_____
Calcium	_____	_____	_____	_____

2. Reactions Between Atoms

a. Using Table 4.2 and Figure 4.1, draw shell models of an atom of these elements.

Calcium   **Carbon**

b. Define:

Ionic bond _____

Covalent bond _____

c. Indicate the number of electrons in the outer shell of these neutral atoms and ions:

Na _____ K^+ _____   Mg _____ H^+ _____   Ca _____ Cl^- _____

d. Using Figure 4.1 as a guide, draw a shell model of these molecules:

Potassium Chloride (KCl)   **Calcium Chloride (CaCl$_2$)**

e. Using Table 4.3 as a guide, draw the structural formula of these molecules:

Methane (CH$_4$)   **Ammonia (NH$_3$)**

Carbon Dioxide (CO$_2$)   **Formaldehyde (CH$_2$O)**

3. pH

a. Define:

Acid _____

Base _____

b. Identify these pH values as either acid or base and rank them in order (left to right) of decreasing strength.

a. 10   b. 3   c. 2   d. 6   e. 8   f. 12   g. 1   h. 7.5

Acid: _____ _____ _____ _____   Base: _____ _____ _____ _____

c. Record the pH of these test solutions.

After-shave lotion _____   Vinegar _____   Household ammonia _____

Alka-Seltzer® _____   Detergent solution _____   Lemon juice _____

d. Indicate the number of drops of 1.0% HCl required to decrease the pH of these solutions from 7 to 6.

Distilled water _____ Buffer solution _____

Explain your results. _____

_____

4. Biological Molecules

a. Prepare a set of standards for the chemical tests and record your results in the chart below.

Standards

Test	Organic Compound Testing for	Substance Tested	Results
Iodine	Starch	Starch	
		Water	
Benedict's	Reducing sugars	Glucose	
		Water	
Paper spot	Lipids	Corn oil	
		Water	
Sudan IV	Lipids	Corn oil	
		Water	
Biuret	Protein	Albumin	
		Water	

b. Indicate the test or tests used to detect:

Protein _____ Starch _____

Reducing sugar _____ Lipids _____

c. Indicate the characteristics of positive tests for these organic compounds as determined by the preparation of standards.

Starch _____

Reducing sugar _____

Lipids _____

Protein _____

d. Perform the chemical tests on the substances provided. Compare your results with the standards and record your results in the chart below.

Unknowns

Substance Tested	Test Results*					Organic Compounds Present
	Iodine	Benedict's	Paper spot	Sudan IV	Biuret	
Onion						
Potato						
Apple						
Egg white						

*+ = positive; − = negative

Student _____

Lab Instructor _____

Diffusion and Osmosis

1. Scientific Method

a. Write in the answer column the terms that correctly complete the following paragraph.

Scientific study begins by making careful systematic __(1)__ . A testable __(2)__ is made that may explain the data. Then, a __(3)__ experiment is designed and conducted to test the __(4)__ . The results are analyzed and a __(5)__ is formed that usually leads to the development of a new __(6)__ .

1. _____
2. _____
3. _____
4. _____
5. _____
6. _____

b. Constrast these terms:

A scientific theory _____

_____

A nonscientific theory _____

_____

c. Is Brownian movement a living process? _____ Describe the cause of Brownian movement.

_____

_____

2. Diffusion and Temperature

a. Define diffusion. _____

_____

What natural phenomenon enables diffusion to occur? _____

_____

b. State the hypothesis to be tested. _____

_____

c. What is the independent variable? _____

d. Record the water temperature for each beaker. A _____°C   B _____°C

e. Compare the rate of diffusion in each beaker.

A _____ B _____

f. Do your results support the hypothesis? _____ If not, what relationship seems to exist between temperature and the rate of diffusion? _____

_____

_____

g. Describe how you would test this new hypothesis. _____

_____

_____

_____

_____

3. Diffusion and Molecular Weight

    a. What is the independent variable in the experiment? _____

    b. Record the diameter of the colored circles after 1 hr:

        methylene blue _____ mm   potassium permanganate _____ mm

    c. Do the results support the hypothesis? _____ If not, what relationship seems to exist between molecular weight and the rate of diffusion? _____

    _____

    _____

    d. Describe how you would test this new hypothesis. _____

    _____

    _____

4. Molecular Size and Diffusion Through a Cellulose Membrane

    a. Describe any color change in your test tube setup after 20 min. _____

    _____

    _____

    b. Explain the color change. _____

    _____

    c. Did starch or glucose diffuse from the sac?

        Starch _____   Glucose _____

    d. Considering the molecules of starch, water, iodine, and glucose, which are small enough to diffuse through the pores in the sac? _____

    e. If you wanted to remove essentially all of the glucose from a glucose-water solution, how would you do it using a cellulose bag, a large beaker, and lots of water? Diagram your setup and describe its operation.

5. Selective Permeability and Living Cell Membranes

a. State the hypothesis to be tested. _____

_____

b. Indicate the presence (+) or absence (0) of anthocyanin in the fluid of:

Tube 1 _____   Tube 2 _____   Tube 3 _____

Explain what happened to the cells to give these results.

Tube 1 _____

Tube 2 _____

Tube 3 _____

c. Do the results support the hypothesis? _____

d. State a conclusion from your results. _____

_____

e. Describe the meaning of selective permeability of cell membranes. _____

_____

_____

6. Osmosis and Concentration Gradients

a. State the null hypothesis to be tested. _____

_____

b. What is the independent variable? _____

c. Indicate the sucrose concentration and the distance the fluid rose in each osmometer tube during a 30-min interval.

Tube 1: % sucrose _____   Distance _____ mm

Tube 2: % sucrose _____   Distance _____ mm

d. Do the results support the hypothesis? _____ If not, what seems to be the relationship between the magnitude of the concentration gradient and the rate of osmosis?

_____

_____

_____

e. Describe how you would test this new hypothesis. _____

_____

_____

7. Osmosis and Living Cells

a. Record the degree of crispness in the celery sticks in:

Distilled water _____   Salt water _____

b. In relation to the protoplasm of the celery cells, which solution is:

Hypotonic? _____   Hypertonic? _____

c. Draw an optical midsection of an *Elodea* cell when mounted in:

Water   **Salt Solution**

d. Explain what happened to *Elodea* cells mounted in salt solution. _____

_____

e. Plants wilt when deprived of water. What happens at the cellular level to cause the wilted appearance? _____

_____

f. Why are plants killed when exposed to concentrated salt water? _____

_____

_____

g. Explain the basis of preserving food with salt or sugar. _____

_____

_____

h. Explain what happened to the *Elodea* cells when the leaf was remounted in tap water._____

_____

i. Based on the knowledge you have learned about osmosis, what movement of water would you expect if the intercellular fluid around your body cells, in contrast to the intracellular fluid, was: Hypotonic? _____

_____

Hypertonic? _____

_____

Explain the basis for your answers. _____

_____

_____

_____

What effect would the concentration gradient have on the rate of osmosis? _____

_____

_____

j. Considering osmosis at the cellular level, why is it necessary to drink adequate amounts of water each day? _____

_____

_____

_____

_____

LABORATORY REPORT 6

Student _____

Lab Instructor _____

Monerans and Protists

1. Monera
   a. Draw a few bacteria cells of each morphological type as observed on the bacterial type slide.

   Bacillus   **Coccus**   **Spirillum**

   b. Can you distinguish cellular contents of bacteria at 400×? _____
      How are bacterial species identified? _____
      _____

   c. Indicate the shape of each of the following and the disease caused by each:
      Clostridium botulinum _____
      Staphylococcus aureus _____
      Treponema pallidum _____
   d. Record the diameter of the no-growth areas.

Antibiotic	Diameter of No-growth Areas (mm)	
	S. epidermidis	*E. coli B*
Ampicillin 10 μg		
Erythromycin 10 μg		
Cephalothin 10 μg		
Chloromycetin 10 μg		

   e. Which antibiotic was most effective against both organisms? _____
   f. Which species was more susceptible to the antibiotics? _____
   g. What geometric shape best describes the cells of these organisms?
         Gloeocapsa _____ *Oscillatoria* _____
   h. Draw a few cells of each organism at 400× to show their shape and arrangement. Label the chromoplasm and the gelatinous sheath, if present.

         Gloeocapsa   ***Oscillatoria***

2. Protista: Protozoans

a. Make a drawing of these zooflagellates from your slides and label the flagellum (-a) and nucleus. Draw a few red blood cells with *T. brucei* to show their relative size.

        Peranema   **Trypanosoma**   **Trichonympha**

b. Use a series of diagrams showing the formation of a pseudopod by *Pelomyxa*.

c. Draw a few "shells" of foraminifera and radiolaria from your slides.

d. Is the anterior end of *Paramecium* rounded or pointed? _____

e. Describe the movement of *Paramecium*. _____
_____

f. Do the cilia beat in unison or in small groups? _____

g. What happens when *Paramecium* bumps into an object? _____
_____

h. Describe the function of each:

    Oral groove _____

    Contractile vacuoles _____

    Food vacuoles _____

i. Describe any pattern that exists in the location of food vacuoles and the progression of digestion.
_____
_____

3. Plant-like Protists

a. List the animal-like and plant-like characteristics of *Euglena*.

    Animal-like _____

    Plant-like _____

b. Explain the distribution of *Euglena* in the partially shaded dish.

_____

_____

_____

c. What is the adaptive advantage of this behavior? _____

_____

d. Draw a few dinoflagellates from your slide and label pertinent parts.

e. Draw the siliceous cell walls of a few diatoms from these sources.

Prepared Slides   **Diatomaceous Earth**

4. **Protista: Slime Molds**

Make drawings of your observations.

Dictyostelium   *Physarum*

5. Examination of Pond Water Organisms

a. Draw a few of the monerans and protists observed in slides of pond water. Indicate the magnification used and the division or phylum of each organism, and label the nucleus, chloroplasts, cell wall, cilia, flagella, and vacuoles, if present.

b. Matching
   1. Bacteria   2. Cyanobacteria   3. Prochlorophytes   4. Flagellated protozoans
   5. Amoeboid protozoans   6. Ciliates   7. Sporozoans   8. Euglenoids
   9. Dinoflagellates   10. Diatoms   11. Slime molds

   _____ Prokaryotic cells   _____ Chloroplasts
   _____ Cell wall absent   _____ Silica cell walls
   _____ Form sporangia   _____ Chlorophylls a and b
   _____ Chlorophyll a only   _____ At least some are parasites
   _____ Animal-like protists   _____ Funguslike protists
   _____ Most are saprotrophs   _____ Plantlike protists
   _____ Nonmotile parasites   _____ Most complex protozoans

c. Unknowns
   Write the numbers of the "unknown" specimens in the correct spaces.

   _____ Bacteria   _____ Euglenoids
   _____ Cyanobacteria   _____ Dinoflagellates
   _____ Flagellated protozoans   _____ Diatoms
   _____ Amoeboid protozoans   _____ Slime molds
   _____ Ciliated protozoans
   _____ Sporozoans

LABORATORY REPORT 7

Student _____

Lab Instructor _____

Fungi

1. Introduction
a. Write the term that matches the phrase.
1. Threadlike filaments composing a fungus _____
2. Dormant reproductive cells dispersed by the wind _____
3. Nonreproductive body of a fungus _____
4. Fleshy reproductive body of a fungus containing spore-forming hyphae _____
5. Compose the hyphae _____
6. Specialized structures forming spores _____

b. Distinguish between saprotrophic and parasitic modes of nutrition.
Saprotrophic _____
Parasitic _____

c. Describe how fungi obtain nutrients from organic substrates. _____
_____

2. Conjugating Molds: *Rhizopus*
a. Contrast the functions of the three types of hyphae in *Rhizopus*.
Stolon hyphae _____
_____

Rhizoid hyphae _____
_____

Sporangiophores _____
_____

b. Diagram the appearance of a *Rhizopus* colony showing the location of mature and immature sporangia and the structures that form the outermost portion of the colony. Label pertinent parts.

c. From your slides, draw a few (a) sporangiophores with immature and mature sporangia, (b) spores, and (c) hyphae, showing relative size. Indicate the color of each and show the septate or nonseptate condition of the hyphae.

3. Sac Fungi

a. Draw an outline of a *Penicillium* colony. Label the region of mature conidiospores and immature conidiospores and the color of each. Draw the appearance of a few hyphae with conidiospores to show how they are formed and the septate or nonseptate condition of the hyphae.

Colony   **Hyphae with Conidiospores**

b. Draw a few yeast cells with buds at 400×. Label the buds.

c. Diagram a small portion of a *Peziza* cup showing the spore-forming hyphae, asci, and ascospores as they appear on your prepared slide.

d. *Peziza* (True/False)

_____ Hyphae are septate.

_____ Hyphal cells forming asci are uninucleate.

_____ Spores in asci vary in number.

_____ The cup is the fruiting body.

_____ Hyphal cells forming cup are binucleate.

_____ Eight spores occur in each ascus.

4. Club Fungi: Mushrooms

a. Draw a few (a) hyphae from a stipe to show their septate or nonseptate condition and (b) basidia and spores.

Hyphae from Stipe　　　　　　　　　　**Basidia and Spores**

b. Mushroom (True/False)

_____ Hyphae are septate.

_____ Hyphal cells forming gills are binucleate.

_____ Each basidium forms two spores.

_____ Each spore has one nucleus.

_____ Basidia form on cap as well as on gills.

_____ Hyphae of stipe are uninucleate.

c. Examine the representative club fungi. List them and indicate their distinguishing characteristics by diagram or statement.

5. Lichens

a. Show by diagram the distribution of algal cells and fungal cells observed on the slide of lichen, x.s.

b. In a lichen, describe the role of the:

Fungus _____

_____

Alga _____

_____

c. Show by diagram the differences between the three types of lichens.

     Crustose   **Foliose**   **Fruticose**

6. Review

Write the numbers of the "unknown fungi" in the correct spaces below.

_____Conjugating fungi _____ Sac fungi _____ Club fungi _____ Lichens

LABORATORY REPORT 8

Student _____

Lab Instructor _____

Algae

1. Introduction

Write the term that matches the phrase.

1. Type of nutrition in algae _____
2. Material forming the cell wall _____
3. Organelle containing photosynthetic pigments _____
4. Generation forming spores _____
5. Generation forming gametes _____
6. Chlorophylls in red algae _____
7. Chloropylls in green algae _____
8. Chlorophylls in brown algae _____
9. Brown pigment in brown algae _____
10. Red pigment in red algae _____

2. Green Algae

a. Describe the distribution of *Chlamydomonas* in the partially shaded petri dish. _____
_____

b. Describe the involvement of the flagella and stigma in this behavior. _____
_____
_____

c. Explain the adaptive advantage of this behavior. _____
_____
_____

d. Describe the movement of the living *Pandorina*. _____
_____

What enables this movement? _____

Do *Pandorina* cells possess an eyespot? _____

e. Do you think *Pandorina* can distinguish light and dark? _____

Design and perform a controlled experiment to test your hypothesis. Describe your experiment and results below. Use diagrams if you wish.

Experiment _____
_____
_____
_____
_____

Results _____
_____

Conclusion _____
_____
_____

f. Describe the movement of *Volvox*. _____

How is this movement possible? _____

_____

g. Draw a few adjacent cells from (1) a *Spirogyra* filament and (2) an *Oedogonium* filament to show the similarities or differences of the cells in each filament. Label pertinent parts.

Spirogyra   **Oedogonium**

h. What common characteristics of green algae and multicellular land plants suggest that green algae are ancestors of land plants? _____

_____

_____

3. Brown and Red Algae

a. Name and give the function of the major parts of a brown alga:

A root-like structure _____

A stemlike structure _____

A leaflike structure _____

b. What part of a brown alga is most important in photosynthesis? _____

c. What is the function of air-filled sacs in large kelp? _____

_____

d. Feel the brown algae. Is the texture rough and sticky or smooth and slippery? _____

_____

e. Considering their habitat, what advantage does this surface texture and a robust, but flexible, body provide for brown algae? _____

_____

_____

f. Considering their habitat, explain the delicate body of most red algae as opposed to the robust body of brown algae. _____

_____

g. What is the function of phycobilins in red algae? _____

_____

4. Unknowns

Write the number of the "unknown" specimens in the correct spaces.

_____Green algae   _____ Brown algae   _____ Red algae

Student _____

Lab Instructor _____

Terrestrial Plants

1. Moss Plants

a. Write the term that matches the phrase.
   1. Generation producing spores   _____
   2. Generation producing sperm and eggs   _____
   3. Dominant generation   _____
   4. Attaches gametophyte to soil   _____
   5. Structure containing spores   _____
   6. Substance required for sperm transport   _____
   7. Primary photosynthetic structures   _____

b. Diagram from your slides moss stemlike and leaflike structures. Draw a few cells in each region to show their appearance and relative size.

         Stemlike Structure   **Leaflike Structure**

2. Ferns

a. List three distinguishing characteristics of vascular plants.
   1. _____
   2. _____
   3. _____

b. Indicate the function of:

Xylem _____

Phloem _____

c. Record the distinguishing characteristics of a fern sporophyte. _____

_____

_____

d. Draw a fern sporophyte and gametophyte and label the rhizome, roots, leaves, and rhizoids.

Sporophyte **Gametophyte**

3. Conifers

a. Describe the distinguishing characteristics of a conifer sporophyte.

_____

_____

b. Draw examples of scalelike and needlelike leaves of conifers.

Scalelike Leaves **Needlelike Leaves**

c. Which is larger, male or female cones? _____

Which type of cone produces pollen? _____ Seeds? _____

4. Flowering Plants

a. List the labels for these figures.

Figure 9.3

1. _____ 4. _____ 7. _____
2. _____ 5. _____ 8. _____
3. _____ 6. _____ 9. _____

Figure 9.4

1. _____ 5. _____ 9. _____
2. _____ 6. _____ 10. _____
3. _____ 7. _____ 11. _____
4. _____ 8. _____

b. Describe the distinguishing characteristics of a flowering plant sporophyte. _____

_____

_____

c. In flowering plants, what is the function of:

Flowers _____

Fruits _____

d. Are *Coleus* leaves parallel veined or net veined? _____

e. Check the statement that best describes the arrangement of vascular bundles in *Coleus*.

_____ Scattered but more abundant near the outer edge of the stem

_____ Form a broken ring near the edge of the stem

_____ Form a solid disk in the center of the stem

f. Is *Coleus* a monocot or a dicot? _____

g. What type of venation occurs in corn leaves? _____

Do the leaves have a petiole? _____

h. Check the statement that best describes the appearance of corn roots.

_____ One major root from which other roots branch

_____ Many roots of the same size originating from the base of the stem

i. Is the corn stem active in photosynthesis? _____

j. Check the statement that best describes the arrangement of vascular bundles in the corn stem.

_____ Scattered but more abundant near the outer edge of the stem

_____ Form a broken ring near the edge of the stem

_____ Form a solid disk in the center of the stem

k. Is corn a monocot or a dicot? _____

l. Indicate the function of the:

Petals _____

Sepals _____

Anthers _____

Ovary _____

m. Diagram a cross section of the *Gladiolus* ovary as observed on your slide. Show the divisions of the ovary and the ovules.

n. Is a *Gladiolus* plant a monocot or dicot? _____

o. Diagram the fruits that have flower parts attached and label those parts.

5. Unknowns

Place the numbers of the "unknown" plants in the correct spaces below.

_____ Moss plants   _____ Ferns   _____ Conifers

_____ Monocots   _____ Dicots

6. Roots of Flowering Plants

a. List the three functions of roots. _____

_____

b. Draw the young root of a germinated grass seed at 40× showing the root hairs and an epidermal cell with its root hair at 100×.

Young Root   **Root Hair**

c. Draw a few cells from these regions of the *Allium* root tip to indicate their relative size.

Region of Cell Division   **Region of Differentiation**

d. Draw the basic arrangement of roots in tap and fibrous root systems.

Tap Roots   **Fibrous Roots**

Which system is best for reaching deep-water sources? _____

Which is best for holding the surface soil to prevent erosion? _____

e. What are adventitious roots? _____

_____

f. Write the term that matches the phrase.
1. Region forming new cells   _____
2. Region where greatest growth occurs   _____
3. Region of greatest water absorption   _____
4. Protects the dividing cells   _____
5. Tissue transporting water and minerals   _____
6. Tissue transporting organic nutrients   _____
7. Tissue containing water-impermeable cells   _____
8. Composed of xylem, phloem, and pericycle   _____
9. Composed of sieve tubes and companion cells   _____
10. Tissue forming branch roots   _____

7. Stems of Flowering Plants

a. Describe the functions of stems. _____

_____

b. Are the vascular bundles in a corn stem more abundant near the center or periphery of the stem? _____ What is the advantage of this arrangement? _____

_____

c. Write the term that matches the phrase regarding dicot stems.
1. Primary supporting tissue   _____
2. Outermost tissue in young stem   _____
3. Outermost tissue in old stem   _____
4. Region of cell division   _____
5. Tissue forming wood of stem   _____
6. Tissue just exterior to vascular cambium   _____
7. Formed of spring and summer wood   _____

d. What is the age of the demonstration section of tree stem? _____

e. Show by diagram the arrangement of vascular tissues in herbaceous (*Medicago*) and young woody dicot (*Quercus*) stems as observed in the prepared slides. Label xylem, phloem, and vascular cambium.

Herbaceous Dicot   **Woody Dicot**

8. Leaves of Flowering Plants

a. Write the term(s) that match(es) the phrase.

1. Photosynthetic tissue of leaf _____

2. Compose a vein _____

3. Waterproof coating on leaf surface _____

4. Control size of stomata _____

b. Draw a small section of *Syringa* leaf, x.s., at 40× from the prepared slide including a vein in cross section. Label the pertinent parts.

c. Desert plants often have very small leaves with a thick cuticle. Explain the value of such adaptations. _____
_____
_____

d. Palisade cells are the most important photosynthetic cells. Explain the adaptive value in having the closely packed palisade layer near the upper surface of the leaf. _____
_____
_____

e. Explain the adaptive value of shade plants having large leaves. _____
_____
_____
_____

Student _____

Lab Instructor _____

Simple Animals

1. Sponges

  a. Write the term that matches the phrase.

    1. Level of organization   _____

    2. Type of symmetry   _____

    3. Cells that create the water current   _____

    4. Cells that engulf and digest food   _____

    5. Skeletal elements of sponges   _____

  b. Trace the path of water flowing through *Grantia*.

    In → _____ → _____ → _____ → _____ → Out

  c. Describe the manner of feeding and the distribution of nutrients. _____

    _____

    _____

    _____

2. Coelenterates

  a. Write the term that matches the phrase.

    1. Type of symmetry   _____

    2. Type of body plan   _____

    3. Embryonic tissues present   _____

    4. Level of organization   _____

    5. Contraction enables movement   _____

    6. Coordinates movement   _____

    7. Special cells used to capture prey   _____

    8. Method of nutrient dispersal   _____

  b. Distinguish the following stages by structure and function.

    Polyp _____

    _____

    Medusa _____

    _____

  c. Show by diagram the feeding process as observed in *Hydra*.

    Feeding Position   **Capture of *Daphnia***   **Engulfment**

d. Digestion in coelenterates is a two-step process. Describe it.
   1. _____
   2. _____
e. Diagram a short length of *Hydra* tentacle showing the stinging cells.

f. In *Hydra*, is extension or retraction more rapid? _____

3. Flatworms
a. Write the term that matches the phrase.
   1. Type of symmetry   _____
   2. Type of body plan   _____
   3. Level of organization   _____
   4. Embryonic tissues present _____
   5. Method of nutrient dispersal _____
   6. Enable gliding movement   _____
b. Digestion is a two-step process. Describe it.
   1. _____
   2. _____
c. What is the advantage of the highly branched gastrovascular cavity? _____
   _____
d. What is the advantage of cephalization? _____
   _____
e. Describe the distribution of *Dugesia* in the partially covered dish. _____
   _____
   Interpret this pattern. _____
   _____
f. Describe the orientation of *Dugesia* to the direction of water movement after swirling the water in the dish. _____
   _____
   Interpret this reaction. _____
   _____
g. Considering the life cycle of the beef tapeworm, state two ways to prevent infestation in beef-eating humans.
   1. _____
   2. _____
h. List the names and estimated lengths of the demonstration tapeworms.
   _____   _____

i. Draw a scolex, gravid proglottid with eggs, and an encysted larva of a beef tapeworm from your slides. Label pertinent structures.

Scolex   **Gravid Proglottid**   **Encysted Larva**

j. Sketch the flukes on the prepared slides. Indicate their names and where the adults live in the host.

4. Roundworms

a. Write the term that matches the phrase.
   1. Type of symmetry _____
   2. Type of body plan _____
   3. Level of organization _____
   4. Type of coelom _____
   5. Type of digestive tract _____

b. Where does the adult *Ascaris* live in the host? _____
   What is the function of the cuticle? _____

c. Describe the developmental pathway of *Ascaris* after a human swallows an egg until the adult lies in the small intestine. _____
   _____
   _____
   _____
   _____

d. Considering the life cycle of *Ascaris,* list two ways to prevent human infestation.
   1. _____
   2. _____

e. Draw a few eggs and larvae (in lung tissue) of *Ascaris* from your slides.

f. Draw the outline of one to two specimens of these roundworms from your slides to show relative size. Indicate the magnification used in your observations.

      Trichinella spiralis   ***Wuchereria bancrofti***   ***Turbatrix aceti***

g. For each roundworm parasite observed, indicate: (1) where the adult lives in the human host, and (2) how humans are infected.

Ascaris	1.	_____
	2.	_____
Trichinella	1.	_____
	2.	_____
Wuchereria	1.	_____
	2.	_____
_____	1.	_____
	2.	_____

5. Unknowns

Place the numbers of the "unknown" specimens in the correct spaces below.

    _____ Sponge   _____ Tapeworm larva

    _____ Coelenterate polyp   _____ Fluke

    _____ Coelenterate medusa   _____ Roundworm adult

    _____ Tapeworm adult   _____ None of the above

LABORATORY REPORT 11

Protostomate Animals

1. Protostomates

Write the term that matches the phrase.

1. Embryonic opening forming mouth _____
2. Type of coelom _____
3. Level of organization _____
4. Body plan _____
5. Type of symmetry _____
6. Embryonic tissues present _____

2. Mollusks

a. List the four distinguishing characteristics of mollusks.

1. _____   3. _____
2. _____   4. _____

b. Matching

1. Monoplacophorans   2. Chitons   3. Snails and slugs   4. Tooth shells
5. Clams and mussels   6. Octopi and squids

_____ Image-forming eyes   _____ Conical shell open at each end
_____ Radula present   _____ Shell of eight plates
_____ Remnants of segmentation   _____ Shell of two valves
_____ All predaceous   _____ Filter-feeders
_____ Digging foot   _____ Crawl on flattened foot
_____ Marine only   _____ Foot modified into tentacles

c. Draw a radula from a prepared slide showing the filelike teeth.

d. Place the number of the "unknown" mollusks in the correct spaces.

_____ Monoplacophorans   _____ Snails and slugs
_____ Chitons   _____ Clams, oysters, and mussels
_____ Tooth shells   _____ Octopi and squids

e. Clams

   1. Write the term that matches the phrase.

     a. Covers visceral mass and secretes shell   _____

     b. Muscles closing valves of shell   _____

     c. Brings water into mantle cavity   _____

     d. Used to dig into mud or sand   _____

     e. Organs of gas exchange   _____

   2. Describe the feeding process. _____

   _____

   _____

f. Squid

   List the adaptations for a predaceous, swimming life-style. _____

   _____

   _____

3. Annelids

a. What is the primary distinguishing characteristics of annelids? _____

b. Matching

   1. Oligochaetes   2. Polychaetes   3. Leeches

   _____ Head with simple eyes   _____ Blood-sucking parasites

   _____ Few setae   _____ Parapodia and many setae

   _____ Hermaphroditic   _____ Superficial rings on somites

   _____ Suckers for attachment   _____ Marine only

c. Place the numbers of the "unknown" annelids in the correct spaces.

   _____ Polychaetes   _____ Oligochaetes   _____ Leeches

d. Earthworms

   1. Write the term that matches the phrase.

     a. Membranes dividing segments   _____

     b. Secretes the egg case   _____

     c. Grinds food into small pieces   _____

     d. Receives sperm during copulation   _____

     e. Removes metabolic wastes   _____

     f. Provide traction for movement   _____

     g. Transports nutrients to body cells   _____

   2. Considering that the intestine digests food and absorbs nutrients into the blood, what is the typhlosole and what is its value? _____

   _____

   _____

3. As observed in your dissection:
   a. Draw a dorsal view of a short length of the ventral nerve cord, including two segmental ganglia and the attached lateral nerves.

   b. Does the crop or gizzard have a thicker wall? _____
      Relate this to the function of each:
      Crop _____
      Gizzard _____

4. Arthropods
   a. List the major characteristics of arthropods. _____
      _____
      _____

   b. Matching
      1. Arachnids   2. Crustaceans   3. Centipedes   4. Millipedes   5. Insects
      _____ three pairs of legs   _____ compound eyes
      _____ two pairs of antennae   _____ four pairs of legs
      _____ cephalothorax and abdomen   _____ head, thorax, and abdomen
      _____ elongate flattened body   _____ no antennae
      _____ body segments fused in pairs   _____ compound and simple eyes
      _____ elongate, dorsally convex body _____ five pairs of legs

   c. Place the numbers of the "unknown" arthropods in the correct spaces.
      _____ Arachnids   _____ Centipedes   _____ Insects
      _____ Crustaceans   _____ Millipedes

   d. List the stages in an insect life cycle with:
      Incomplete metamorphosis _____
      Complete metamorphosis _____

   e. Considering the complete metamorphosis of a butterfly life cycle as an example, what is the adaptive advantage of:
      1. the different feeding patterns of larvae and adult? _____
         _____

      2. the pupal stage? _____

f. Crayfish dissection
   1. Write the term that matches the phrase.
      a. Form paddle for swimming _____
      b. Body divisions _____
      c. Excretory organs _____
      d. Number of legs _____
      e. Shell-like covering of cephalothorax _____
      f. Number of appendages per body segment _____
      g. Type of eyes _____
   2. Indicate the number of longitudinal rows of gills. _____   To what are gills in the outer row
      attached? _____
      How many gills are on each side? _____
      What protects the delicate gills from injury? _____
   3. What are the yellowish organs with a granular texture that fill most of the cephalothorax?
      _____
   4. Draw a dorsal view of the supraesophageal ganglion ("brain") and the subesophageal gan-
      glion (first ventral ganglion) showing how they are connected and any lateral nerves present.

   5. Draw a cross section of the abdomen and label the muscles, intestine, and nerve cord.

g. Grasshopper dissection
  1. Write the term that matches the phrase.
    a. Body divisions   _____
    b. Pairs of legs   _____
    c. Excretory organs   _____
    d. Respiratory system   _____
    e. Type of eyes   _____
  2. How many body segments compose the thorax? _____
  3. What is the function of spiracles? _____
  4. How many gastric caeca are present? _____
  5. Draw several contiguous individual lenses of the compound eye of a crayfish and grasshopper to show their differences in shape.

  Crayfish   **Grasshopper**

Student _____

Lab Instructor _____

Deuterostomate Animals

1. **Echinoderms**
   a. Write the term that matches the phrase.
      1. Fate of the blastopore _____
      2. Formed from second embryonic opening _____
      3. Type of skeleton _____
      4. Major distinctive characteristic _____
      5. Structures involved in gas exchange _____
      6. Type of symmetry _____
      7. Surface on which mouth is located _____
   b. Matching
      1. Brittle stars   2. Sea cucumbers   3. Sea lilies   4. Sea stars   5. Sea urchins

      _____ Star-shaped with broad-based arms

      _____ Star-shaped with narrow-based arms

      _____ Attached by stalk from aboral surface

      _____ Arms used for locomotion

      _____ Elongate forms lacking arms and spines

      _____ Globose forms with movable spines; mouth with five teeth

      _____ Ambulacral grooves with tube feet

      _____ Ambulacral grooves without tube feet

   c. Write the numbers of the "unknown" echinoderms in the correct spaces.

      _____ Brittle stars   _____ Sea cucumbers   _____ Sea lilies

      _____ Sea stars   _____ Sea urchins

   d. Sea star dissection
      1. How many arms are present? _____ How many digestive glands are in each arm? _____
         How many gonads are in each arm? _____
      2. What part of the water vascular system extends into each arm? _____;
         terminates in a tube foot? _____
      3. Estimate the number of tube feet in one arm. _____

2. **Chordates: General**
   a. List the four distinguishing characteristics of chordates. _____
      _____
      _____
   b. In tunicates and cephalochordates, describe the function of:
      Pharyngeal gills _____
      Notochord _____
   c. Indicate the following characteristics of chordates:
      Type of symmetry _____   Fate of the blastopore _____

d. Matching

    1. Tunicates   2. Cephalochordates   3. Vertebrates

    _____ Filter feeders   _____ Motile larva and nonmotile adult

    _____ Head, trunk, and tail   _____ Marine only

    _____ Chordate characteristics, usually in   _____ Gills for feeding and gas exchange

             embryo only   _____ Fishlike, headless adult

    _____ Endoskeleton   _____ Notochord is only support

3. Fishes

a. Describe the trend from tunicates to bony fish in the:

Number of gills _____

Function of gills _____

b. Draw the cut surface of a cross section through the tail of a lamprey, shark, and perch to show the vertebra, nerve cord, and notochord, if present. Label pertinent parts.

             Lamprey   **Shark**   **Perch**

c. In what ways is a vertebral column an improvement over a notochord? _____

_____

_____

d. In what ways are bony fish better adapted to a swimming way of life than cartilaginous fish?

_____

_____

e. Based on your observations, what evidence of segmentation is found in bony fish? _____

_____

_____

f. Based on your observation of fish in the aquarium, describe the respiratory movements of bony fish. _____

_____

_____

4. Amphibians to Mammals

a. List two reasons why amphibians are poorly adapted for terrestrial life.

    1. _____

    2. _____

b. What evidence of a fish ancestry is found in amphibians? _____

_____

c. List three ways that reptiles are adapted to terrestrial life.

    1. _____

    2. _____

    3. _____

d. Explain the presence of pharyngeal slits in the chick embryo. _____

_____

e. List three adaptations for flight observed in birds.

    1. _____

    2. _____

    3. _____

f. Draw the beaks of seed-eating, insect-eating, and flesh-eating birds to show their adaptations.

          Seed Eating   **Insect Eating**   **Flesh Eating**

g. Compare walking movements in a salamander, lizard, and rat by indicating which organism is described by the phrase.

    1. Greatest lateral movement of backbone   _____

    2. Least lateral movement of backbone   _____

    3. Greatest lateral extension of legs   _____

    4. Least lateral extension of legs   _____

h. Describe the advantage of homeothermy. _____

_____

    Describe any disadvantages. _____

_____

i. Explain the value of hair and feathers in homeothermic animals. _____

_____

j. List two evidences of reptilian ancestry found in the rat.

    1. _____   2. _____

5. Chordate Summary

a. Matching

1. Jawless fish (modern) 2. Cartilaginous fish 3. Bony fish 4. Amphibians
5. Reptiles 6. Birds 7. Mammals

_____ Diaphragm and hair   _____ Feathers, horny beak

_____ Notochord in adult   _____ Cartilaginous endoskeleton

_____ Leathery-shelled eggs   _____ Intrauterine development

_____ No scales, jaws, or operculum   _____ Lungs, paired limbs, no scales

_____ Epidermal scales   _____ Dermal scales

_____ Operculum, swim bladder   _____ Subterminal mouth, paired fins

_____ Poikilothermic   _____ Homeothermic

_____ Two-chambered heart   _____ Four-chambered heart

_____ Seven pairs of gill slits   _____ Bony endoskeleton

_____ Milk-secreting glands   _____ Claws or nails

_____ Hard-shelled eggs   _____ Leightweight dermal scales

_____ Ribcage   _____ Three-chambered heart

b. Write the numbers of the "unknown" vertebrates in the correct spaces.

_____ Jawless fish   _____ Amphibians   _____ Birds

_____ Cartilaginous fish _____ Reptiles   _____ Mammals

_____ Bony fish

6. Summary of the Animal Kingdom

Compare the animal groups by placing the correct number(s) in the spaces provided.

1. Sponges 2. Coelenterates 3. Flatworms 4. Roundworms 5. Mollusks
6. Annelids 7. Arthropods 8. Echinoderms 9. Chordates

_____ Bilateral symmetry in adult   _____ Radial symmetry in adult

_____ Saclike body plan   _____ Endoskeleton

_____ Tube-within-a-tube body plan   _____ Organ level

_____ Ectoderm, endoderm, mesoderm   _____ Exoskeleton

_____ Dorsal tubular nerve cord   _____ Ventral nerve cord(s)

_____ Organ system level   _____ Tissue level

_____ Pseudocoelomate   _____ Protostomate

_____ Cellular-tissue level   _____ Pharyngeal gill slits

_____ Gas-exchange organs   _____ Circulatory system

_____ Excretory organs   _____ Eucoelomate

_____ Notochord   _____ Acoelomate

_____ Water vascular system   _____ Ectoderm and endoderm only

_____ Exoskeleton, jointed appendages   _____ No skeleton, segmentation

Student _____

Lab Instructor _____

Photosynthesis

1. Introduction

a. Write the term that matches the phrase.
   1. Organisms synthesizing organic nutrients   _____
   2. Form of energy captured by photosynthesis   _____
   3. Form of energy formed by photosynthesis   _____
   4. End product of photosynthesis   _____
   5. Storage form of carbohydrates in plants   _____
   6. Source of O_2 formed by photosynthesis   _____
   7. Source of oxygen atoms in glucose formed by photosynthesis   _____
   8. Source of hydrogen atoms in glucose formed by photosynthesis   _____
   9. Source of carbon atoms in glucose formed by photosynthesis   _____

b. List the key events in each reaction:

Light reaction _____

_____

Dark reaction _____

_____

c. Write the term that matches the phrase regarding leaf tissues.
   1. Main photosynthesizing tissue   _____
   2. Coating that prevents excessive water loss   _____
   3. Openings for gas exchange   _____
   4. Transports water to leaf cells   _____
   5. Provides supporting framework for blade   _____

d. Diagram a portion of the internal structure of a *Syringa* leaf, x.s. Label the upper and lower epidermis, palisade mesophyll layer, spongy mesophyll, and a vein.

2. Carbon Dioxide and Photosynthesis

a. Indicate the independent variable. _____

b. Indicate the presence or absence of starch.

Plant A _____ Plant B _____

c. Was the hypothesis supported? _____ State a conclusion from your results. _____
_____
_____

3. Light and Photosynthesis

a. Indicate the independent variable. _____

b. Indicate the presence or absence of starch in the:

Exposed leaf tissue _____

Shielded leaf tissue _____

c. Was the hypothesis supported? _____ State a conclusion from your results. _____
_____
_____

4. Chlorophyll and Photosynthesis

a. Indicate the independent variable. _____

b. Indicate the presence or absence of starch and sugar in the:

	Starch	**Sugar**
Green *Coleus* leaf tissue	_____	_____
Nongreen *Coleus* leaf tissue	_____	_____
Green corn leaf	_____	_____
Nongreen corn leaf	_____	_____

c. What is the source of the sugar in the nongreen plant tissues?

Coleus _____

Corn _____

d. State a conclusion from your results regarding chlorophyll and photosynthesis. _____
_____
_____

5. Chloroplast Pigments

a. Attach your chromatogram and label the pigment lines of carotene, chlorophyll a, chlorophyll b, and xanthophyll.

b. Indicate the chlorophyll type that was:
   Most soluble _____ Least soluble _____

c. Explain why all leaf pigments are not visible in a healthy leaf.
   _____
   _____

d. How do you explain the appearance of yellow pigments in many plant leaves in autumn?
   _____
   _____

e. Examine Figure 15.5 and determine:
   1. The three parts of the spectrum responsible for most of photosynthesis.
      _____ _____ _____

   2. Colors of light absorbed primarily by chlorophyll a.
      _____ _____

   3. Colors of light absorbed primarily by chlorophyll b.
      _____ _____

   4. Explain why most leaves appear green. _____
      _____

6. Light Intensity and the Rate of Photosynthesis
   a. Indicate the independent variable. _____
   b. What is used to indicate the rate of photosynthesis? _____
      _____

      Why is this a good indicator of photosynthetic action? _____
      _____

   c. An assumption in the experiment is that light intensity decreases as distance from the light
      source increases. Is this a valid assumption? _____ If you wished to verify it, how would you
      do it? _____
      _____

   d. Explain why a heat filter must be used when using a spot lamp. _____
      _____

   e. State the null hypothesis to be tested. _____
      _____
      _____

f. Record your data in the chart below.

Light Intensity and the Rate of Photosynthesis

Distance from Light Source (cm)	Readings (ml)		ml O_2/3 min	ml O_2/min
	Start	Stop		
30	———	———	———	———
	———	———	———	———
	———	———	———	———
	Average		———	———
60	———	———	———	———
	———	———	———	———
	———	———	———	———
	Average		———	———
90	———	———	———	———
	———	———	———	———
	———	———	———	———
	Average		———	———
120	———	———	———	———
	———	———	———	———
	———	———	———	———
	Average		———	———

g. Was your hypothesis supported? _____ State a conclusion from your results. _____
_____
_____

h. Explain why your conclusion should be modified to "within the range of values tested"?
_____
_____

LABORATORY REPORT 16

Cellular Respiration

1. Introduction

a. Write the summary equation for the aerobic respiration of glucose. Underline the reactants and circle the products.

_____

b. Write the term that matches the phrase.
   1. Serves as final hydrogen acceptor _____
   2. Source of carbon in carbon dioxide formed _____
   3. Uncontrolled oxidation _____
   4. Molecules enabling a controlled oxidation _____
   5. Provides energy for immediate cellular work _____
   6. Combines with ~P to form ATP _____
   7. Produces 36 ATP units, net _____
   8. Produces 2 ATP units, net _____

2. Respiration and Carbon Dioxide Production

a. Record the results of your experiments in the table below.

Carbon Dioxide Production in Animals and Germinating Seeds

Tube	Contents plus Bromthymol Blue	Color of Bromthymol Blue After Test Interval	CO_2 Concentration Increase	No Change
Exp. 1: Humans				
1	Exhaled air	_____	_____	_____
2	Atmospheric air	_____	_____	_____
Exp. 2: Germinating Seeds and Invertebrates				
1	Germinating seeds	_____	_____	_____
2	Live invertebrates	_____	_____	_____
3	Nothing	_____	_____	_____

b. What is the independent variable in these experiments? _____

c. What is the purpose of tube 2 in experiment 1 and tube 3 in experiment 2? _____

d. State a conclusion from experiment 1. _____

e. State a conclusion from experiment 2. _____

3. Respiration and Photosynthesis

a. Record the results of your experiment with photosynthesizing and nonphotosynthesizing leaves in the table below.

Carbon Dioxide Production and Photosynthesis

Tube	Contents plus Bromthymol Blue	Color of Bromthymol Blue after Test Interval	CO_2 Concentration Increase	No Change
1	Nonphotosynthesizing leaves	_____	_____	_____
2	Photosynthesizing leaves	_____	_____	_____
3	Nothing	_____	_____	_____

b. Explain your results. _____

_____

_____

_____

4. Respiration and Heat Production

a. Indicate the temperature of each vacuum bottle.

Bottle 1 _____   Bottle 2 _____   Bottle 3 _____

b. State a conclusion from the results. _____

_____

_____

_____

5. Temperature and Respiration Rate

a. State the null hypothesis to be tested regarding the effect of temperature on respiration rate of the organisms. _____

_____

_____

_____

b. Record the data collected for the rate of respiration in germinating pea seeds and invertebrates at 10°C, room temperature, and 40°C. *Add your room temperature to the chart.

Oxygen Consumption in Pea Seeds and Invertebrates

Organism	Temp. (°C)	ml O_2/3 min 5 Replicates					Average ml O_2/3 min	Combined Weight of Organisms (g)	Average ml O_2/hr/g
		1	*2*	*3*	*4*	*5*			
Peas	10								
	*								
	40								
Invert.	10								
	*								
	40								
Control	10								
	*								
	40								

c. Do the results support your hypothesis? _____

d. State a conclusion from the results. _____

_____

_____

e. Do you think your conclusion applies to all organisms? _____

Explain. _____

_____

_____

_____

6. **Respiration Rate in a Frog and a Mouse**

a. State the null hypothesis to be tested regarding the effect of temperature on the respiration rate of a frog and mouse. _____

_____

_____

b. Record the data collected for the rate of respiration in a frog and a mouse at 10°C, room temperature, and 40°C. *Add your room temperature to the chart.

Oxygen Consumption in a Frog and Mouse

Organism	Temp. (°C)	*ml O₂/3 min 5 Replicates*					Average ml O₂/3 min	Weight (g)	Average ml O₂/hr/g
		1	*2*	*3*	*4*	*5*			
Frog	10								
	*								
	40								
Mouse	10								
	*								
	40								

c. Do the results support your hypothesis? _____

d. State a conclusion from the results. _____

_____

_____

e. Explain the different effects of 10°C on respiration rate in the frog and mouse. _____

_____

_____

_____

f. Plot the average respiration rate (ml O₂/hr/g) of each organism tested in the graph below. Select and record a different symbol for each organism so your graph may be easily read.

Symbols

Peas: Frog:

Invertebrates: Mouse:

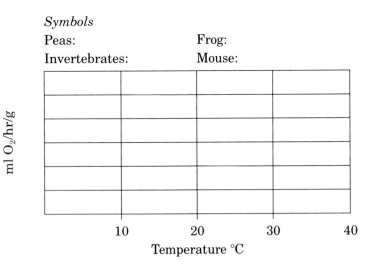

ml O₂/hr/g

Temperature °C

LABORATORY REPORT 17

Student _____

Lab Instructor _____

Digestion

1. Introduction
    a. Write the term that matches the phrase.
        1. Catalyze hydrolysis of foods   _____
        2. Digestion within cells   _____
        3. Determines specificity of enzyme   _____
        4. Enzymatic hydrolysis   _____
        5. Digestion outside of cells   _____
    b. Summarize the action of digestive enzymes. _____
_____
_____

2. Digestion in *Paramecium*
    Show by diagram the location and color of food vacuoles in your specimen.

3. Human Digestive System
    a. Write the term that matches the phrase.
        1. Produces bile   _____
        2. Digestive fluid formed by stomach mucosa   _____
        3. End product of protein digestion   _____
        4. Where carbohydrate digestion begins   _____
        5. Site of decay of nondigestible materials   _____
        6. Absorption of nutrients is completed   _____
        7. Wavelike contractions of digestive tract   _____
        8. Separate oral and nasal cavities   _____
        9. Carries food from pharynx to stomach   _____
      10. Where digestion of food is completed   _____
      11. Break food into smaller pieces   _____
      12. Pushes food into pharynx in swallowing   _____
      13. End product of fat digestion   _____
      14. End product of carbohydrate digestion   _____
      15. Reabsorbs water from nondigestible material   _____

b. List the layers of the intestinal wall from outside in.

_____   _____   _____   _____

c. List the labels for Figure 17.3.

1. _____   4. _____   7. _____
2. _____   5. _____   8. _____
3. _____   6. _____

d. List the labels for Figure 17.4.

1. _____   7. _____   13. _____
2. _____   8. _____   14. _____
3. _____   9. _____   15. _____
4. _____   10. _____   16. _____
5. _____   11. _____   17. _____
6. _____   12. _____   18. _____

4. Digestion of Starch

a. Record your results by placing an "S" for a positive test for starch or an "M" for a positive test for maltose in the appropriate squares.

	Tube Numbers and pH																	
	pH 5						pH 8						pH 11					
Temp.	1	1B	2	2B	3	3B	4	4B	5	5B	6	6B	7	7B	8	8B	9	9B
5° C																		
37°C																		
70°C																		

b. List the control tubes expected to give starch tests that are:

positive _____   negative _____

What should be done if these results are not attained? _____
_____

c. Indicate the optimum pH and temperature, among those tested, for amylase activity.

pH _____   Temp. _____°C

Do you think amylase may be active at another pH and temperature? _____

How would you test your hypothesis? _____
_____

d. How do some pH values and temperatures inactive amylase? _____
_____

e. Would you expect the enzymes of all organisms to work equally well at the optimum temperature determined by this experiment? _____

Explain. _____
_____
_____

f. Pancreatic amylase is a component of pancreatic juice that is released into the small intestine. What can you hypothesize about the pH of the digestive fluids in the small intestine? _____
_____

Gas Exchange

1. **Introduction**
   a. What is the physical process by which this exchange occurs? _____
   b. What is the relationship between the area of the respiratory surface and the rate of gas exchange? _____
      _____
   c. Describe the function of tracheae in insects. _____
      _____
   d. From your slide, draw part of a tracheal trunk and its smaller tracheal branches.

   e. Diagram a small portion of each type of gill and show how the surface area is increased in each.
      Clam   **Crayfish**   **Fish**

2. **Mammalian Respiratory System**
   a. List the labels for Figure 18.3.
      1. _____   5. _____   9. _____
      2. _____   6. _____   10. _____
      3. _____   7. _____   11. _____
      4. _____   8. _____
   b. List in sequence the air passages of the human respiratory system.
      Nasal cavity → _____ → _____ →
      _____ → _____ → lungs
   c. Write the term that matches the phrase.
      1. Site of gas exchange   _____
      2. Air passage that enters a lung   _____
      3. Lining that removes foreign particles   _____
      4. Filters, humidifies, and warms air   _____
      5. Tubules carrying air into alveoli   _____

d. Draw the microscopic appearance of each:

Normal Lung Tissue

Emphysematous Lung Tissue

3. Mechanics of Breathing

a. When the rubber sheet is pulled down, the volume of the bell jar is _____
and the pressure in the bell jar is _____, which causes air to flow
_____ the balloons.

b. When the rubber sheet is pushed up, the volume of the bell jar is _____
and the pressure in the bell jar is _____, which causes air to flow
_____ the balloons.

4. Lung Volumes

Record your:

Tidal volume _____ ml   Vital capacity _____ ml

Expiratory reserve volume _____ ml   Inspiratory reserve volume _____ ml

5. Control of Breathing

a. List three hypotheses to explain the increase in breathing rate.

1. _____

2. _____

3. _____

b. In experiment 2, what changes occurred in the concentration of gases in the container?

O_2 _____ CO_2 _____

How would these changes affect the blood concentrations of:

O_2 _____ CO_2 _____

c. In experiment 3, what changes occurred in the blood concentrations of:

O_2 _____ CO_2 _____

Which hypothesis has been eliminated by experiment 3? _____

_____

d. State the correct hypothesis based on the results of experiment 4. _____

_____

6. Gas Exchange in Leaves

a. Describe the pathway for gas exchange in leaves. _____

_____

b. Indicate the net directional flow of O_2 and CO_2 between the leaf and the atmosphere when the leaf is:

	O_2	CO_2
Photosynthesizing	_____	_____
Not photosynthesizing	_____	_____

c. What adaptations of a leaf facilitate gas exchange? _____

_____

_____

Student _____

Lab Instructor _____

Internal Transport in Plants

1. **Vascular Tissue**
   a. Write the term that matches the phrase.
      1. Tissue transporting water and minerals _____
      2. Tissue transporting organic nutrients _____
      3. Composes xylem in conifers _____
      4. Composes xylem in flowering plants _____
      5. Larger cells of phloem _____
      6. Smaller cells of phloem _____
      7. Composes wood of woody plants _____
      8. Vascular tissue formed by dead cells _____
      9. Direction of fluid movement in xylem _____
      10. Direction of nutrient movement in phloem _____
   b. Draw a few cells of these cell types from your slides.

   Vessels   **Tracheids**   **Sieve Tubes and Companion Cells**

   c. Based on your microscopic study, write the term that matches the phrase.
      1. Vascular tissue uppermost in a leaf vein _____
      2. Vascular tissue innermost in alfalfa stem _____
      3. Vascular tissue outermost in oak stem _____
      4. Vascular tissue innermost in vascular bundles of corn stem _____

2. **Root Pressure**
   a. Record the movement of fluid by root pressure in the table below.

	Time (min)				
	0	*30*	*60*	*90*	*120*
Reading (ml)					

   b. Record the average rate of fluid movement by root pressure. _____ ml/min

   c. Describe the importance of root hairs. _____

_____

   d. Show by diagram the distribution of root hairs on the young root of a germinated grass seed. Draw a root hair showing its attachment to the root and its nucleus.

             Root Hair Distribution (40×)   **Root Hair (400×)**

3. Transpiration

   a. What process causes water loss from the leaves in transpiration? _____

   b. Record the movement of fluid by transpiration in the table below.

	Time (min)					
	0	*2*	*4*	*6*	*8*	*10*
Reading (ml) (normal)						
Reading (ml) (wind)						

   c. Record the average transpiration rate (ml/min).

      No wind _____ Wind _____

   d. Place numbers in front of the following items to indicate the sequence of the water movement from potometer bottle to atmosphere.

      _____ Bottle   _____ Stomata   _____ Atmosphere

      _____ Substomatal space   _____ Xylem   _____ Mesophyll

   e. Name two other environmental factors that will affect the rate of transpiration. _____

_____

If time permits, conduct an experiment to test the effect of one of these factors. Record the results and your conclusion.

f. Which process, root pressure or transpiration, contributes more to the movement of water up the stem? _____

g. What causes plants to wilt? _____
_____

4. Stomatal Density

a. Calculate the density of stomata on your leaf based on your observations.

Calculation of Stomata Density

Leaf Surface	Area of Field (mm²)	Stomata per Field	Stomata per cm²	Leaf Surface Area (cm²)	Stomata per Leaf Surface
Upper					
Lower					

Use this space for your calculations.

b. Indicate the average density of stomata (stomata/cm^2) on each surface.

Upper-leaf surface _____   Lower-leaf surface _____

Explain the advantage of this distribution. _____

_____

_____

How many stomata are on the leaf? _____

c. What adaptations would you expect in leaves of desert plants in regard to:

Size? _____

Number of stomata per leaf? _____

Explain. _____

_____

_____

d. What adaptations would you expect in leaves of plants in a shaded, moist environment in regard to:

Size? _____

Number of stomata per leaf? _____

Explain. _____

_____

_____

LABORATORY REPORT 20

Internal Transport in Animals

1. **Introduction**
   a. Write the term that matches the phrase.
      1. Basic function of circulatory systems   _____
      2. Circulatory system with hemolymph   _____
      3. Circulatory system with blood   _____
      4. Respiratory pigment in annelids   _____
      5. Blue-green pigment in certain arthropods   _____
   b. Contrast open and closed circulatory systems.
      Open system _____

      _____

      Closed system _____

      _____

   c. Draw a few red cells from slides of amphibian, reptilian, and mammalian blood.

Amphibian	**Reptilian**	**Mammalian**

2. **Human Blood**
   a. Indicate the characteristic(s) that most easily distinguish(es):
      Erythrocytes from leukocytes _____

      _____

      Granulocytes from agranulocytes _____

      _____

   b. Write the term that matches the phrase.
      1. Most abundant blood cell   _____
      2. Most abundant white cell   _____
      3. Liquid portion of the blood   _____
      4. Cells lacking nuclei   _____
      5. Cells with lobed nuclei and red cytoplasmic
         granules   _____
      6. Cells with a large, kidney-shaped nucleus and no
         cytoplasmic granules   _____

7. Cell fragments that initiate clotting _____
8. Blood cells that fight infection _____
9. The smallest white blood cells _____
10. The rarest white blood cells _____

3. Blood Typing

a. Indicate your ABO blood group. _____ Rh type _____

b. Indicate compatibility (C) and incompatibility (I) of possible blood transfusions shown in the chart below.

Blood Type and Antigen of Donor	Blood Type (and Antibodies) of Recipient			
	O (a,b)	A (b)	B (a)	AB (none)
O				
A				
B				
AB				

c. Which ABO blood type may receive blood from the other three types in emergencies? _____
This is the **universal recipient**.

d. Which ABO blood type may donate blood to the other three types in emergencies? _____
This is the **universal donor**.

e. Considering *both* ABO and Rh blood types, indicate:
Universal donor _____ Universal recipient _____

f. Infants suffering from erythroblastosis fetalis may require a massive blood transfusion. What blood type or types should be given to a baby with a blood type of A,Rh$^+$? _____
Explain your answer. _____
_____

4. Mammalian Heart

a. How is the mammalian heart more efficient than the frog heart? _____
_____
_____
_____
_____

b. List the labels for Figure 20.4.

1. _____   8. _____
2. _____   9. _____
3. _____   10. _____
4. _____   11. _____
5. _____   12. _____
6. _____   13. _____
7. _____   14. _____

c. Write the term that matches the phrase regarding mammalian heart structure and function.
   1. Returns blood to right atrium   _____
   2. Returns blood to left atrium   _____
   3. Separates ventricles   _____
   4. Pumps blood into pulmonary arteries   _____
   5. Pumps blood into aorta   _____
   6. Prevents backflow of blood into right atrium   _____
   7. Prevents backflow of blood into left atrium   _____
   8. Prevents backflow of blood into right ventricle _____
   9. Prevents backflow of blood into left ventricle   _____
   10. Contraction phase of heart cycle   _____
   11. Relaxation phase of heart cycle   _____

5. **Sheep Heart Dissection**
   a. Explain the difference in size and wall thickness of the ventricles. _____
   _____
   _____

   b. Describe the function of the chordae tendineae. _____
   _____

   c. Circle the terms that describe the chordae tendineae.
        elastic   nonelastic   thick   thin   pliable   stiff   opaque   transparent
   d. Circle the terms that apply to the semilunar valves.
        elastic   nonelastic   thick   thin   pliable   stiff   opaque   transparent
   e. Circle the terms that apply to the atrioventricular valves.
        elastic   nonelastic   thick   thin   pliable   stiff   opaque   transparent

6. **Cardiac Muscle**
   Draw a small portion of cardiac muscle from your slide. Label a nucleus and an intercalated disc.

7. Pattern of Circulation

a. Write the term that matches the phrase.

1. Vessels carrying blood from the heart   _____
2. Vessels carrying blood to the heart   _____
3. Smallest blood vessels   _____
4. Controls flow of blood into capillary   _____
5. Smallest arteries   _____

b. By what process does O_2 pass from capillary blood into body cells? _____
_____

c. Describe the flow of blood through a capillary in the frog's foot. _____
_____

d. Draw the following from your observations to show the relative (a) diameter of the vessel lumen and (b) wall thickness.

Normal Vein	Normal Artery	Atherosclerotic Artery

Explain the hazards of atherosclerosis. _____
_____
_____
_____
_____

e. Trace the flow of blood from the left ventricle to the intestine and back to the left ventricle.
_____
_____
_____
_____
_____

f. What arteries carry deoxygenated blood? _____

g. What veins carry oxygenated blood? _____

h. List the labels for Figure 20.10.

1. _____   4. _____   7. _____
2. _____   5. _____   8. _____
3. _____   6. _____   9. _____

LABORATORY REPORT 21

Excretion

1. **Nitrogenous Wastes**

   a. Indicate the excretory organs removing nitrogenous wastes in each:

   Annelids _____ Crustaceans _____

   Insects _____ Mammals _____

   b. List three forms of nitrogenous wastes excreted by animals.

   1. _____ 2. _____ 3. _____

   Which compound is most effective in water conservation? _____

   Which compound is most toxic? _____

2. **Mammalian Urinary System**

   a. List the labels for Figures 21.1 and 21.2.

Figure 21.1	**Figure 21.2**
1. _____	1. _____	8. _____
2. _____	2. _____	9. _____
3. _____	3. _____	10. _____
4. _____	4. _____	11. _____
5. _____	5. _____	12. _____
6. _____	6. _____	13. _____
7. _____	7. _____	14. _____
8. _____		

   b. Describe the function of each.

   Kidney _____

   Urethra _____

   Urinary bladder _____

   Ureter _____

   c. Diagram the structural relationship between the Bowman's capsule, the glomerulus, afferent arteriole, and efferent arteriole.

d. List in sequence the parts of a nephron.

1. _____   3. _____

2. _____   4. _____

3. Urine Formation

a. Explain the adaptive value of increased blood pressure in the glomerulus. _____

_____

_____

b. Describe what happens in each:

Glomerular filtration _____

_____

Tubular reabsorption _____

_____

Tubular secretion _____

_____

c. Using the data in Table 21.1, what percentage of the blood passing through the kidneys becomes filtrate? _____ filtrate becomes urine? _____

What happens to the rest of the filtrate? _____

d. Why is a low volume of urine production important in terrestrial animals but unimportant in aquatic animals? _____

e. Considering Table 21.2, explain how the high concentration of urea in urine (column 4) is attained. _____

_____

f. Does the kidney simply maintain the concentration of urea at a safe level or remove all of it from the blood? _____

Explain. _____

g. Comparing filtrate and urine, how many times is urea concentrated in the urine (columns 3 and 4)? _____

How is that accomplished? _____

h. Comparing the concentration of mineral salts in columns 1 and 5, very little is removed from the blood. What happens to most of the salts in the filtrate? _____

i. Explain the protein concentrations in each column:

Column 2 _____

Column 5 _____

j. Does glucose easily diffuse into Bowman's capsule? _____

How do you know this? _____

What happens to the glucose in the filtrate? _____

Explain the slightly lower concentration of glucose in column 5 as compared to column 1.

_____

_____

4. Urinalysis

a. Record the results of your analysis of the simulated urine samples in the table below. Use an X to indicate the presence of a component and record the color and specific gravity.

Characteristic	Tubes					
	1	*2*	*3*	*4*	*5*	*6*
Glucose						
Bilirubin						
Ketone						
Specific gravity						
Blood						
pH						
Protein						
Nitrite						
Color						

b. Indicate for each urine sample the health condition of the "patient" based on your urinalysis. Select the health conditions from the list that follows.

Tube 1 _____

Tube 2 _____

Tube 3 _____

Tube 4 _____

Tube 5 _____

Tube 6 _____

Diabetes insipidis

This disorder is caused by a deficiency of antidiuretic hormone which results in the inability of kidney tubules to reabsorb water. Symptoms are constant thirst, weight loss, weakness, and production of 4 to 10 liters of urine each day. The urine is very dilute (a low specific gravity) and is nearly colorless.

Diabetes mellitus

This disorder is caused by a deficiency of insulin which results in a decreased ability for glucose to enter cells. Therefore, glucose accumulates in the blood and fats are used excessively in cellular respiration producing ketones as a waste product. Symptoms include weakness, fatigue, weight loss, and excessive blood glucose levels. In uncontrolled diabetes, a urine sample contains excess glucose, ketones, and a low pH.

Hepatitis

Hepatitis is usually caused by a viral infection. In type A hepatitis, symptoms may include fatigue, fever, generalized aching, abdominal pain, and jaundice (yellowish tinge to the whites of the eyes due to excessive blood levels of bilirubin). Urine is dark amber in color due to excessive bilirubin.

Glomerulonephritis

In this kidney disease, excessive permeability of Bowman's capsule allows proteins and red blood cells to enter the filtrate. They are present in the urine since they cannot be reabsorbed.

Hemolytic anemia

A number of conditions cause the destruction of red blood cells releasing hemoglobin into the blood plasma, for example, malaria and incompatible blood transfusions. The urine of such patients contains hemoglobin and may be red-brown or smoky in color.

Normal

See Table 21.3 for the composition of normal urine.

Strenuous exercise and a high-protein diet

Athletes in excellent health may produce urine with an acid pH, presence of a small amount of protein, and an elevated specific gravity due to the presence of the proteins.

Urinary tract infection

Bacterial urinary tract infections may involve the urethra (urethritis), urinary bladder (cystitis), ureters, or the kidney. Production of alkaline urine promotes the growth of infectious bacteria. Urethritis often causes a burning pain during urination. Symptoms may include a low-grade fever and discomfort of the affected region. Urine will show a positive test for nitrite because bacteria in the urine convert nitrate, a normal component, into nitrite. Severe infections may also cause the urine to contain blood and pus (leukocytes) producing a cloudy urine.

c. Which characteristics would you expect to find in a urine sample if:

There was an acute kidney infection? _____

_____

_____

The glomerular blood pressure was chronically high? _____

_____

_____

d. Explain the relationship between the volume of urine, its specific gravity, and its color. _____

_____

_____

_____

_____

LABORATORY REPORT 22

Student _____

Lab Instructor _____

Chemical Control

1. **Introduction**

   Write the term that matches the phrase.

   1. Chemical messengers _____
   2. Chemicals released into synaptic junctions _____
   3. Chemical messengers formed by endocrine glands _____
   4. Chemicals formed by neurons and affecting nonadjacent cells _____
   5. Chemicals formed by neurons and affecting adjacent cells _____
   6. Chemical messengers in vascular plants _____
   7. Aggregations of endocrine cells in mollusks and higher animals _____

2. **IAA and Plant Growth**

   a. Write the term that matches the phrase.

   1. Growth of shoot toward light _____
   2. Growth of root toward the pull of gravity _____
   3. Growth of shoot away from the pull of gravity _____
   4. Growth of root away from light _____
   5. Chemical causing tropisms _____

   b. Considering experiments 1 and 2 and that IAA promotes growth:

   1. Where is IAA produced in the coleoptile? _____
   2. What portion of the coleoptile responds to the IAA? _____
      _____
   3. How does IAA move from the site of production to the site of action? _____
   4. Is IAA soluble in water? _____Why do you think so? _____

   c. When an agar block containing coleoptile-tip extract was placed on the right side of a tipless coleoptile in experiment 3:

   1. Was the IAA concentration greater in the region of elongation on the right or left side of the coleoptile? _____
   2. Why do you think so? _____
   3. Which side of the coleoptile grew more rapidly? _____
   4. What is the action of IAA? _____

   d. In experiment 4:

   1. What is the role of group A? _____
   2. How do you explain the growth of group B? _____
      _____

e. In experiment 5:
  1. What is the role of groups A and C? _____
  2. On which side of the coleoptile in group B was the IAA concentration:
     Higher? _____ Lower? _____
  3. Why do you think so? _____
  4. State two possible explanations for the effect of light on IAA.
     a. _____
     _____
     b. _____
     _____

f. Explain the growth pattern of the plants in the geotropism demonstration.
   Plant A _____
   _____
   _____
   Plant B _____
   _____
   _____

g. Explain the growth pattern of plants in the phototropism demonstration.
   Plant A _____
   _____
   _____
   Plant B _____
   _____
   _____

h. Explain the growth pattern of plants that received these applications?
   Lanolin only _____
   _____
   Lanolin and IAA _____
   _____

3. **Thyroxine and Metabolic Rate**
   a. Record the data regarding food consumed and weight gained in the table below.

Food Consumption and Weight Gained (grams) in Mice

Mouse	Starting Weight	Weight Today	Weight Gained	Food Consumed	Weight Gained / Food Consumed
A					
B					
C					

   b. Which mouse used the food most efficiently? _____
   c. Summarize the effect of thyroxine on weight gain in young mice. _____
   _____
   _____

d. Record the oxygen consumption (milliliters) in the table below.

Oxygen Consumption (millimeters) in Mice

Mouse	Replicates (3 min each)					Average ml O_2/min	Average ml O_2/g/hr
	1	*2*	*3*	*4*	*5*		
A							
B							
C							

e. Summarize the effect of thyroxine on the metabolic rate of young mice. _____

_____

_____

_____

_____

4. Chemical Control of Heart Rate

a. Record the data from the control of heart rate experiment in the table below

Heart Rate of the Frog

Beats per Minute	Condition
_____	Before injection
_____	30 sec after injection of 0.05 ml epinephrine
_____	3 min after injection of 0.05 ml epinephrine
_____	30 sec after injection of 0.05 ml acetylcholine
_____	3 min after injection of 0.05 ml acetylcholine
_____	30 sec after injection of 0.25 ml epinephrine
_____	3 min after injection of 0.25 ml epinephrine
_____	30 sec after injection of 0.25 ml acetylcholine
_____	3 min after injection of 0.25 ml acetylcholine

b. Indicate how long it took for the effect of the chemicals to be recognized after injection.
0.05 ml of epinephrine _____ 0.05 ml of acetylcholine _____
Did the effect last 3 min or longer for the injection of:
0.25-ml epinephrine? _____ 0.25-ml acetylcholine? _____

c. Summarize the effect of the chemicals on the heart rate.
Epinephrine _____
Acetylcholine _____

d. What is the advantage of epinephrine secretion by the adrenal gland during stress? _____

_____

_____

_____

_____

5. Review

Matching

1. IAA   2. Thryoxine   3. Epinephrine   4. Acetylcholine   5. Norepinephrine

_____ Neurotransmitter   _____ Plant hormone

_____ True hormone   _____ Secreted by an endocrine gland

_____ Acts on adjacent cells   _____ Acts on nonadjacent cells

LABORATORY REPORT 23

Neural Control

1. **Neurons**
   a. Write the term that matches the phrase.
      1. Nerve cell _____
      2. Neuron carrying impulses to CNS _____
      3. Neuron carrying impulses from CNS _____
      4. Composed of cranial and spinal nerves _____
      5. Composed of brain and spinal cord _____
      6. Cells providing support for nerve cells _____
      7. Part of neuron containing nucleus _____
      8. Process carrying impulses from cell body _____
      9. Process carrying impulses toward cell body _____
      10. Neuron with all parts located in the CNS _____
      11. Junction of axon tip and adjacent neuron _____
      12. Secretes neurotransmitter _____
      13. Receives neurotransmitter _____
      14. Forms myelin sheath in PNS neurons _____
      15. Forms myelin sheath in CNS neurons _____
      16. Necessary for regrowth of neuron processes _____
      17. Spaces between Schwann cells _____
      18. A stimulatory neurotransmitter _____
      19. An inhibitory neurotransmitter _____
      20. Cells located between neurons in the CNS _____
   b. Explain the mechanism of one-way transmission of impulses at a synapse. _____
   _____
   _____

   c. Draw a giant multipolar neuron and a nerve, x.s., as observed on the prepared slides. Label
      pertinent parts.

   Neuron Cell Body   **Nerve, x.s.**

2. The Brain

a. Write the term that matches the phrase.

1. Fissure between cerebral hemispheres _____

2. Cerebral lobe containing visual center _____

3. Outer portion of cerebrum _____

4. Small ridges on surface of cerebrum _____

5. Cerebral lobe containing hearing center _____

6. Protective membranes covering CNS _____

7. Remnant of a third eye _____

8. Endocrine gland attached to hypothalamus _____

b. List the labels for Figure 23.4.

1. _____ 4. _____ 6. _____

2. _____ 5. _____ 7. _____

3. _____

c. List the labels for Figure 23.7.

1. _____ 5. _____ 9. _____

2. _____ 6. _____ 10. _____

3. _____ 7. _____ 11. _____

4. _____ 8. _____ 12. _____

d. Matching

1. Cerebrum   2. Cerebellum   3. Thalamus   4. Hypothalamus   5. Medulla oblongata
6. Corpus callosum   7. Ventricles

_____ Uncritical awareness of sensations   _____ Muscular coordination

_____ Contains cerebrospinal fluid   _____ Controls water balance

_____ Critical interpretation of sensations   _____ Largest part of mammalian brain

_____ Regulates heart rate   _____ Controls body temperature

_____ Connects cerebral hemispheres   _____ Intelligence, will

_____ Breathing control center

e. Is the relative size of the following larger in humans or sheep?

Cerebrum _____   Cerebellum _____

f. Matching

1. White matter   2. Gray matter

_____ Exterior of cerebrum   _____ Exterior of spinal cord

_____ Myelinated fibers   _____ Nonmyelinated fibers

_____ Neuron cell bodies   _____ Axon tips in spinal cord

3. Spinal Cord

a. List the labels for Figure 23.8.

1. _____ 5. _____ 8. _____

2. _____ 6. _____ 9. _____

3. _____ 7. _____ 10. _____

4. _____

b. Write the term that matches the phrase.
   1. Neuron carrying impulses to effector _____
   2. Neuron receiving impulses from receptor _____
   3. Neuron in dorsal root _____
   4. Neuron in ventral root _____
   5. Location of sensory neuron cell body _____
   6. Connects sensory and motor neurons _____
c. Describe the effect of an injury that cuts the:
   Dorsal root of a spinal nerve _____
   _____

   Ventral root of a spinal nerve _____
   _____

   Spinal cord _____
   _____

d. Diagram from your slide (1) a cat spinal cord, x.s., labeling gray and white matter, ventral fissure, and central canal, and (2) a motor neuron cell body and associated processes labeling the nucleus, nucleolus, and neurofibrils at the base of the processes.

Cat Spinal Cord, x.s.   **Motor Neuron Cell Body**

4. Human Reflexes

a. Rate the strength of your patellar response as strong (+ + +), moderate (+ +), weak (+), or none (0):

   Left leg: without diversion _____   with diversion _____
   Right leg: without diversion _____   with diversion _____
   Explain any differences in response. _____
   _____
   _____

b. If you touch a hot stove, you will reflexively jerk back your hand a split second before you feel the pain. Explain the delayed pain response. _____
   _____
   _____
   _____

c. Describe the effect on the size of the pupil when:
   The subject is looking into a darkened area. _____
   The subject is looking into a bright light. _____

d. When a penlight is shined into the left eye, what response is seen:

In the left eye? _____

In the right eye? _____

Explain your observations. _____

_____

_____

_____

5. Spinal Reflex in the Frog

a. Record the response (leg movement) to the acetic acid stimulus before the sciatic nerve is cut.

Site of Stimulation	Right Hind Leg	Left Hind Leg	Right Front Leg	Left Front Leg
Calf of right hind leg				
Calf of left hind leg				

b. Describe the impulse pathways in this response. _____

_____

_____

c. Record the response (leg movement) after the right sciatic nerve was cut.

Site of Stimulation	Right Hind Leg	Left Hind Leg	Right Front Leg	Left Front Leg
Calf of right hind leg				
Calf of left hind leg				

d. Explain the response when the right hind leg was stimulated. _____

_____

_____

e. These experiments have been performed on a frog that is brain dead.

Could the frog be kept alive on a frog respirator? _____

Would the frog ever be able to recover to a normal life? _____

Do your responses also apply to brain dead humans? _____

What does this tell you about the progressive nature of death? _____

_____

_____

_____

LABORATORY REPORT 24

Sensory Perception

1. The Eye

a. List the labels for Figure 24.1.

1. _____ 6. _____ 11. _____
2. _____ 7. _____ 12. _____
3. _____ 8. _____ 13. _____
4. _____ 9. _____ 14. _____
5. _____ 10. _____ 15. _____

b. Write the term that matches each meaning.

1. Outer white fibrous layer _____
2. Fills space in front of lens _____
3. Controls amount of light entering eye _____
4. Transparent window in front of eye _____
5. Layer with photoreceptors _____
6. Controls shape of lens _____
7. Opening in center of iris _____
8. Fills space behind lens _____
9. Site of sharp, direct vision _____
10. Receptors for color vision _____
11. Carries impulses from eye to brain _____
12. Receptors for black and white vision _____
13. Membrane covering front of eye _____
14. Focuses light on retina _____
15. Junction of retina and optic nerve _____

2. Visual Tests

a. Record the distance from your eye to Figure 24.2 at which the blind spot is detected.

Left eye _____ Right eye _____

Explain any difference in the distances. _____

_____

b. Record the near-point distance for your eyes.

Left eye: without glasses _____ with glasses _____

Right eye: without glasses _____ with glasses _____

Explain any differences in the distances. _____

_____

How do your near-point values compare with your age? See Table 24.1.

_____

c. Is astigmatism present?
   Left eye: without glasses _____   with glasses _____
   Right eye: without glasses _____   with glasses _____
d. Record the acuity for each eye.
   Left eye: without glasses _20/_____   with glasses _20/_____
   Right eye: without glasses _20/_____   with glasses _20/_____
e. Record your responses to the color-blindness test in the table.

Responses to the Ishihara Color-Blindness Test Plates

Plate No.	Subject	Normal	Red-Green Color Blind		Totally Color Blind
1		12	12		12
2		8	3		0
3		5	2		0
4		29	70		0
5		74	21		0
6		7	0		0
7		45	0		0
8		2	0		0
9		0	2		0
10		16	0		0
11		Traceable	0		0
			Red Cones Absent	Green Cones Absent	
12		35	5	3	0
13		96	6	9	0
14		Can trace two lines	Purple	Red	0

*0 means the subject sees no pattern in the test plate.

Are you color blind? _____   If so, describe your type of color blindness. _____
_____

3. The Ear
a. List the labels for Figure 24.4.
   1. _____ 6. _____ 11. _____
   2. _____ 7. _____ 12. _____
   3. _____ 8. _____ 13. _____
   4. _____ 9. _____
   5. _____ 10. _____

b. Write the term that matches each meaning.
   1. Contains sound receptors _____
   2. Bone in which inner ear is embedded _____
   3. Contains receptors for static equilibrium _____
   4. Connects middle ear to pharynx _____
   5. Interprets impulses as sound _____
   6. Contain receptors for dynamic equilibrium _____
   7. Conduct vibrations across middle ear _____
   8. Membrane separating outer and middle ear _____

4. **Ear Physiology**
   a. Record the average distance for the watch-tick test.
      Left ear _____ Right ear _____ Class average _____
   b. State your conclusion from the Rinne test.
      Left ear _____
      Right ear _____
   c. Indicate your degree of wavering in the static equilibrium test. Use the terms *slight, moderate,* and *great* to indicate variations.
      Standing on both feet with arms at side
      Eyes open _____ Eyes closed _____
      Standing on both feet with arms extended
      Eyes open _____ Eyes closed _____
   d. How important are visual clues in maintaining balance? _____
      _____

5. **Skin Receptors**
   a. Indicate the minimal distances that provided a two-point sensation.
      Forearm _____ Back of neck _____
      Palm of hand _____ Fingertip _____
      What is the value of this distribution? _____
      _____
      _____

   b. Indicate the adaptation time when using
      A single coin _____ Several coins _____
      What is the value of adaptation to stimuli? _____
      _____
      _____

   c. Indicate the smallest temperature differential that you could detect by immersing your finger in water. _____°C
      Indicate the range and average of the class. Range _____°C Average _____°C
   d. Was the sensation stronger or weaker when your entire hand was immersed in the water? _____
      _____ Explain the sensation. _____
      _____
      _____

e. Did the sensation change with time when your hands were immersed in
Ice water? _____ Warm water? _____
Explain. _____
_____
_____

f. Describe the sensation when both hands were placed in tap water after immersion in ice and
warm water, respectively. _____
_____
How do you account for this? _____
_____
_____

g. What is the physiological basis for a boy looking for his hat only to discover that he is wearing
it? _____
_____
_____

h. What is the physiological basis for a girl believing that the temperature of water in a swimming
pool is "fine" after testing it with her foot but complaining that it is "freezing cold" after jumping
in? _____
_____

i. Indicate the role of the following in sensory perception.
Receptors _____
_____
Brain _____
_____
Nerves _____
_____

LABORATORY REPORT 25

The Skeletal System

1. Skeletal Types

   a. Write the term that matches the phrase.

     1. Skeleton within the body   _____

     2. Skeleton of body fluids and body wall   _____

     3. Skeleton exterior to the body   _____

     4. Most common skeleton among invertebrates   _____

     5. Type of skeleton in vertebrates   _____

     6. Join bones together at articulations   _____

     7. Secretes exoskeleton in arthropods   _____

     8. Hardens arthropod exoskeletons   _____

   b. Matching

     1. Calcareous exoskeleton   2. Bony endoskeleton   3. Chitinous exoskeleton

     4. Cartilaginous endoskeleton   5. Endoskeleton of calcareous plates

     _____ Mollusks   _____ Sharks and rays

     _____ Echinoderms   _____ Amphibians

     _____ Arthropods   _____ Mammals

   c. Explain the relationship between molting and growth in arthropods. _____

     _____

     _____

   d. Name three functions that arthropod exoskeletons and vertebrate endoskeletons have in common.

     1. _____

     2. _____

     3. _____

2. Macroscopic Bone Structure

   a. List the labels for Figure 25.1.

     1. _____   4. _____   7. _____

     2. _____   5. _____   8. _____

     3. _____   6. _____

b. Write the term that matches the phrase.
    1. Location of yellow marrow   _____
    2. Fills spaces in cancellous bone   _____
    3. Type of bone forming the diaphysis   _____
    4. Type of bone forming the epiphyses _____
    5. Fibrous membrane covering bone   _____
    6. Forms blood cells   _____
    7. Cartilage between diaphysis and epiphyses _____
    8. Protects articular surface of the bone   _____
    9. Site of growth in length   _____
c. Distinguish between the epiphyseal disc and epiphyseal line. _____

_____

_____

d. What advantage is provided by the articular cartilage? _____

_____

e. Why is adequate dietary calcium necessary for the development of strong bones? _____

_____

f. Describe the appearance, feel, and function of the articular cartilage.
    1. Appearance _____
    2. Feel _____
    3. Function _____
g. Circle the following terms that describe the ligaments.
       pliable   stiff   weak   strong   elastic   nonelastic

3. Microscopic Study
  a. Matching
     1. Hyaline cartilage   2. Dense fibrous connective tissue   3. Fibrocartilage   4. Bone
     _____ Tendons, ligaments   _____ Connective tissue
     _____ Matrix of calcium salts   _____ Many white fibers in matrix
     _____ Articular cartilage   _____ Intervertebral discs
  b. Does the bone seem to have a good supply of blood? _____
  c. Does cancellous bone decrease the weight of a bone? _____
    Does it significantly decrease the strength of a bone? _____
  d. Draw a portion of these tissues from your slides. Label pertinent parts.
            Bone, x.s.   **Hyaline cartilage**

Fibrocartilage   **Dense Fibrous Connective Tissue**

e. Describe the relationship between lacunae and osteocytes. _____

_____

f. What part of bone tissue is nonliving? _____

g. What is the function of canaliculi? _____

4. The Human Skeleton

a. List the labels for Figure 25.6.

1. _____	9. _____	16. _____
2. _____	10. _____	17. _____
3. _____	11. _____	18. _____
4. _____	12. _____	19. _____
5. _____	13. _____	20. _____
6. _____	14. _____	21. _____
7. _____	15. _____	22. _____
8. _____		

b. Matching

   1. Clavicle 2. Humerus 3. Femur 4. Os coxae 5. Scapula 6. Ulna and radius
   7. Tibia and fibula 8. Mandible 9. Skull 10. Sacrum 11. Sternum 12. Vertebrae
   13. Tarsals 14. Carpals 15. Phalanges

_____ Ankle bones	_____ Form backbone
_____ Bones of lower arm	_____ Bone of upper arm
_____ Finger bones	_____ Bones of lower leg
_____ Shoulder blade	_____ Collarbone
_____ Wrist bones	_____ Breastbone
_____ Thighbone	_____ Lower jawbone
_____ Five fused vertebrae	_____ Hipbones

c. Matching

   1. Immovable 2. Slightly movable 3. Hinge 4. Gliding 5. Pivot 6. Ball and Socket

_____ Vertebrae 1 and 2	_____ Humerus/ulna
_____ Humerus/scapula	_____ Finger joints
_____ Joints between vertebrae	_____ Joints of cranial bones
_____ Femur/os coxae	_____ Wrist and ankle
_____ Femur/tibia	_____ Lower jaw/skull

5. Sexual Differences of the Pelvis

a. Using Table 25.1, record your observations of a male and female pelvis and of the "unknown" pelvis.

Characteristic	Male	Female	Unknown
General structure			
Acetabula			
Pubic angle			
Sacrum			
Coccyx			
Pelvic brim			
Ischial spines			
Pelvic ratio			

b. What is the gender of the "unknown" pelvis? _____

c. What is the advantage of the adaptations of the female pelvis? _____

_____

6. Fetal Skeleton

a. What portions of the:

Long bones are formed of bone? _____

Cranial bones are formed of bone? _____

b. What part of the fetal skeleton seems to have the largest circumference? _____

_____

c. What is the advantage of the incomplete ossification of the cranial bones prior to birth? _____

_____

7. Skeletal Adaptations

a. In fast-running animals, the legs are:

_____ Extended laterally   _____ Located under the body

b. Matching: Compare the monkey and human skeletons.

1. Monkey   2. Human

_____ Forward-facing eye orbits   _____ Grasping hands

_____ Heavier, thicker vertebrae   _____ Grasping feet

_____ Broader, heavier pelvis   _____ Narrower, lighter pelvis

_____ More posterior skull attachment   _____ More central skull attachment

_____ Vertebral curvatures   _____ Hind legs shorter than front legs

_____ Many tail vertebrae

LABORATORY REPORT 26

Student _____

Lab Instructor _____

Muscles and Movement

1. Vertebrate Muscle Tissues

  a. What characteristic of muscle tissue enables movement? _____

  b. Matching

    1. Smooth muscle   2. Skeletal muscle   3. Cardiac muscle

    _____ Attached to bones   _____ Voluntary control

    _____ Heart muscle   _____ Sarcolemma present

    _____ Multinucleate fibers   _____ Striations present

    _____ Involuntary control   _____ In walls of blood vessels

    _____ Rapid contractions, but quickly tiring   _____ Rhythmic, untiring contractions

  c. Draw a few teased muscle fibers from your slides. Label pertinent parts and indicate the magnification.

    Smooth   **Skeletal**   **Cardiac**

2. Skeletal Muscle Action

  a. Indicate the action of the antagonists to muscles that cause:

    Flexion _____   Adduction _____

  b. The stationary end of a muscle is the: _____; the action end of a muscle is the: _____

  c. Considering Figure 26.4, indicate the site of insertion for each:

    Biceps _____   Triceps _____

  d. Are most muscle insertions arranged to provide maximum mechanical advantage? _____
    Explain your answer. _____

    _____

    _____

    _____

443

e. Is the vertebrate body adapted for speed of movement or great strength? _____
Explain your answer. _____
_____
_____

f. Indicate the directional movement and lever class for each illustration in Figure 26.6.

Illustration	**Movement**	**Lever Class**
A	_____	_____
B	_____	_____
C	_____	_____
D	_____	_____
E	_____	_____
F	_____	_____
G	_____	_____

3. Frog Muscles

a. Do skeletal muscles extend across a joint or are they located only between joints? _____

b. Indicate the actions of these muscles in the frog's leg:
1. Triceps femoris _____
2. Semimembranosus _____
3. Gastrocnemius _____
4. Peroneus _____
5. Sartorius _____
6. Gracilis major _____
7. Extensor cruris _____

4. Ultrastructure and Contraction

a. Write the term that matches the phrase.
1. Thick myofilament with cross-bridges _____
2. Thin myofilaments attached to Z lines _____
3. Form boundaries of sarcomere _____
4. Myofilaments composing I band _____

b. Record the length of the fibers before and after adding ATP, K^+, and Mg^{++}.
Before _____ mm   After _____ mm

c. Draw the appearance of striations of the fibers before and after adding ATP, K^+, and Mg^{++}.

Before   **After**

LABORATORY REPORT 27

Cell Division

1. Introduction

Write the type of cell division that matches each phrase.

1. Occurs in prokaryote cells _____
2. Occurs in haploid eukaryote cells _____
3. Occurs in diploid eukaryote cells _____
4. Forms cells with identical genetic composition _____
5. Forms cells with half the chromosome number of the parent cell _____

2. Mitotic Division

a. Write the term that matches each phrase.

1. Compose a replicated chromosome _____
2. Division of the cytoplasm _____
3. 5% to 10% of the cell cycle _____
4. 90% to 95% of the cell cycle _____
5. Stage of interphase where chromosomes replicate _____
6. Members of a chromosome pair _____
7. Diploid chromosome number in humans _____
8. Haploid chromosome number in humans _____

b. Write the mitotic stage that matches each phrase.

1. Nuclear membrane and nucleolus disappear _____
2. Spindle is formed _____
3. Separation of sister chromatids _____
4. Daughter cells are formed _____
5. Rod-shaped chromosomes are first visible _____
6. Chromosomes line up on equatorial plane _____
7. New nuclei are formed _____
8. Daughter chromosomes migrate to opposite poles of the cell _____

c. Draw whitefish blastula cells in the stages noted below as they appeared on your slide. Label pertinent structures.

| **Interphase** | **Prophase** | **Metaphase** |

| **Early Anaphase** | **Late Anaphase** | **Telophase** |

d. Draw onion root tip cells in the following stages as they appeared on your slide. Label pertinent structures.

| **Prophase** | **Anaphase** | **Telophase** |

e. Describe how mitotic cell division in plants differs from that in animal cells. _____
_____

f. In the onion root tip, do the daughter cells occupy the same column of cells as the parent cell?
_____ Is it the same for all cells in the root tip? _____

g. Describe the process of mitotic cell division in your own words without using the names of the phases. _____
_____
_____
_____
_____

3. **Meiotic Division**
    a. Considering meiosis in humans, indicate the:
        1. Number of chromosomes in the parent cell _____
        2. Number of chromatids in daughter cells formed by first division _____
        3. Number of chromosomes in daughter cells formed by first division _____
        4. Number of chromosomes in daughter cells of second division _____
        5. Number of haploid cells formed by meiotic division of parent cell _____
    b. Diagram the arrangement of the chromosomes in mitosis and meiosis as they would appear in the phases noted below where the diploid (2n) chromosome number is 4.
    Mitosis
    Metaphase   **Anaphase**

    Meiosis I
    Metaphase   **Anaphase**

    Meiosis II
    Metaphase   **Anaphase**

c. Indicate the number of chromosomes in human cells formed by:

Mitotic cell division _____ Meiotic cell division _____

d. Summarize in your own words the process of meiosis without using the names of the phases.

_____

_____

_____

_____

_____

_____

e. Meiotic division is the process leading to the formation of specific cells in animals and plants. Name these cells.

Animals _____ Plants _____

f. Indicate the number of chromosomes in a human:

Egg cell _____ Sperm _____

g. (True/False)

The zygote formed by the union of haploid gametes receives:

A random number of chromosomes from
    each parent   _____

Equal numbers of chromosomes from
    each parent   _____

One member of each chromosome pair from
    each parent   _____

Both members of each chromosome pair from
    each parent   _____

4. Summary

    a. Summarize the primary role of mitotic cell division in:

       Unicellular organisms _____

       Infant humans _____

       Adult humans _____

    b. Summarize the role of meiotic cell division in:

       Animals _____

       _____

       Plants _____

       _____

Student _____

LABORATORY REPORT 28

Lab Instructor _____

Reproduction in Simple Organisms

1. Reproductive Patterns

Write the type of reproduction that matches each phrase.

Asexual sporulative   Sporic sexual   Zygotic sexual   Gametic sexual   Vegetative

1. Meiotic division of diploid zygote _____
2. Haploid sporangia produce mitospores _____
3. Budding and fragmentation _____
4. Diploid gonads form haploid gametes _____
5. Diploid sporangia produce meiospores _____
6. Haploid gametes formed by mitotic division _____
7. Cell division of unicellular organisms _____

2. Monerans and Protists

Write the term that matches each phrase.

1. Asexual reproductive method in monerans _____
2. Asexual reproductive method in protists _____
3. Sexual reproductive method in monerans _____
4. Sexual reproductive method in *Paramecium* _____
6. Attachment site in conjugating *Paramecium* _____
7. Process forming gamete nuclei in *Paramecium* _____
8. Process forming zygote nuclei in *Paramecium* _____
9. Process forming offspring in *Paramecium* _____

3. Fungi

a. Write the term that matches each phrase regarding *Rhizopus*.

1. Diploid portion of life cycle _____
2. Asexual reproductive type _____
3. Sexual reproductive type _____
4. Cell division forming spores _____

b. Draw the appearance of budding yeast cells and mature sporangia and zygospores of *Rhizopus*.

| **Budding Yeast** | ***Rhizopus*** | ***Rhizopus*** |
| **Cells** | **Sporangia** | **Zygospores** |

c. Write the term that matches each phrase regarding *Coprinus*.
1. Cytoplasmic fusion of gametelike cells _____
2. Site of gamete nuclei fusion _____
3. Type of hyphae forming the basidiocarp _____
4. Cell division forming basidiospores _____
5. Location of basidia on the basidiocarp _____

d. Draw a few dikaryotic hyphae with basidia and basidiospores.

4. Green Algae

a. Write the term that matches each phrase.
1. Vegetative reproduction in *Spirogyra* and *Oedogonium* _____
2. Asexual reproduction in *Oedogonium* _____
3. Sexual reproduction in *Spirogyra* _____
4. Sexual reproduction in *Ulva* _____
5. Sexual reproduction in *Oedogonium* _____
6. Gamete-forming generation in *Ulva* _____
7. Diploid generation in *Ulva* _____
8. Fate of zygote in *Spirogyra* _____
9. Fate of zygote in *Oedogonium* _____
10. Fate of zygote in *Ulva* _____

b. Draw from your slides of *Oedogonium* the following structures. Label pertinent parts.

Antheridium   **Oogonium**   **Zoospores**

c. Indicate a major advantage and disadvantage of asexual reproduction.
Advantage _____

Disadvantage _____

d. Indicate a major advantage and disadvantage of sexual reproduction.

Advantage _____

_____

Disadvantage _____

_____

Student _____

Lab Instructor _____

Reproduction in Land Plants

1. **Alternation of Generations**

   Write the term that matches each phrase.

   1. Haploid adult generation _____
   2. Diploid adult generation _____
   3. Union of gametes (process) _____
   4. Cell division forming gametes _____
   5. Generation formed by growth of meiospores _____
   6. Spore-forming generation _____
   7. Gamete-forming generation _____
   8. Cell formed by fusion of gametes _____
   9. Type of cell division forming spores _____
   10. Generation formed from zygote _____

2. **Mosses**

   a. Write the term that matches each phrase.

   1. Dominant generation _____
   2. Organ producing sperm _____
   3. Organ producing egg _____
   4. Formed by growth of protonema _____
   5. Organ producing meiospores _____
   6. Partially parasitic generation _____
   7. Provide motility for sperm _____
   8. Haploid generation _____
   9. Formed by meiotic cell division _____

   b. Explain why the process of fertilization usually restricts mosses to moist habitats. _____
   _____

   c. From your slides, make drawing of the following structures. Label pertinent parts.

Antheridium with Sperm	Archegonium with Egg	Capsule with Spores

3. Ferns

a. Write the term that matches each phrase.

1. Dominant generation   _____

2. Clusters of sporangia   _____

3. Formed by meiotic cell division   _____

4. Haploid generation   _____

5. Provide motility for sperm   _____

6. Plant part containing sporangia   _____

7. Organ containing egg   _____

8. Organ producing sperm   _____

9. Organ producing spores   _____

b. What environmental conditions are required for sperm transport? _____

_____

c. What advantage is gained by the antheridia and archegonia being located on the underside of the gametophyte? _____

_____

d. From a fresh specimen, draw the appearance of a sorus at 40× and a single sporangium at 100×. Label pertinent parts.

              Sorus   **Sporangium**

e. From a slide of fern gametophyte, draw the appearance of the underside of the gametophyte and label pertinent parts. Draw enlargements of an antheridium and archegonium.

4. Seed Plants

Write the term that matches each phrase.

1. Organ forming microspores   _____
2. Organ forming megaspores   _____
3. Develop from microspores   _____
4. Develop from megaspores   _____
5. Generate sperm nuclei   _____
6. Transfer of pollen to megasporangia   _____
7. Composed of female gametophyte, embryo sporophyte, and associated tissues   _____

5. Cone-Bearing Plants

a. Write the term that matches each phrase.

1. Dominant generation   _____
2. Cones forming pollen grains   _____
3. Cones with megasporangia   _____
4. Cones with microsporangia   _____
5. Cones forming seeds   _____
6. Develops from functional megaspore   _____
7. Develops from microspores   _____
8. Method of pollination   _____
9. Male gametophyte   _____
10. Method of seed dispersal   _____

b. From your slides and fresh specimens, draw the following structures and label pertinent parts.

**Microsporangium
with
Pollen Grains**

**Megasporangium
with
Archegonium**

**Pine Seed
(whole)**

**Pine Seed
(sectioned
showing embryo)**

6. Flowering Plant

Write the term that matches each phrase.

1. Anthers _____
2. Ovules _____
3. Develops from microspores _____
4. Develops from functional megaspore _____
5. Number of eggs in female gametophyte _____
6. Transfer of pollen from anther to pistil _____
7. Composed of ovule, embryo, and endosperm _____
8. A ripened or mature ovary _____
9. Common pollinating agent in grasses _____
10. Pollinating agent in most flowering plants _____
11. Used to attract insects _____
12. Formed by union of polar and sperm nuclei _____

7. Flowers, Fruits and Seeds

a. List the labels for Figure 29.6.

1. _____   5. _____   9. _____
2. _____   6. _____   10. _____
3. _____   7. _____   11. _____
4. _____   8. _____

b. Draw the appearance of these structures of a lily flower as observed with a dissecting microscope. Label pertinent parts.

Anther with Pollen   **Ovary, x.s.**

c. Why is a corn kernel considered a fruit and not a seed? _____
_____

d. What is the stored nutrient in the bean seed and corn kernel? _____
Where is it concentrated in the:
Bean seed? _____   Corn kernel? _____

e. Indicate the basic function of a:
Seed _____
Fruit _____

f. List the names and types of the representative fruits.

Name	Type
_____	_____
_____	_____
_____	_____
_____	_____
_____	_____

Student _____

LABORATORY REPORT 30

Lab Instructor _____

Reproduction in Vertebrates

1. **Types and Patterns of Reproduction**
   a. Write the term that matches each phrase.
      1. Type of sexual reproduction in animals _____
      2. Two types of vegetative reproduction in coelenterates _____
      3. Animals with asexual sporulative reproduction _____
      4. Sperm and eggs released into water _____
      5. Sperm placed in female reproductive tract _____
      6. Refers to the presence of male and female sex organs in same individual _____
   b. Define:
      Oviparous reproduction _____
      _____
      _____

      Viviparous reproduction _____
      _____
      _____

      Ovoviviparous reproduction _____
      _____
      _____

   c. Matching
      1. Oviparous reproduction   2. Viviparous reproduction   3. Ovoviviparous reproduction
      _____ Mammals   _____ Internal fertilization
      _____ External fertilization   _____ Embryo nourished by yolk
      _____ Embryo nourished by mother   _____ Fish and amphibians
      _____ Few sharks and snakes   _____ Birds and reptiles
   d. Indicate the usual reproductive habitat of organisms that use:
      External fertilization _____   Internal fertilization _____
   e. What reproductive adaptations of amphibians restrict their distribution? _____
      _____
   f. What reproductive adaptations enabled reptiles to colonize the land? _____
      _____
      _____
      _____

g. Matching
   1. Chorion   2. Allantois   3. Yolk sac   4. Amnion

   _____ Envelops embryo   _____ Envelops yolk

   _____ Embryonic urinary bladder   _____ Part of umbilical cord

   _____ Outermost membrane   _____ Provides "private pond" for embryo

   _____ Forms placenta

h. The number of eggs produced by vertebrates decreases progressively from fish to mammals.
How do you explain this? _____

i. Describe the function of the placenta and umbilical cord in animals.
Placenta _____

Umbilical cord _____

2. Human Reproductive Systems

a. List the labels for Figure 30.2.

   1. _____   6. _____   11. _____

   2. _____   7. _____   12. _____

   3. _____   8. _____   13. _____

   4. _____   9. _____   14. _____

   5. _____   10. _____   15. _____

b. Trace the path of sperm from the seminiferous tubules to the urethra. _____

c. List the labels for Figure 30.3.

   1. _____   5. _____   9. _____

   2. _____   6. _____   10. _____

   3. _____   7. _____   11. _____

   4. _____   8. _____

d. Contrast the function of the urethra in males and females.
Males _____
Females _____

e. Spermatozoa of all animals require a fluid medium in which to swim to the egg. How is this
provided in mammals? _____

f. Matching
   1. Testes   2. Ovaries   3. Oviduct   4. Vas deferens   5. Penis   6. Prostate gland
   7. Prepuce   8. Vagina   9. Uterus   10. Ejaculatory duct

   _____ Secretion that activates sperm   _____ Male copulatory organ

   _____ Produces eggs   _____ Carries eggs to uterus

   _____ Site of embryo development   _____ Female copulatory organ

   _____ Carries sperm to urethra   _____ Removed in circumcision

   _____ Produces spermatozoa   _____ Lined with ciliated cells

g. Is there an advantage in the sperm swimming against the current in the oviduct to reach the egg? _____ Explain your answer. _____
_____

h. Indicate the site of production and the action of each:
Testosterone _____
Estrogen _____
Progesterone _____

i. Examine models or preserved human embryos and fetuses and compare their ages with the degree of development. When born prematurely, a fetus has about a 15% chance of survival at 24 weeks and nearly 100% at 30 weeks. A fetus is full term at 40 weeks.
   1. At what age is the fetus recognizable as human? _____
   2. At 8 weeks:
      What is the crown-buttocks length of the fetus? _____
      What percentage of the length consists of the head? _____
      Are the fingers and toes developed? _____
      Are the eyes, ears and nose well developed? _____
   3. What trend occurs in head-body proportions during development?
      _____
   4. Considering the role of the placenta, why is it important for the prospective mother to avoid drugs of all types? _____
      _____
      _____
   5. Why does a fetus assume a "fetal position" in the uterus?
      _____

3. Microscopic Study

a. From your slides, draw a small portion of (1) a seminiferous tubule showing the cells involved in spermatogenesis and (2) the appearance of human sperm. Label pertinent parts.

Seminiferous Tubule   **Human Sperm**

b. Matching
   1. Diploid   2. Haploid   3. Formed by first meiotic division
   4. Formed by second meiotic division

   _____ Spermatogonium   _____ Oogonium
   _____ Spermatid   _____ Secondary oocyte
   _____ First polar body   _____ Spermatozoa
   _____ Secondary spermatocyte   _____ Primary oocyte

c. From your slides, draw (1) the appearance of the ovary section at 40×, (2) a portion of the ovary showing a Graafian follicle with its secondary oocyte, and (3) a portion of an ovary showing a corpus luteum. Label pertinent parts.

Ovary Section	**Graafian Follicle**	**Corpus Luteum**

4. Birth Control

a. Contraceptives are birth control methods that prevent the union of sperm and a secondary oocyte. Which of the methods in Table 30.2 are *not* contraceptives? _____

_____

b. Which contraceptives provide a barrier to the entrance of sperm into the uterus? _____

_____

c. Which contraceptives are chemicals that kill sperm? _____

_____

d. Which contraceptive uses hormones to prevent ovulation? _____

_____

e. Which birth control method prevents the implantation of an early embryo? _____

_____

f. Explain why the rhythm method is not very effective. _____

_____

_____

g. Does using a condom guarantee that you will not be infected with HIV or pathogens of other STDs? _____ Explain. _____

_____

_____

h. Induced abortion is the removal of a developing embryo or fetus from the uterus prior to term. Is it better to prevent unwanted pregnancies by birth control or rely on induced abortion? _____

Explain. _____

_____

_____

Student _____

Lab Instructor _____

Fertilization and Early Development

1. Introduction

Write the term that matches each phrase.

1. Penetration of egg by a sperm _____
2. Mitotic division of fertilized egg _____
3. Fusion of egg and sperm nuclei _____
4. Solid ball of cells (embryonic stage) _____
5. Hollow ball of cells (embryonic stage) _____
6. Cell formed by fertilization _____
7. Embryonic stage formed by gastrulation _____
8. Color of secretion containing sperm _____
9. Color of secretion containing eggs _____

2. Activation and Early Embryology

a. Write the term that matches each phrase.

1. Chemicals that attract sperm to eggs _____
2. Hemisphere of egg containing most yolk _____
3. Hemisphere of egg penetrated by sperm _____
4. Extension of egg engulfing sperm head _____
5. Prevents penetration by additional sperm _____
6. Time until most eggs are activated _____

b. From your slide, draw a few nonactivated and activated eggs. Show the distribution of sperm around the activated eggs. Label pertinent parts.

Nonactivated Eggs   **Activated Eggs**

c. Explain the distribution of the sperm. _____

_____

d. List the labels for Figure 31.5.

1. _____   4. _____   7. _____
2. _____   5. _____   8. _____
3. _____   6. _____   9. _____

e. Record the time required to reach each stage.

Two-cell stage _____   Four-cell stage _____   Eight-cell stage _____

f. Indicate the most abundant stage after each time period.

12 hour _____   24 hr _____   48 hr _____

3. **Experiments**

a. In experiment 1, do the four cells have identical capabilities? _____

Does the chemical differences among the cells extend along a polar or equatorial gradient?

_____

b. In experiment 2, do the chemical differences within the zygote extend along a polar or equatorial gradient? _____

What embryonic tissue is normally formed by the:

Animal hemisphere _____   Vegetal hemisphere _____

4. **Chordate Development**

a. Indicate the plane (polar or equatorial) of the cleavage divisions.

   1. First division _____

   2. Second division _____

   3. Third division _____

b. Are the planes the same as in the sea urchin? _____

c. Why are the cells formed in the vegetal hemisphere larger than those of the animal hemisphere?

_____

d. In which hemisphere does gastrulation occur? _____

Is this the same site as in the sea urchin? _____

e. Indicate the embryonic tissue that forms these structures.

   1. Embryonic gut _____

   2. Mesodermal pouches _____

   3. Notochord _____

   4. Neural tube _____

   5. Lining of the coelom _____

f. From your slide, draw a frog neurula in cross section. Label the pertinent parts. Color your drawing as follows: ectoderm—blue; neural tube—green; mesoderm—red; endoderm—yellow.

g. What similarities have you observed in embryos of the sea urchin, amphioxus, and frog?

_____

_____

_____

LABORATORY REPORT 32

Early Embryology of the Chick

1. Egg Structure
a. Write the term that matches each phrase.
   1. Envelops the true egg _____
   2. Contains the egg nucleus _____
   3. Fat-rich nutrient in true egg _____
   4. Protein-rich nutrient exterior to true egg _____
   5. Suspends true egg _____
   6. Prevent evaporative water loss _____
   7. Protects true egg from mechanical injury _____

b. Describe the egg structure that enables the blastodisc to remain upward. _____
_____
_____
_____

c. Describe the advantage provided by the greater amount of yolk in a bird egg in contrast to the sea urchin egg. _____
_____
_____

2. Development Before Incubation
a. Write the term that matches each phrase.
   1. Site of blastodisc penetration by sperm _____
   2. Embryonic stage at time egg is laid _____
   3. Formed by divisions of the blastodisc _____
   4. Homologous with the sea urchin blastula _____
   5. Cavity between yolk and blastoderm _____
   6. Outer layer of cells of blastoderm _____
   7. Inner layer of cells of blastoderm _____

b. Explain why cleavage of a bird egg is so different from that of the sea urchin. _____
_____
_____

c. List the labels for Figures 32.4 and 32.5.

Figure 32.4	**Figure 32.5**
1. _____	1. _____
2. _____	2. _____
3. _____	3. _____
4. _____	

3. Development After Incubation Begins

a. Write the term that matches each phrase.

1. Direction of endoderm growth _____
2. Elongated depression bordered by ridges _____
3. Homologous with blastopore of sea urchin _____
4. Fate of primitive streak and primitive pit _____
5. First part of tubular gut to form _____
6. Carry blood from the yolk to the embryo _____
7. Form body muscles _____
8. Most rapidly developing part of 96-hr chick _____
9. Fate of otic vesicle _____
10. Fate of optic cup _____

b. Describe the formation of the mesoderm in the 24-hr chick. _____
_____
_____
_____

4. Living 48-hour Embryo

a. In the living 48-hr embryo, does the:

Head turn to the right or left? _____
Heart bulge to the right or left? _____
Was the same pattern observed on the prepared slide? _____
Does this pattern seem random or determined? _____

b. In the living 48-hr embryo:

Was blood present? _____
Was the heart beating? _____

c. Indicate the optimum temperature for:

Chick development _____
Sea urchin development _____

d. What is the ultimate control of embryonic development? _____

LABORATORY REPORT 33

Student _____

Lab Instructor _____

Heredity

1. Fundamentals

Write the term that matches each phrase.

1. Traits passed from parents to progeny _____
2. Part of DNA coding for a specific protein _____
3. Contain homologous genes _____
4. Alternate forms of a gene _____
5. Alleles of a gene pair are identical _____
6. Alleles of a gene pair are different _____
7. Observable form of a trait _____
8. Genetic composition determining a trait _____
9. Allele expressed in heterozygote _____
10. Allele not expressed in heterozygote _____

2. Monohybrid Crosses with Dominance

a. In corn plants, plant height is controlled by one gene with two alleles. How many alleles that control the trait for height are present in each cell nucleus of a corn plant? _____
How many alleles that control the trait for height are present in the nucleus of each gamete formed by a corn plant? _____

b. Indicate the genotypes of these corn plants.

homozygous tall _____   heterozygous tall _____   dwarf _____

c. Indicate the genotypes of possible gametes of these corn plants.

homozygous tall _____   heterozygous tall _____   dwarf _____

d. Determine the predicted phenotype ratio in progeny of a cross between two monohybrid tall corn plants.

Parent phenotypes   _____   _____
Parent genotypes   _____   _____
Gametes   _____ _____   _____ _____

Punnett square

Genotype ratio   _____
Phenotype ratio   _____

e. Examine the corn seedlings in tray 1 that are progeny of a cross of two monohybrid tall corn plants. Record the number of:

Tall plants _____ Dwarf plants _____

Determine the observed phenotype ratio of tall plants to dwarf plants by dividing the number of plants of each type by the number of dwarf plants. Round your answers to whole numbers.

_____ Tall ÷ _____ dwarf = _____   _____ Dwarf ÷ _____ dwarf = 1

Record the observed phenotype ratio: _____ Tall : 1 dwarf

Is the observed ratio different from the predicted ratio? _____

If so, explain why this may have happened. _____

_____

f. Examine the corn seedings in tray 2. Record the number of:

Green plants _____ Albino plants _____

Determine the phenotype ratio of green plants to albino plants by dividing the number of plants of each type by the number of albino plants. Round your answers to whole numbers.

_____ Green ÷ _____ albino = _____   _____ Albino ÷ _____ albino = 1

Record the observed phenotype ratio: _____ green : 1 albino

Indicate the type of simple dominant/recessive cross that produces this kind of ratio: _____

_____

Based on this observed ratio and using "G" to represent the allele for green leaves and "g" to represent the allele for albino leaves, record the phenotypes and genotypes of the parent plants:

Phenotypes   _____   _____

Genotypes   _____   _____

g. Determine the expected progeny of a cross between:

1. Heterozygous tall and dwarf corn plants

Parent phenotypes   _____   _____

Parent genotypes   _____   _____

Gametes   _____ _____   _____ _____

Punnett square

Genotype ratio   _____

Phenotype ratio   _____

2. Homozygous green and albino corn plants (set up your own Punnett square.)

Genotype ratio   _____

Phenotype ratio   _____

h. Indicate the possible genotypes of parent pea plants in crosses yielding the following ratios.

Phenotype Ratio	Parent Genotypes	
1. 3 round seeds : 1 wrinkled seeds	_____ X	_____
2. all white flowers	_____ X	_____
3. all purple flowers (3 possibilities)	_____ X	_____
	_____ X	_____
	_____ X	_____
4. 1 round seeds : 1 wrinkled seeds	_____ X	_____

i. Is it possible to observe differences between homozygous and heterozygous individuals possessing the dominant allele? _____

j. To determine the genotype of a tall pea plant by a test cross, it should be crossed with a: _____ pea plant. In such a test cross, indicate the genotype of the tall pea plant if:

All the progeny are tall _____

Half the progeny are tall _____

k. The table below lists several human traits that are determined by a simple dominant/recessive mode of inheritance. Record your phenotypes and the frequency of the phenotypes among members of your class.

Human Dominant/Recessive Traits Determined by a Single Gene*

Trait	Phenotype	Your Phenotype	Number in Class with Phenotype
Handedness	Right handed*		
	Left handed		
Ear lobes	Free*		
	Attached		
Freckles	Freckled*		
	Nonfreckled		
Hair line	Widow's peak*		
	Straight		
Little finger	Bent*		
	Straight		
Tongue roll	Yes*		
	No		
Rh factor	Rh$^+$*		
	Rh$^-$		

*Indicates dominant traits.

l. For each trait, is the dominant phenotype always more abundant among class members?

_____

m. Assuming your class to be representative of the general population, what can you conclude about the relationship between dominant and recessive phenotypes and their frequencies in the population? _____

_____

n. Newlyweds Bill and Sue are nonfreckled. Since each had one parent who had freckles, they wonder what the probability is of their children having freckles. What would you tell them?

_____

_____

o. Mary has freckles, but her husband Dick does not. Mary's father has freckles but her mother does not. What is the probability that Mary and Dick's child will have freckles? _____

p. Linda is pregnant and Rh⁻. Her husband Tom is Rh⁺ like both of his parents. What is Linda's genotype? _____What are the two possible genotypes for Tom? _____
What is the probability of their baby being Rh⁺:
if Tom is homozygous for Rh⁺? _____
if Tom is heterozygous for Rh⁺? _____

3. Monohybrid Crosses with Codominance

a. Explain the meaning of codominance. _____
_____

b. Indicate the genotype and phenotype ratios of these crosses in snapdragons:
1. Pink flowers × white flowers
Genotype ratio: _____
Phenotype ratio: _____
2. Pink flowers × pink flowers
Genotype ratio: _____
Phenotype ratio: _____

c. How would you produce seeds that yield 100% pink-flowering snapdragons? _____
_____

d. What is the probability that parents, who are heterozygous for sickle-cell hemoglobin, will have a child with sickle-cell anemia? _____

4. Multiple Alleles: ABO Blood Types

a. Indicate the expected genotype and phenotype ratios for these matings:
1. $I^A I^B \times ii$
Genotype ratio: _____
Phenotype ratio: _____
2. $I^A i \times I^B i$
Genotype ratio: _____
Phenotype ratio: _____

b. (True/False)
_____ A type O child may have two type A parents.
_____ A type O child may have one type AB parent.
_____ A type B child may have two type AB parents.
_____ A type AB child may have one type O parent.

c. Ann (type A, Rh⁺ blood) is suing Joe (type AB, Rh⁻ blood) for child support claiming that he is the father of her child (type B, Rh⁺ blood). On the basis of blood type, is it possible that Joe is the father? _____ What ABO blood type(s) would Joe have to possess to *prove* that he is *not* the father? _____

5. Dihybrid Crosses with Dominance

a. Determine the phenotype ratio in progeny of a cross between two yellow-round dihybrid pea plants.

Parent phenotypes _____   _____

Parent genotypes _____   _____

Gametes _____ _____ _____ _____   _____ _____ _____ _____

Punnett square

_____ _____ _____ _____

Phenotype ratio _____ yellow-round _____ green-round

_____ yellow-wrinkled _____ green-wrinkled

b. List the possible gametes of a dihybrid with a genotype of AaBb.

Possible gametes _____ _____ _____ _____

c. If two tall-green dihybrid corn plants were crossed, each with a genotype of TtGg, what is the *expected phenotype ratio* in the progeny? (T = tall, t = dwarf, G = green, g = albino) Recall that tall and green are dominant.

Phenotype ratio _____ tall-green _____ dwarf-green

_____ tall-albino _____ dwarf-albino

d. In the lab, there are several trays of corn seedlings that are progeny of a cross between tall-green dihybrids. Each tray is labeled tray 3. Count and record the phenotypes in *all* these trays. Determine the *observed phenotype ratio* by dividing the number of each phenotype by the number of dwarf-albino plants and rounding to the nearest whole number.

Phenotype ratio _____ tall-green _____ dwarf-green

_____ tall-albino _____ dwarf-albino

Explain any difference between the expected and observed phenotype ratios. _____

_____

6. Linked Traits

a. A dihybrid was crossed with a homozygous recessive for each trait and the cross produced only two classes of progeny. How do you explain this? _____

_____

b. Why are more males color blind than females? _____

_____

c. Determine the expected phenotype ratio of children from these matings:

1. Normal female (XX) × color-blind male (X^CY)

_____

2. Carrier female (XX^C) × normal male (XY)

_____

7. Pedigree Analysis

Examine the pedigrees shown below, and determine whether the inherited trait (■ or ●) is dominant, recessive, or sex-linked recessive; □ = male; ○ = female.

a.

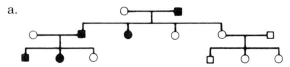

_____

b.

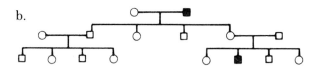

_____

c.

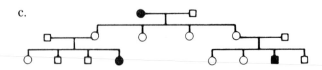

_____

8. Polygenic Inheritance

Determine the expected phenotype ratio in progeny produced by the cross of a wheat plant with medium red seeds ($R_1r_1R_2r_2$) and a wheat plant with white seeds ($r_1r_1r_2r_2$). _____
_____
_____

9. Chi-Square Analysis

a. In the example of chi-square determination in Table 33.6, what is the hypothesis being tested?
_____

Would the hypothesis be supported by a chi square of 3.952? _____

b. Indicate the value of chi square determined in Table 33.8. _____
Is the hypothesis supported? _____

c. Indicate the value of chi square determined in Table 33.9. _____
Is the hypothesis supported? _____

d. Biologists like to have a large number of progeny to analyze when doing genetic crosses. Why is this of value? _____
_____

e. Indicate the chi-square values and parental genotypes for the corn ears from experimental crosses.

Ear No.	Chi Square	Parental Genotypes
_____	_____	_____
_____	_____	_____
_____	_____	_____

LABORATORY REPORT 34

Molecular and Chromosomal Genetics

1. DNA

    a. Write the term that matches each phrase.

       1. Molecule containing genetic information   _____

       2. Sugar in DNA nucleotides   _____

       3. Pyrimidine that pairs with adenine   _____

       4. Purine that pairs with cytosine   _____

       5. Chemical bonds joining complementary nitrogen
          bases   _____

       6. Number of nucleotide strands in DNA   _____

       7. Two molecules forming sides of the DNA "ladder"   _____

    b. In DNA replication, what determines the sequence of nucleotides in the new strands of nucleotides that are formed? _____

_____

    c. After determining the pairing pattern of the nitrogenous bases in Figure 34.2, add the missing bases to this hypothetical strand of DNA.
       A = adenine, C = cytosine, G = guanine, T = thymine

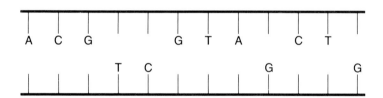

    d. At the left is a hypothetical segment of DNA. After replication it forms two strands as shown at the right. Add the missing bases in the replicated strands.

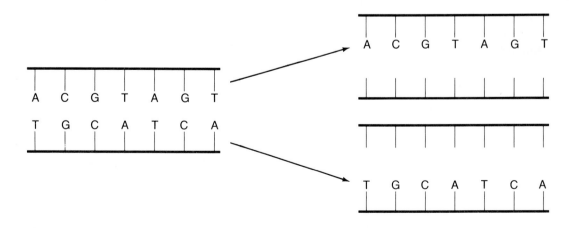

2. RNA Synthesis

a. Write the term that matches each meaning.

1. Sugar in an RNA nucleotide   _____

2. Number of nucleotide strands in RNA   _____

3. RNA base that pairs with adenine   _____

4. Template for RNA synthesis   _____

b. The figure below shows a portion of DNA molecule whose strands have separated to synthesize mRNA. The bases are shown for the strand that is *not* serving as the template. Add the bases of the DNA strand serving as the template and the bases of the newly formed RNA molecule.

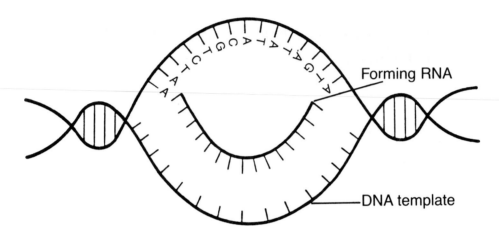

Forming RNA

DNA template

3. Protein Synthesis

a. Write the term that matches each meaning.

1. Carries the genetic code to ribosomes   _____

2. Formed of a mRNA base triplet   _____

3. Specifies a particular amino acid   _____

4. Sites of protein synthesis   _____

5. Carries amino acids to ribosome   _____

6. tRNA triplet that pairs with codon   _____

7. Codon that starts polypeptide synthesis   _____

8. Codons that stop polypeptide synthesis   _____

b. Indicate the DNA base triplet and tRNA anticodons for the mRNA codons in Figure 34.4.

Amino Acid	DNA Base Triplet	Anticodon
Tryptophan	_____	_____
Lysine	_____	_____
Valine	_____	_____
Methionine	_____	_____

c. Using Table 34.1, indicate the possible sequences of mRNA codons that code for the polypeptides below.

methionine - lysine - histidine - tryptophan - glutamine - tryptophan - (stop)

_____ _____ _____ _____ _____ _____ _____

_____ _____ _____ _____ _____ _____

_____

d. Molecular biologists use probes of mRNA to locate specific gene loci on DNA segments. Use the codon sequences determined in 3c as probes to locate the gene (complimentary base sequence) in the segment of template (single-strand) DNA shown below. Circle the located gene and indicate the probe that was successful in locating it.

-A-G-T⁞T-C-T⁞T-A-C⁞C-C-T⁞G-A-A⁞C-G-G⁞C-A-T⁞T-C-A⁞

-G-A-C⁞A-T-T⁞C-C-G⁞C-T-A⁞G-A-C⁞C-A-T⁞T-A-C⁞T-T-C⁞

-G-T-A⁞A-C-C⁞G-T-T⁞A-C-C⁞A-T-C⁞G-T-A⁞T-C-A-

Successful probe _____

e. Consider the following hypothetical gene. Indicate the mRNA codons that it forms and the sequence of amino acids produced in the polypeptide chain.

-T-A-C⁞T-T-A⁞G-A-A⁞A-T-A⁞C-C-G⁞A-A-G⁞A-C-T-

mRNA codons

_____

Amino acid sequence

_____

f. For the following mutations, show the effect by indicating the mRNA codons and the amino acid sequence in the polypeptide.

1. Base substitution

-T-A-C⁞T-T-A⁞G-A-G⁞A-T-A⁞C-C-G⁞A-A-G⁞A-C-T-

mRNA codons

_____

Amino acid sequence

_____

2. Triplet addition

-T-A-C⁞T-T-A⁞G-G-T⁞G-A-A⁞A-T-A⁞C-C-G⁞A-A-G⁞A-C-T-

mRNA codons

_____

Amino acid sequence

_____

3. Base addition

-T-A-C┤T-T-A┤G-A-A┤A-T-C-A┤C-C-G┤A-A-G┤A-C-T-

mRNA codons

_____

Amino acid sequence

_____

4. Base deletion (site indicated by arrow)

↓

-T-A-C┤T-T-A┤G-A-A┤A-T┤C-C-G┤A-A-G┤A-C-T-

mRNA codons

_____

Amino acid sequence

_____

g. What seems to determine the magnitude of a mutation's effect? _____

_____

_____

_____

h. Summarize how DNA controls cellular functions. _____

_____

_____

_____

_____

_____

4. Chromosomal Abnormalities

a. Prepare karyotypes for patients A, B, and C on the following three pages.

b. Indicate the sex and chromosomal abnormality, if any, for each patient whose karyotype you analyzed.

Patient	**Sex**	**Abnormality**
A	_____	_____
B	_____	_____
C	_____	_____

c. Amniocentesis is used to obtain fetal cells from the amniotic fluid for chromosomal analysis. If you were a prospective parent of a fetus with a chromosomal abnormality, what factors would you consider in deciding whether to abort the fetus?

_____

_____

_____

_____

_____

Human Karyotype Analysis Form
Patient A

A

| 1 | 2 | 3 |

B

| 4 | 5 |

C

| 6 | 7 | 8 | 9 | 10 | 11 | 12 | X |

D

| 13 | 14 | 15 |

E

| 16 | 17 | 18 |

F

| 19 | 20 |

G

| 21 | 22 | Y |

Sex of subject _____   Number of chromosomes _____

Chromosomal disorder _____

Human Karyotype Analysis Form
Patient B

A

B

1   2   3   4   5

C

6   7   8   9   10   11   12   X

D

E

13   14   15   16   17   18

F

G

19   20   21   22   Y

Sex of subject _____   Number of chromosomes _____

Chromosomal disorder _____

Human Karyotype Analysis Form
Patient C

A

_____   _____   _____
  1   2   3

B

_____   _____
  4   5

C

_____   _____   _____   _____   _____   _____   _____   _____
  6   7   8   9   10   11   12   X

D

_____   _____   _____
 13   14   15

E

_____   _____   _____
 16   17   18

F

_____   _____
 19   20

G

_____   _____   _____
 21   22   Y

Sex of subject _____   Number of chromosomes _____

Chromosomal disorder _____

LABORATORY REPORT 35

Student _____

Lab Instructor _____

Evolution

1. The Fossil Record

a. What groups of organisms were present during the Proterozoic era?

_____

What groups were most abundant? _____

What groups are found in the oldest fossils? _____

b. Indicate the period of first appearance for these groups:

Chordates _____ Angiosperms _____

Psilopsids _____ Arthropods _____

Reptiles _____ Bryophytes _____

Gymnosperms _____ Ferns _____

Jawed fish _____ Amphibians _____

Birds _____ Mammals _____

c. Examine the fossils available and complete the chart below.

Fossils

Number	Name	Period of First Appearance	Period of Greatest Abundance	Period of Extinction
1				
2				
3				
4				
5				
6				
7				
8				
9				
10				

d. Indicate the period when these groups invaded the land:

Vascular plants _____ Arthropods _____

Vertebrates _____

e. What groups show a decrease in diversity in the Permo-Triassic Crisis? _____

_____

481

f. Use data from Figures 35.4 and 35.5 to indicate the probable immediate ancestor of these groups:

Ferns _____   Amphibians _____

Birds _____   Psilopids _____

Jawed fish _____   Mammals _____

Angiosperms _____   Gymnosperms _____

Reptiles _____   Bryophytes _____

g. Indicate the vertebrate group that was:

First to appear _____   Last to appear _____

h. What vertebrate group declined in the Cretaceous during the Laramide Revolution? _____

2. Evidence from Vertebrate Embryology

a. Do early vertebrate embryos have a similar structure and appearance? _____

b. Does similarity of embryological development indicate relationship? _____

c. How do you explain the pharyngeal pouches in chick and pig embryos? _____

_____

_____

3. Evidence from Vertebrate Morphology

a. Does homology of limb bone structure indicate relationship? _____

b. What are the three basic limb bones of the anterior appendage in vertebrates? _____

_____

c. How many digits are in the basic foot plan of tetrapods? _____

d. What bones are used to support the wing membrane in bats? _____

e. How do the bones of bird wings differ from those of bat wings? _____

_____

_____

f. Which digit forms the foot of the modern horse? _____

What human structure is homologous with the hoof? _____

g. How many years were required for the dog-sized, four-toed browsing animal to evolve into the modern horse? _____

h. Considering Figure 35.8, is extinction or survival the trend in the evolution of the horse? _____ Explain your response. _____

_____

i. Why do organisms become extinct? _____

_____

_____

_____

4. Evidence from Biochemistry

a. What three biochemical molecules occur in all organisms? _____

_____

b. Does similarity of molecules in living organisms indicate relationship? _____

_____

c. Record your results and conclusions in the table below. Rate the degree of precipitation on a 1-to-5 scale, 5 being the greatest amount of precipitation.

Test Serum	Degree of Precipitation	Animal Source of the Serum
I		
II		
III		
IV		
V		
VI	5	Human

d. Do your results agree with the data in Table 35.1? _____
   If not, explain the differences. _____
   _____

e. Based on Table 35.1, are chimpanzees more closely related to:
   Gorillas or humans? _____
   Gorillas or orangutans? _____
   Explain the basis for your answers. _____
   _____
   _____

f. Based on Table 35.1, which primate group would you expect to appear:
   Earliest in the fossil record? _____
   Latest in the fossil record? _____

5. Human Evolution

a. List the labels for Figure 35.9.

1. _____   6. _____
2. _____   7. _____
3. _____   8. _____
4. _____   9. _____
5. _____

b. Write the term that matches each phrase.
   1. Earliest hominid   _____
   2. First human   _____
   3. Continent where humans evolved   _____
   4. First modern man   _____
   5. First human living on three continents _____

6. Skull Analysis

a. Plot the cephalic indexes of members of the class.

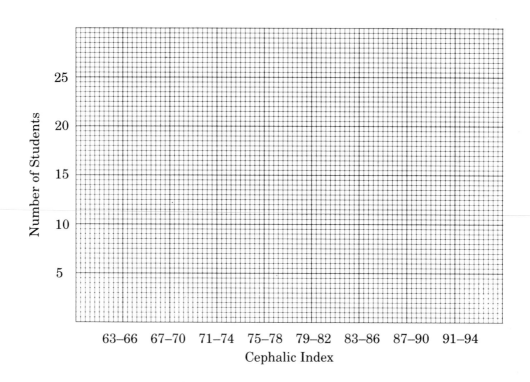

b. Indicate the following the data from your class.

Average cephalic index _____   Ranges of indexes _____

c. Considering the data in item 6b, what difficulties are encountered in trying to make firm conclusions about a single fossil specimen? _____

_____

_____

d. Why are indexes better than simple measurements for comparing fossil specimens? _____

_____

_____

e. Record the results of your skull analyses in the table below.

Skull Comparison

Characteristics	Chimpanzee	A. africanus	H. erectus	H. sapiens neanderthalensis	H. sapiens sapiens (Cro-Magnon)	H. sapiens sapiens (modern)
Cranial breadth						
Cranial length						
Cranial index						
Cranial breadth						
Facial breadth						
Skull proportion index						
Facial projection length						
Skull length						
Facial projection index						
Browridges						
Forehead						
Canine length						
Skull and vertebra attachment						
Chin						

f. Which of the fossil hominid skulls has a cranial (cephalic) index that falls within the range of indexes for the class? _____

_____

g. Have there been any major skull changes in humans during the last 40,000 years? _____
Explain your response. _____

_____

_____

h. For each hominid skull, plot the skull proportion index and the facial projection index against each other on the graph. Label each plot as to organism.

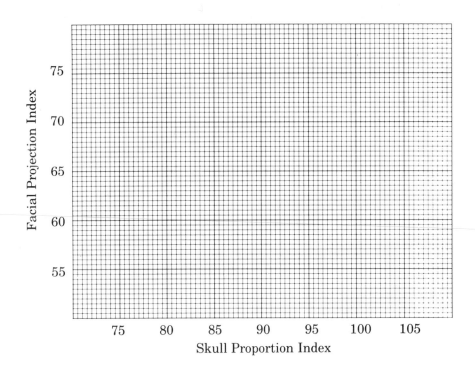

i. Do the indexes plotted in item 6h seem to vary independently or in association with each other?

_____

Explain the basis of your answer. _____

_____

_____

_____

j. Summarize the human evolutionary trends based on your analysis.
Cranium size _____

_____

_____

Face _____

_____

_____

Browridges _____

_____

_____

Skull attachment _____

_____

LABORATORY REPORT 36

Evolutionary Mechanisms

1. CHW Equilibrium

a. Define:

Population _____

Gene pool _____

Evolution _____

Gene mutation _____

Natural selection _____

_____

Gene flow _____

Genetic drift _____

b. Evolution depends on genetic variability within populations and at least one of three additional factors. List them. _____

c. Under what conditions does the CHW equilibrium operate? _____

_____

_____

Do these exist in natural populations? _____

d. Indicate the allele and genotype frequencies in these populations:

1. 91% tasters and 9% nontasters

Allele frequencies: T _____   t _____

Genotype frequencies: TT _____   Tt _____   tt _____

2. 36% tasters and 64% nontasters

Allele frequencies: T _____   t _____

Genotype frequencies: TT _____   Tt _____   tt _____

e. If the conditions of the CHW law are met, what will be the allele frequencies in a population of 75% tasters and 25% nontasters after 10 generations: T _____   t _____

f. Indicate the frequencies of the following in your class:

Tasters _____   T _____   t _____   TT _____   Tt _____   tt _____

2. Natural Selection

a. In the selective breeding of horses, who or what does the selecting? _____

In natural selection, who or what does the selecting? _____

b. With 100% selection against a dominant phenotype, how long does it take to remove the dominant allele from the population? _____

How will the dominant allele return to the population, if ever? _____

_____

c. With 20% selection against the 64% tasters in the population, indicate the frequency of tasters after five generations. _____

d. With 100% selection against the 36% nontasters in the population, indicate the frequency of nontasters after five generations. _____
Explain the presence of nontasters. _____
_____

e. Are dominant or recessive alleles easier to remove by selection? _____
Explain. _____
_____

f. Is selection against the genotype or phenotype? _____

g. Is selection more effective against recessive alleles in haploid or diploid organisms? _____
Explain your response. _____
_____
_____

h. Would a lethal recessive mutant tend to spread through a population of:
haploid organisms? _____ diploid organisms? _____
Explain your responses. _____
_____
_____

3. Genetic Drift

a. Record the allele frequencies from Table 36.2.

Generations	Frequency of T	Frequency of t
0		
1		
3		
5		
6		
7		
8		
9		
10		

b. Is there a pattern to the changes in allele frequencies? _____
What is the cause of the changes in allele frequency? _____
_____

c. Why is genetic drift unimportant in large populations? _____
_____
_____
_____

LABORATORY REPORT 37

Ecological Relationships

1. **Energy Flow**
    a. Write the term that matches each meaning.
        1. Sum of all ecosystems in the world   _____
        2. All populations within an ecosystem   _____
        3. Interbreeding members of a species in an ecosystem   _____
        4. A defined area consisting of biotic and abiotic components   _____
        5. All parts of earth where organisms exist   _____
        6. Provide energy and organic nutrients for all members of the community   _____
        7. Extract last energy from organic matter   _____
        8. Animals feeding on plants only   _____
        9. Composed mainly of fungi and bacteria   _____
        10. Animals feeding on other animals   _____
        11. Animals feeding on both plants and animals   _____
        12. Animal group not preyed upon by other animals   _____
        13. Convert light energy into chemical energy   _____
        14. Convert organic matter into inorganic matter   _____
        15. Ultimate source of energy for the earth   _____
    b. Show the flow of energy through a community consisting of decomposers, herbivores, producers, and carnivores.

        _____ → _____ → _____ → _____
    c. Explain why only about 10% of the energy stored in one tropic level is transferred to the next higher level. _____
        _____
        _____
        _____

    d. In dry years, the biomass of producers is significantly decreased. What would be the effect on:
        A rodent population? _____
        A hawk population? _____
    e. What group of organisms is not shown in Figure 37.1? _____
    f. Circle the simplified food chain that would support a larger human population.
        Cereals and legumes → beef → humans
        Cereals and legumes → humans
        Explain your answer. _____
        _____

g. Categorize the organisms (by number) in Figure 37.2 according to their ecological roles.
   Producers _____ Herbivores _____
   Primary carnivores _____ Secondary carnivores _____
   Tertiary carnivores _____
h. In which categories would humans be placed? _____
_____

i. What would result from a significant reduction of the insect population by insecticides?
_____
_____

j. Coyotes feed primarily on rodents, but occasionally may kill a lamb in sheep-ranching areas. Ranchers sometimes conduct coyote-killing campaigns to protect their flocks. Evaluate the advisability of such practices. _____
_____
_____

k. Evaluate the effect on the ecosystem of farmers killing rodents with poison. _____
_____

2. Biogeochemical Cycles
a. List the processes (by number) involved in the carbon cycle shown in Figure 37.3.
   _____ Combustion   _____ Consumption   _____ Death
   _____ Photosynthesis _____ Respiration
   Which of these processes is a major contributor to air pollution in our industrialized society?
   _____

b. Scientists are concerned that an increase of CO_2 in the atmosphere may result in an increase in the average global temperature. What would be the source of this additional CO_2? _____
_____

c. How is the carbon in dead organisms returned to circulation? _____
_____

d. What is the origin of the energy available in fossil fuels? _____
_____

e. What is the global abiotic reservoir for nitrogen? _____
   What organisms are able to use this nitrogen source? _____
f. Indicate the form of nitrogen required by:
   Animals _____ Plants _____
g. What advantage does symbiotic nitrogen-fixing bacteria provide legumes? _____
_____

h. Diagram (1) a portion of a clover root nodule showing the *Rhizobium* bacteria and (2) *Rhizobium* bacteria in a smear on your slide.
   In a Root Nodule   **In a Smear of a Root Nodule**

3. Ecosystem Analysis

a. Based on macroscopic observations, which microecosystem seems to have:

The greatest density of autotrophs? _____

The lowest density of autotrophs? _____

The greatest density of heterotrophs? _____

The lowest density of heterotrophs? _____

b. Based on macroscopic observations, record the relative abundance of each kind of crustacean in the microecosystems in the table below. Use this scale: 0 = none; 5 = most abundant.

Based on examination of your slides, record the presence of autotrophs in the microecosystems by placing an "X" in the appropriate spaces in the table below. Compare your data with that of the entire class.

Organism	Acid Pollution	Normal	Organic Pollution
Cyanobacteria *Oscillatoria*			
Protista *Euglena*			
Peridinium			
Green Algae *Microspora*			
Pandorina			
Spirogyra			
Crustacea *Cyclops*			
Daphnia			
Gammarus			

c. Indicate the autotroph and heterotroph most tolerant of:

	Autotroph	Heterotroph
Acid pollution	_____	_____
Organic pollution	_____	_____

d. Indicate the ecosystem with the:

Greatest diversity of autotrophs. _____

Least diversity of autotrophs. _____

e. Indicate the ecosystem with the:

Greatest diversity of heterotrophs. _____

Least diversity of heterotrophs. _____

f. Which ecosystem (in nature) would have the:

Greatest density of decomposers? _____

Least density of decomposers? _____

g. If the action of decomposers is curtailed, what would be the effect on other organisms in the community? _____

Explain your answer. _____

_____

_____

_____

h. An organic polluted lake typically has a deficient concentration of dissolved oxygen. Explain why this happens. _____

_____

_____

i. Desirable fish species, such as trout, cannot live in either an acid polluted lake or an organic polluted lake. Explain why.

Acid polluted lake. _____

_____

_____

Organic polluted lake. _____

_____

_____

j. What steps should be taken to reduce water pollution by:

Acid rain? _____

_____

_____

Organic pollution? _____

_____

_____

k. Most large U.S. cities have an air pollution problem. What are the major causes of air pollution?

_____

If you were mayor of a city plagued by air pollution and considering economic, health and ecological factors, how would you try to reduce air pollution? _____

_____

_____

_____

l. List three environmental problems in your area, their causes, and possible solutions.

1. _____

_____

2. _____

_____

3. _____

_____

LABORATORY REPORT 38

Student _____

Lab Instructor _____

Population Growth

1. Introduction

Write the term that matches each meaning.

1. Sum of all limiting factors _____

2. Maximum theoretical reproductive capacity _____

3. Limiting factors whose effect is unchanged by population size _____

4. Limiting factors whose effect changes proportionately with population size _____

2. Growth Curves

a. Plot the theoretical and realized growth and growth rate curves below.

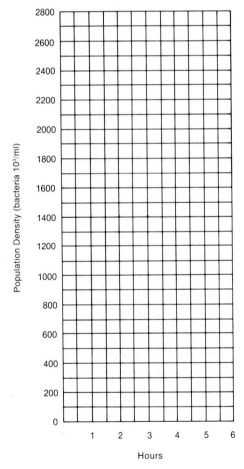

Theoretical and realized growth curves of bacterial populations

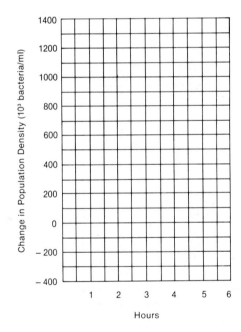

Theoretical and realized growth rate curves of bacterial populations

b. Indicate the time interval when the growth rate was greatest.
   Theoretical growth _____ Realized growth _____

c. In the realized growth curve:
   When did cell deaths most nearly equal cell "births?" _____
   Why did the realized bacterial population die out? _____
   _____

   What limiting factors determined the realized population growth pattern? _____
   _____

   Were density-dependent or density-independent factors involved? _____
   _____

d. In item 2a, show the shape the realized growth curve would have taken if additional nutrient broth was added at the fifth hour. Use a red pencil to draw the new curve.

e. Write the term that matches each meaning regarding Figure 38.1.
   1. Growth curve with a J shape _____
   2. Growth curve with an S shape _____
   3. Phase of explosive growth _____
   4. Point where limiting factors take effect _____
   5. Phase of equilibrium between biotic potential and environmental resistance _____
   6. Where birth equals death _____
   7. Where resources per capita are least _____

3. Human Population Growth

a. Plot the human population growth curve below using data in Table 38.3.

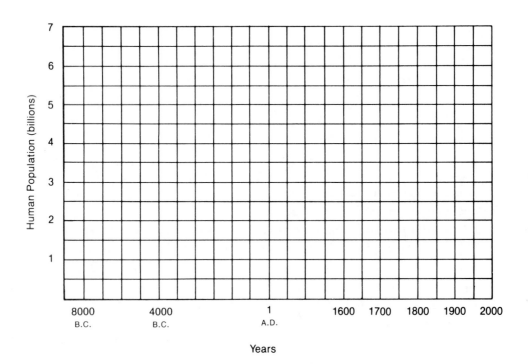

Human population growth curve

b. Does the human population growth curve resemble the theoretical growth curve or the realized growth curve? _____

c. In what phase of the realized growth curve does the human population curve appear to be in? _____

d. What events in human history have enabled the exponential growth of the human population? _____ _____

e. Are human populations subject to the interaction of biotic potential and environmental resistance like natural populations? _____

f. List any limiting factors controlling the size of natural populations that do not affect human populations. _____

g. Humans are subject to certain limiting factors that are not observed in other animal populations. Name two of these. _____ _____

h. If nothing is done to check the growth of the human population, will deaths ultimately equal births? _____ What will be the shape of the population growth curve? _____

i. Ultimately, there are only two ways to decrease population growth. Name them. _____ _____

   Which is preferable? _____

j. Record the growth rate and doubling time as calculated for Table 38.4.

	Growth Rate **(%/yr)**	**Doubling Time** **(yr)**
Developing countries	_____	_____
Developed countries	_____	_____
Africa	_____	_____
Asia	_____	_____
Europe	_____	_____
Latin America	_____	_____
North America	_____	_____
Oceania	_____	_____
CIS	_____	_____

k. What region has the:

   Fastest population growth? _____

   Slowest population growth? _____

l. The rapid growth rate in developing countries has resulted from a decreased death rate without a proportionate decrease in birthrate. What has brought this about? _____ _____

m. What relationship exists between population size and resources per person? _____

n. Are the earth's resources infinite? _____ Does it seem likely that resources can increase to keep pace with the projected population growth? _____

o. The standard of living is based on available resources per person. If a country's population is increasing at a rate of 2.0% per year, what rate of increase in productivity is required to maintain the same standard of living? _____

p. At the carrying capacity, deaths equal births and resources are just adequate to maintain the existing population size. Would it be advantageous to curtail the human population (zero growth) before it levels off at the carrying capacity due to natural causes? _____ Explain your response. _____

_____

_____

_____

q. Are developing countries likely to improve their standard of living without decreasing their population growth? _____ Explain your answer. _____

_____

_____

_____

r. Famine has been a serious problem in Africa in recent years. What has caused the famine?

_____

_____

Food was provided by developed countries to help the people faced with starvation. Was the donated food a real or artificial increase in the carrying capacity of the environment?

_____

What is the long-range solution to this problem? _____

_____

_____

_____

_____

_____

s. If you were a policymaker in the United States government, what policies would you propose to stimulate a decrease in population growth:
In developing countries receiving foreign aid? _____

_____

_____

_____

_____

In the United States? _____

_____

_____

_____

_____

_____

LABORATORY REPORT 39

Animal Behavior

1. **Introduction**
   a. Define:
   Innate behavior _____

   _____

   Learned behavior _____

   _____

   b. Provide the terms that match the statements.
      1. A rapid, stereotyped movement of the body or
         body part in response to a stimulus   _____
      2. Movement toward or away from a stimulus   _____
      3. Predictable, stereotyped, complex movements
         triggered by a releaser   _____
      4. Slower or faster movement of an animal in
         response to a stimulus   _____

2. **Taxes in Planaria**
   a. Are the planaria randomly dispersed over the bottom of the dish? _____
   Describe the distribution. _____
   How do you explain their distribution? _____

   _____

   b. When the planaria moved into the shaded area did they:
   Stay there? _____ Become motionless? _____
   After 10 min indicate the number of planaria in the:
   Lighted half _____ Shaded half _____
   Do the results support the hypothesis? _____
   Explain. _____

   _____

   How is this behavior beneficial to planaria? _____

   c. Did the planaria exhibit rheotaxis? _____ What type? _____
   How is this behavior beneficial to planaria? _____

   _____

   d. What physical process disperses the chemicals in the water? _____
   How long did it take to get a response from the planaria? _____
   Explain the time lag. _____
   Do your observations support the hypothesis? _____
   Explain. _____

   _____

How is this response beneficial to planaria? _____

_____

e. Why was the light shield necessary? _____

_____

f. Record your results and place the class totals in parentheses.

| Time | Location of Planarian | |
	Bottom Half	Top Half

g. Did the results support the hypothesis? _____
   Explain. _____

_____

How is this response beneficial to planaria? _____

_____

3. Behavior in Sowbugs

a. What behavior do you expect to observe if sowbugs exhibit negative kinesis to light. _____

_____

_____

What would indicate negative phototaxis? _____

_____

b. What behavior do you expect to observe if sowbugs exhibit positive kinesis to moisture? _____

_____

_____

c. How did sowbugs respond to being touched? _____

_____

What type of innate response is this? _____
How is this response beneficial to sowbugs? _____

_____

d. How do moving sowbugs respond when you tap the dish? _____

_____

What type of innate response is this? _____
How is this response beneficial to sowbugs? _____

_____

e. Do the sowbugs move randomly over the bottom of the dish? _____
   If not, describe your observations. _____

_____

_____

Do they seem to use their antennae to establish their position by touch? _____

Explain. _____

_____

_____

f. Record your observations and place the class totals in parentheses.

Time	Number in the Shade		Number in the Light	
	Moving	Not Moving	Moving	Not Moving

g. Do the results support the hypothesis? _____

If not, explain. _____

_____

h. Record your observations and place the class totals in parentheses.

Time	Number in Moist Area		Number in Dry Area	
	Moving	Not Moving	Moving	Not Moving

i. Do the results support the hypothesis? _____

If not, explain. _____

_____

j. Record your observations and place the class totals in parentheses.

Time	Number in the Dry Shade		Number in the Moist Light	
	Moving	Not Moving	Moving	Not Moving

k. Which stimulus produced a stronger response? _____

l. Is there a tendency for sowbugs to form aggregations when inactive?

_____

Can you explain this behavior solely on the basis of preferences for darkness and moisture?

_____

If not, what other factors may be involved? _____

_____

_____

4. Taxes in Fruit Flies

a. Record the distribution of fruit flies in the vial.

Condition	Number of Flies	
	Top Half	Bottom Half
After 3 min with stopper up		
After 3 min with bottom up		

b. State three hypotheses to explain these observations.

1. _____

_____

2. _____

_____

3. _____

c. Use a separate sheet of paper to explain how you determined the response of fruit flies to light and gravity. For each hypothesis tested provide the following:

1. State the hypothesis.

2. Describe your methods.

3. Record your results.

4. State your conclusion.

APPENDIX A

COMMON PREFIXES, SUFFIXES, AND ROOT WORDS

Prefixes of Quantity

amphi-, diplo-	both, double, two
bi-, di-	two
centi-	one hundredth
equi-, iso-	equal
haplo-	single, simple
hemi-, semi-	one-half
hex-	six
holo-	whole
quadri-	four
milli-	one thousandth
mono-, uni-	one
multi-, poly-	many
oligo-	few
omni-	all
pento-	five
tri-	three

Prefixes of Direction or Position

ab-, de-, ef-, ex-	away, away from
acro-, apici-	top, highest
ad-, af-	to, toward
antero-, proto-	front
anti-, contra-	against, opposite

archi-, primi-, proto-	first
circum-, peri-	around
dia-, trans-	across
deutero-	second
ecto-, exo-, extra-	outside
endo-, ento-, intra-	inner, within
epi-	upon, over
hypo-, infra-, sub-	under, below
hyper-, supra-	above, over
inter-, meta-	between
medi-, meso-	middle
pre-, pro-	before, in front of
post-, postero-	behind
retro-	backward
ultra-	beyond

Miscellaneous Prefixes

a-, an-, e-	without, lack of
chloro-	green
con-	together, with
contra-	against
dis-	apart, away
dys-	difficult, painful

erythro-	red
eu-	true, good
hetero-	different, other
homo-, homeo-	same, similar
leuco-, leuko-	white
meta-	change, after
micro-	small
neo-	new
necro-	dead, corpse
pseudo-	false
re-	again
sym-, syn-	together, with
tachy-	quick, rapid

Miscellaneous Suffixes

-ac, -al, -alis, -an, -ar, -ary	pertaining to
-asis, -asia, -esis	condition of
-blast	bud, sprout
-cide	killer
-clast	break down, broken
-elle, -il	little, small
-emia	condition of blood
-fer	to bear
-ia, -ism	condition of
-ic, -ical, -ine, -ous	pertaining to
-id	member of group
-itis	inflammation
-lysis	loosening, split apart
-oid	like, similar to
-logy	study of
-oma	tumor
-osis	a condition, disease
-pathy	disease
-phore, -phora	bearer
-sect, -tome, -tomy	cut
-some, -soma	body
-stat	stationary, placed
-tropic	change, influence
-vor	to eat

Miscellaneous Root Words and Combining Vowels

andro	man, male
arthr, -i, -o	joint
aut, -o	self
bio	life
blast, -i, -o	bud, sprout
brachi, -o	arm
branch, -i	gill

bronch, -i	windpipe
carcin, -o	cancer
cardi, -a, -o	heart
carn, -i, -o	flesh
cephal, -i, -o	head
chole	bile
chondr, -i, -o	cartilage
chrom, -at, -ato, -o	color
coel, -o	hollow
crani, -o	skull
cuti	skin
cyst, -i, -o	bag, sac, bladder
cyt, -e, -o	cell
derm, -a, -ato	skin
entero	intestine
gastr, -i, -o	stomach
gen	to produce
gyn, -o, gyneco	female
hem, -e, -ato	blood
hist, -io, -o	tissue
hydro	water
lip, -o	fat
mere	segment, body section
morph, -i, -o	shape, form
myo	muscle
neph, -i, -o	kidney
neur, -i, -o	nerve
oculo	eye
odont	tooth
oo, ovi	egg
oss, -eo, osteo	bone
oto	ear
path, -i, -o	disease
phag, -o	to eat
phyll	leaf
phyte	plant
plasm	formative substance
pneumo, -n	lung
pod, -ia	foot
proct, -o	anus
soma, -to	body
stasis, stat, -i, -o	stationary, standing still
sperm, -a, -ato	seed
stoma, -e, -to	mouth, opening
therm, -o	heat
troph	food, nourish
ur, -ia	urine
uro, uran	tail
viscer	internal organ
vita	life
zoo, zoa	animal

APPENDIX B

COMMON METRIC UNITS AND TEMPERATURE CONVERSIONS

Common Metric Units

Category	Symbol	Unit	Value	English Equivalent
Length	km	kilometer	1000 m	0.62 mi
	m	meter*	1 m	39.37 in.
	dm	decimeter	0.1 m	3.94 in.
	cm	centimeter	0.01 m	0.39 in.
	mm	millimeter	0.001 m	0.04 in.
	μm	micrometer	0.000001 m	0.00004 in.
Mass	kg	kilogram	1000 g	2.2 lb
	g	gram*	1 g	0.04 oz
	dg	decigram	0.1 g	0.004 oz
	cg	centigram	0.01 g	0.0004 oz
	mg	milligram	0.001 g	
	μg	microgram	0.000001 g	
Volume	l	liter*	1 l	1.06 qt
	ml	milliliter	0.001 l	0.03 oz
	μl	microliter	0.000001 l	

*Denotes the base unit.

TEMPERATURE CONVERSIONS

Fahrenheit to Celsius

$$°C = \frac{5}{9}(°F - 32)$$

Celsius to Fahrenheit

$$°F = \left(\frac{9}{5}\right)(°C) + 32$$

APPENDIX C

OIL-IMMERSION TECHNIQUE

The oil-immersion objective enables a magnification of $95\times$ to $100\times$ because the oil prevents loss of light rays and permits the resolution of two points as close as 0.2 μm (1/100,000 in.).

The working distance of the oil-immersion objective is extremely small, and care must be exercised to avoid damage to the slide or the objective or both. Although focusing with the low-power and high-dry objectives before using the oil-immersion objective is not essential, it is the preferred technique.

Use the following procedure:
1. Bring the object into focus with the low-power objective.
2. Center the object in the field and rotate the high-dry objective into position.
3. Refocus using the fine-adjustment knob and readjust the iris diaphragm. Also, recenter the object if necessary.
4. Rotate the revolving nosepiece halfway between the high-dry and oil-immersion objectives to allow sufficient room to add the immersion oil.
5. Place a drop of immersion oil on the slide over the center of the stage aperture. Move quickly from bottle to slide to avoid dripping the oil on the stage or table.
6. Rotate the oil-immersion objective into position. Note that the tip of the objective is immersed in the oil.
7. A slight adjustment with the fine-adjustment knob may be needed to bring the object into focus. Remember that the distance between the slide and objective cannot be decreased much without damaging the slide or objective or both.

After using the oil-immersion objective, wipe the oil with lens paper before using the low- and high-dry objectives to avoid getting oil on these objectives.

When finished with the oil-immersion objective, wipe the oil from the objective, slide, and stage.

APPENDIX D

THE CLASSIFICATION OF ORGANISMS

This classification of organisms fits quite well with most introductory biology texts. However, it should be noted that biologists are not in total agreement regarding the classification of all organisms, and this classification may differ from your text or the preferences of your instructor. You should keep in mind that the placement of organisms in a classification system is a dynamic, rather than static, process. A lack of unity is expected in such a process as changes are brought about by new data.

Kingdom MONERA Unicellular prokaryotes without distinct nuclei and membrane-bound organelles.

Division SCHIZOPHYTA*	Bacteria
Division CYANOBACTERIA	Blue-green algae
Division CHLOROXYBACTERIA	Prochlorophytes

Kingdom PROTISTA Mostly unicellular eukaryotes with distinct nuclei and membrane-bound organelles.

Animal-like Protists

Phylum MASTIGOPHORA*	Flagellated protozoans
Phylum SARCODINA	Amoeboid protozoans
Phylum CILIOPHORA	Ciliated protozoans
Phylum SPOROZOA	Nonmotile, parasitic protozoans

*Botanists use the term *division* for this category, while zoologists use the term *phylum*.

Plantlike Protists
Division CHRYSOPHYTA — Diatoms and golden-brown algae
Division EUGLENOPHYTA — Flagellated algae lacking cell walls
Division PYRROPHYTA — Fire algae (dinoflagellates)

Funguslike Protists
Division MYXOMYCOTA — Plasmodial slime molds
Division ACRASIOMYCOTA — Cellular slime molds

Kingdom FUNGI

Plantlike saprotrophic heterotrophs typically with multinucleate cells. Body composed of hyphae.

Division ZYGOMYCOTA — Conjugating or algal fungi
Division ASCOMYCOTA — Sac fungi
Division BASIDIOMYCOTA — Club fungi
Divisoin DEUTEROMYCOTA — Imperfect fungi

Kingdom PLANTAE

Multicellular eukaryotes with rigid cell walls; primarily photosynthetic autotrophs.

Complex Algae
Division RHODOPHYTA — Red algae
Division PHAEOPHYTA — Brown algae
Division CHLOROPHYTA — Green algae

Nonvascular Land Plants
Division BRYOPHYTA — Mosses, liverworts, and hornworts

Vascular Land Plants Without Seeds
Division PSILOPHYTA — Psilopsids
Division LYCOPHYTA — Club mosses
Division SPHENOPHYTA — Horsetails
Division PTEROPHYTA — Ferns

Vascular Plants with Seeds
Gymnosperms
Division CONIFEROPHTYA — Conifers
Division CYCADOPHYTA — Cycads
Division GINKGOPHYTA — Ginkgos
Division GNETOPHYTA — Gnetophytes

Angiosperms
Division ANTHOPHYTA — Flowering plants
Class MONOCOTYLEDONES — Monocots (e.g., grasses, lilies)
Class DICOTYLEDONES — Dicots (e.g., beans, oaks, snapdragons)

*Botanists use the term *division* for this category, while zoologists use the term *phylum*.

Kingdom ANIMALIA

Multicellular eukaryotic heterotrophs; cells without cell walls or chlorophyll; primarily motile.

Phylum PORIFERA	Sponges

Radial Protostomes

Phylum COELENTERATA	Coelenterates (diploblastic)
Class HYDROZOA	Hydroids
Class SCYPHOZOA	True jellyfish
Class ANTHOZOA	Sea anemones and corals

Bilateral Protostomes

Phylum PLATYHELMINTHES	Flatworms (triploblastic)
Class TURBELLARIA	Free-living flatworms
Class TREMATODA	Flukes (parasitic)
Class CESTODA	Tapeworms (parasitic)
Phylum NEMATODA	Roundworms
Phylum ROTIFERA	Rotifers
Phylum MOLLUSCA	Mollusks; soft bodied and usually with a shell
Class AMPHINEURA	Chitons
Class MONOPLACOPHORA	Neopilina (remnants of segmentation)
Class SCAPHOPODA	Tooth shells
Class GASTROPODA	Snails and slugs
Class PELECYPODA	Clams and mussels
Class CEPHALOPODA	Squids and octopi
Phylum ANNELIDA	Segmented worms
Class POLYCHAETA	Sandworms
Class OLIGOCHAETA	Earthworms
Class HIRUDINEA	Leeches
Phylum ONYCHOPHORA	Peripatus
Phylum ARTHROPODA	Animals with an exoskeleton and jointed appendages
Class CRUSTACEA	Crustaceans; crabs, crayfish, barnacles
Class ARACHNIDA	Scorpions, spiders, ticks, mites
Class CHILOPODA	Centipedes
Class DIPLOPIDA	Millipedes
Class INSECTA	Insects

Deuterostomes

Phylum ECHINODERMATA	Spiny-skinned, radially symmetrical animals
Class CRINOIDEA	Sea lilies, feather stars
Class ASTEROIDEA	Sea stars
Class OPHIUROIDEA	Brittle stars
Class ECHINOIDEA	Sea urchins
Class HOLOTHUROIDEA	Sea cucumbers
Phylum HEMICHORDATA	Acorn worms
Phylum CHORDATA	Chordates
Subphylum UROCHORDATA	Tunicates
Subphylum CEPHALOCHORDATA	Lancelets
Subphylum VERTEBRATA	Vertebrates
Class AGNATHA	Jawless fishes
Class CHONDRICHTHYES	Cartilaginous fishes

Class OSTEICHTHYES	Bony fishes
Class AMPHIBIA	Salamanders, frogs, toads
Class REPTILIA	Lizards, snakes, crocodiles, turtles
Class AVES	Birds
Class MAMMALIA	Mammals
Subclass PROTOTHERIA	Egg-laying mammals
Subclass METATHERIA	Marsupial mammals
Subclass EUTHERIA	Placental mammals